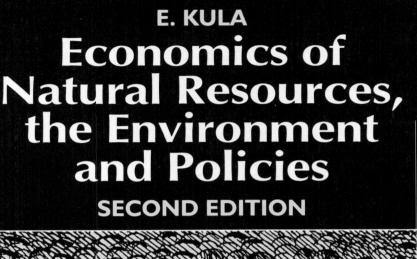

E. KULA
Economics of Natural Resources, the Environment and Policies
SECOND EDITION

GLOBAL WARMING, A FASCINATING PROBLEM...

D1520544

CHAPMAN & HALL

Economics of Natural Resources, the Environment and Policies

In memory of Dr Burhan F. Derbi

Economics of Natural Resources, the Environment and Policies

E. Kula

CHAPMAN & HALL
London · Glasgow · Weinheim · New York · Tokyo · Melbourne · Madras

Published by Chapman & Hall, 2–6 Boundary Row, London SE1 8HN, UK

Chapman & Hall, 2–6 Boundary Row, London SE1 8HN, UK

Blackie Academic & Professional, Wester Cleddens Road, Bishopbriggs, Glasgow G64 2 NZ, UK

Chapman & Hall GmbH, Pappelallee 3, 69469 Weinheim, Germany

Chapman & Hall Inc., One Penn Plaza, 41st Floor, New York, NY 10119, USA

Chapman & Hall Japan, Thomson Publishing Japan, Hirakawacho Nemoto Building, 6F, 1- 7-11 Hirakawa-cho, Chiyoda-ku, Tokyo 102, Japan

Chapman & Hall Australia, Thomas Nelson Australia, 102 Dodds Street, South Melbourne, Victoria 3205, Australia

Chapman & Hall India, R. Seshadri, 32 Second Main Road, CIT East, Madras 600 035, India

First edition 1992

Reprinted 1993

Second edition 1994

©1992, 1994 E. Kula

Typeset in 10/12 pt Times by EXPO Holdings, Malaysia

Printed in Great Britain by St Edmundsbury Press, Bury St Edmunds, Suffolk

ISBN 0 412 57640 6

A catalogue record for this book is available from the British Library

Library of Congress Catalog Card Number: 93-74894

∞Printed on permanent acid-free text paper, manufactured in accordance with ANSI/NISO Z39.48-1992 and ANSI/NISO Z39.48-1984 (Permanence of Paper).

Contents

Preface

The economic activities of humanity, particularly during the last couple of centuries, have had a profound impact on the natural environment. Fast depletion of the world's forest resources, fish stocks, fossil fuels and mine deposits have raised many moral as well as practical questions concerning present and future generations. Furthermore, a number of global environmental problems such as acid rain, the 'greenhouse effect' and depletion of the ozone layer are causing concern throughout the world. What does economics say about the exploitation of nature's scarce resources?

This book, which is a much expanded version of an earlier publication, *Economics of Natural Resources and the Environment,* aims mostly at final-year undergraduates reading subjects such as economics, business studies, environmental science, forestry, marine biology, agriculture and development studies. There is also a good deal of material – especially in the chapters on fisheries, forestry, valuation and discounting – that post-graduate students may find useful as stepping-stones.

The material presented stems from my lectures to final-year students at the University of Ulster during the last 12 years, and some of my research work. When I moved to Northern Ireland in 1982 I was given a course called 'Economics of Exhaustible Resources' to teach. This has changed its title and focus a number of times along with the structure of the University. My early reading lists included a number of journal articles and books written on the subject. However, some of my students asserted that most of the recommended material was unintelligible to undergraduates. This is what led me to decide to write this book.

The book is arranged in 11 chapters. Chapter 1 is about the history of economic thought on natural resources and the environment which starts with classical writers such as R.T. Malthus, D. Ricardo, J.S. Mill and W.S. Jevons, and continues right up to the present time, including twentieth-century thinkers such as Pigou, Boulding, Huxley, Simon, etc. No doubt the reader will notice that there is a good deal of common ground between the classical and modern writers.

Chapter 2, Economics and Policies in Fisheries, begins with the relentless problem of common access which affects many world fisheries. Attention is then focused on the theory of fisheries in which two concepts – the maximum sustainable yield recommended by some marine biologists and the optimum sustainable yield recommended by economists – are explained. These theories are discussed using two models: comparative statistics and dynamics. After that, fishery policies in the European Community and the United States are explained. The chapter ends with a discussion on recent developments in fishery management, including the concept of right based fishing which was put into practice by the New Zealand government in 1985.

Chapter 3 is on the economics of forestry and makes use of the British and Irish experiences. This chapter is based almost entirely on my research work and resulting publications in various journals such as *Journal of Agricultural Economics, Forestry, Resources Policy, The Royal Bank of Scotland Review, The Irish Banking Review* and *Irish Journal of Agricultural Economics*. British and Irish experiences are perhaps the most thought-provoking cases as forest destruction was most reckless in these countries. When governments decided to reverse the trend they came up against a number of difficulties such as discounting, lack of interest by the private sector and the absence of forestry culture in general. Despite over 80 years of effort by the authorities, forestry is still an infant sector in the British Isles. The chapter concludes with aspects of forestry in the European Community, the United States, and forest destruction in the tropics.

Chapter 4, on agriculture and the environment, deals primarily with environmental problems created by farming activities, especially intensive agriculture. The problems in focus are: inorganic nitrate pollution, pesticide pollution, animal waste pollution, which is acute in countries like Holland, soil erosion and salination. The chapter also considers the changing pattern of land use and reports on some predictions about future prospects.

Chapter 5 focuses on the economics of mining, petroleum and natural gas which stems from the works of Gray (1916) and Hotelling (1931). The basic idea is that in exploitation of these resources the net price of deposits should go up in line with the market rate of interest. Various factors which may affect this principle are then discussed along with the work of Slade (1982) who actually tested the theory with some interesting results. The chapter ends with an outline of various trends in fossil fuel use.

Chapter 6, on environmental degradation, starts with a broad discussion of the concept of externalities which is the key to many environmental problems. Then it is explained that zero externality is not the socially optimal level. Various methods, such as the bargaining solution, the common law solution, taxation, direct control, propaganda and public ownership, all of which aim to attain the optimum level of environmental degradation, are outlined. Various analyses are included of the proposed European Carbon Tax which may well be brought into operation throughout the European Community in due course. Then the discussion turns to the public policy in the United Kingdom, the United States, the European Community and the formerly communist East European countries.

Chapter 7 focuses on international environmental problems such as acid rain, destruction of the ozone layer, the 'greenhouse effect' and destruction of biological diversity. Various major international conventions such as the Stockholm Conference, the Tblisi Declaration, the Geneva Convention, the Villach

Conference, the Toronto Conference and the recent United Nations Conference on Environment and Development (known as Rio 1992) are discussed.

Chapter 8 is devoted to valuation methods for environmental attributes. The methods discussed are: cost–benefit and cost-effectiveness analyses, hedonistic price approach, contingent valuation, travel cost analysis, existence value and bequest value. The chapter also includes risk and uncertainty in assessing the value of environmental attributes.

Various economic theories related to natural wonders were constructed in the late 1960s and early 1970s in the United States, mainly in response to development proposals threatening to affect natural features such as Hell's Canyon, the Grand Canyon and the Everglades. Chapter 9 first gives a brief description of a number of natural wonders (a list which should not be regarded as exhaustive) then a simple comparative static theory, similar to that employed by Gannon (1969), is used to emphasize the point that the wisdom of economic theory is strongly in favour of conservation.

Chapter 10 deals with one of the most urgent environmental problems known to humanity, that of nuclear waste storage. After a brief history of nuclear power and types of wastes, there is a report on the currently proposed methods of nuclear waste disposal. The chapter then turns to the costs of building permanent repositories, retrievable storage facilities and waste transportation. The main focus of attention is the Waste Isolation Pilot Plant (WIPP) which is likely to become the United States' first purpose-built nuclear dumping site. The long-term health and monitoring costs of this facility are analysed and the chapter concludes with nuclear waste disposal policies in the United States, the UK and the European Community.

The last chapter describes ordinary and modified discounting methods in natural resource and environmental policies. The chapter also reports on a number of recent debates in the literature regarding the modified discounting method which would have a profound impact on policy-making. As an illustration, the new method is used on two forestry projects, policy-making in fishery management and a nuclear waste storage plant in the United States. The last project emphasizes the brutality of the ordinary discounting on future generations. Indeed the health cost of this project ($6.07 billion) which is to last a million years and may cause suffering and fatalities among numerous generations, is mindlessly reduced to $9940 with a 5% discount rate.

This book would not have been completed without the help and support of many. I am particularly grateful to Dr Ian Hodge, University of Cambridge, for reading chapters on the economics of environmental degradation and natural wonders, and making many valuable comments. The late Professor Bill Carter, University of Ulster, provided scientific information for the 'greenhouse effect' and the depletion of the ozone layer and read the chapters on environmental degradation. Two anonymous marine biologists made useful comments on the fishery management, the right based fishing. The Forest Service of Northern Ireland provided case material which I used for cost–benefit analysis of forestry in Northern Ireland. Mr Robert Scott, Baronscourt, allowed me to use some of his data of private sector forestry. Professor Ronald Cummings, University of New Mexico and Mr Robert Neill of the Environmental Evaluation Group, Albuquerque, and Dr Stanley Logan of Logan & Associates, Santa Fe, helped out on the nuclear waste storage project which I used as case material in Chapter 10. Dr Nick Hanley of Environmental Economics Research Group, University of Sterling, made many valuable comments on the whole manuscript. I am grateful to journals such as *Environment and Planning A*, *Journal of Agricultural Economics*, *Resources Policy*, *Project Appraisal* and *Journal of Environmental Economics and Management,* for granting me permission to reproduce some published material, i.e. tables, figures and various discussions. My thanks extend to my wife for her patient typing of this manuscript and for her tolerance of my workload.

Erhun Kula
Belfast

1

Development of ideas on natural resources and the environment

Is the future of the world system bound to be growth and then collapse into a dismal, depleted existence? Only if we make the assumption that our present way of doing things will not change. There are ample evidence of mankind's ingenuity and social flexibility … Man must explore himself – his goals and values – as much as the world he seeks to change. The dedication to both tasks must be unending. The crux of the matter is not whether the human species will survive, but even more, whether it can survive without falling into a state of worthless existence.

Club of Rome

The history of economic thought on natural resources and the environment is a relatively young subject – the earliest notable discussions go back to the eighteenth century. In this chapter the evolution of some ideas on scarcity and the environment will be discussed in, more or less, chronological order.

1.1 MALTHUSIAN PROBLEM

Most of our thoughts are influenced by the issues and problems of our times and Malthus was no different. His ideas about future prospects were shaped by the events taking place during his lifetime, 1766–1834. In his largely agrarian society some very significant and even turbulent incidents were in progress. The industrial revolution witnessed an unprecedented growth in the population of Britain, whilst structural change in the economy saw an increasing proportion of this population earning its living from non-agricultural pursuits. At the same time the innovation of new technology in both agriculture and industry led to increases in productivity. England's achievements in fighting its long French wars at the end of the eighteenth century and at the same time feeding its growing population were particularly impressive. Alongside all these, profound changes were taking place in many branches of science such as zoology, botany, chemistry and astronomy. Furthermore, advancing medicine coupled with improvements in sanitation achieved a remarkable reduction in mortality.

Some leading philosophers of the time, such as Goodwin and Condercet, were deeply inspired by these events, particularly by the French Revolution and had a vision of society largely free from wars, diseases, crimes and resentments where every man sought the good of all. Unimpressed by Goodwin and Condercet, Malthus published his famous book *An Essay on the Principle of Population as it Affects the Future Improvement of Society* (Malthus, 1798), which came as a shock to those who were optimistic about future prospects.

After observing the growth of population in the United States where food was plenty he came to the conclusion that, if unchecked, population had a tendency to double itself every 25 years. He implied that this was neither the maximum nor the actual growth rate as the available statistics at the time were unreliable but suggested that unchecked population would increase in a geometric fashion, e.g.

$$2, 4, 8, 16, 32, 64, \ldots$$

The food supply, on the other hand, could only be increased in arithmetic progression, e.g.

$$2, 4, 6, 8, 10, 12, \ldots$$

The main reason for this was the law of diminishing returns which simply suggests that as the supply of land is fixed, increasing other inputs on it will increase the food supply at a diminishing rate. Malthus recognized the possibility of opening up new territories but argued that this would be a very slow process and that the quality of lands in the new territories would be inferior to the existing ones. The law of diminishing returns, after all, was only common sense and thus it did not require much empirical proof.

Two conflicting powers are in operation in Malthusian analysis: the power of population and the power in the earth to produce food, a struggle that the latter could not win. The effects of these two unequal powers will, somehow, be kept equal by strong and constantly operating checks on population. The ultimate check on population is the food shortage. Malthus placed other checks in two groups, positive and preventive. The former included wars, famines and pestilence; the latter abortion, contraception and moral restraint. He favoured neither contraception nor abortion as practical means of curtailing the population growth.

The theory which Malthus was trying to construct is a complex one. On the one hand he argued that population would increase when food supply increased. On the other hand, food supply would increase when the population expanded, making more labourers available to work the land. In other words, population depends upon food supply, food supply depends, to some extent, upon population.

The major problem with the Malthusian theory is that it fails to recognize the advance of agricultural technology. As an economic law, diminishing returns, a topic which Malthus explored further in his theory of rent (Malthus, 1815), holds only for a constant state of technology (West, 1815). Furthermore, Malthus failed to see that as income and education levels improve, people may change their attitude by restricting, voluntarily, the size of their family, as occurred in modern Europe.

Almost 200 years after its publication, the Malthusian theory is still a popular debate among economists and political scientists. In some parts of the world it even dominates governments' policies; for a while it may go out of fashion only to come back again. The events which took place in China during the Great Leap Forward are an interesting example to illustrate how the fortune of Malthusian theory can change quite dramatically in a short space of time.

The first Five Year Plan of 1953–57 turned out to be a very successful experiment in China: the average annual growth rate of the gross domestic product was estimated to be at least 12% in real terms. This was followed by a bumper harvest in 1958 which increased the conviction of the Chinese rulers that the country was ripe for a great leap ahead to emancipate her from reliance on the Soviet Union and break out of the vicious circle of backwardness.

In late 1958 the New China News Agency reported that incredibly high yields in grain output had smashed the fragile theory of the diminishing return of the soil and put the final nail in the Malthusian coffin. At the same time the press campaign promoting family planning was reduced substantially and posters, lectures and film shows about birth control began to disappear. The manufacture of contraceptive medicines and devices were also cut back and their display in pharmacies came to an end.

During that time Mao himself stated that the decisive factor for the success of the revolution was the growing Chinese population. The more people there were, the more views and suggestions and the more fervour and energy. Though China's millions were poor, this was not a bad thing because poor people want change, want to do things, and most of all, want revolution. Intoxicated by wild and inaccurate statistics that politically inspired peasants were transforming the countryside, Mao abandoned Malthusianism and became convinced that China's food problems were completely resolved and directed Party workers to reorientate their efforts towards industry. Early harvest 'reports' indicated that a total grain output of 450 million tons was about to be achieved in 1959, which led him to express concern over what to do with the surplus grain.

In the spring of 1959 the truth was beginning to dawn on some Party members that agriculture was failing badly. The Party leaders, on the other hand, were trying desperately to protect the illusion created by the Great Leap Forward. By August 1959, despite renewed attacks against 'counter revolutionaries' who were trying to deny the achievements of the Great Leap, the rising tide of disillusionment had dampened the enthusiasm of the Central Committee of the party. The extent of agricultural failure during the Great Leap years was very substantial indeed. Table 1.1 shows grain output figures reported by seven different agencies, including Chinese officials. There were numerous reasons for the failure: close planting, recommended by the revolutionary cadres against the advice of agricultural experts which resulted in contraction of output; adoption of untested farming methods irrespective of local conditions; overambitious irrigation schemes without proper consideration of water, manpower and material resources; widespread shortages of agricultural labour created by drafting millions to work in industry; failure of industry to provide farm machinery, tools and fertilizer for agriculture; destruction of productivity incentives for farm labourers, which were condemned as capitalist practices; and finally, bad weather conditions. All these contributed to the decline of agriculture. Food shortages lowered peasants physical capabilities. A large number of farm animals died due to lack of feed and care, and weeds overran the fields.

During the Great Leap years the Malthusian ultimate check on population was ruthlessly efficient. Kula (1989b) argues that the famine, which resulted from agricultural failure, and the hardships imposed on the population by the revolutionary leaders killed at least 25 million people. By the end of 1962 a rigorous family planning programme was beginning again, signifying the renaissance of Malthusianism.

Malthus was one of the earliest writers to realize the limitations of our world, which at the time were arable land and its food-growing capacity in the face of a constantly growing population. He believed, sincerely, that eventually humanity will be trapped in a dismal state of existence from which it will not be able to escape. Although the shadow of Malthus is still hanging over a good part of the contemporary world, as far as western Europe is concerned the population is stabilized with enviable results. In effect, this highly crowded area is now pro-

Table 1.1 Grain output estimates for China 1957–61

Reporting body	Grain output (million tons)				
	1957	1958	1959	1960	1961
Chinese officials	185	250	270	150	162
US consulate in Hong Kong	–	194	168	160	167
Taiwanese sources	–	195	160	120	130
Japanese sources	–	200	185	150	160
W. Hoeber and Rockwell	185	175	154	130	140
Dawson	185	204	160	170	180
Jones and Poleman	185	210	192	185	–
Average	185	204	184	152	157

Source: Kula (1989b).

ducing surplus food, a problem which is exactly the opposite to that predicted by Malthus.

Many forecast that with current trends the combined population of India and China alone will be somewhere around 4 billion by the middle of the next century, a case which is likely to make Malthus one of the most talked about thinkers throughout the twenty-first century.

In this respect, Huxley (1983) argues that the coming age will not be the space age or the age of enlightenment or prosperity, but the age of overpopulation with all of its frightening consequences such as poverty, environmental degradation, wars, falling average intelligence and erosion of civil liberties, not only in the developing countries but throughout the world. The fantastic increases in human numbers are taking place not in desirable and productive areas, which are already densely populated, but in regions which are inhospitable, whose soils are being eroded by the frantic efforts of desperate farmers to raise more food and whose easily available mineral and other deposits are being squandered with reckless extravagance. This biological background will advance inexorably towards the front of the historic stage and will remain the central problem for centuries to come throughout the globe.

Clean water, penicillin and simple preventative medicines are relatively cheap commodities which help to reduce death rates; even the poorest governments are rich enough to provide their subjects a substantial measure of death control by making these commodities easily available. Birth control, on the other hand, is a different matter which depends upon the co-operation of an entire population. It must be practised daily by countless individuals from whom it demands more will-power and knowledge than most of the world's illiterate possess. Whereas death control is achieved easily, birth control is achieved with great difficulty. In this century, while death rates have fallen with startling suddenness, birth rates have either remained at their previous levels or if they have fallen, have fallen very little.

In addition to deteriorating environmental quality and dwindling natural deposits, Huxley emphasizes two more problems: creeping dictatorship and falling average intelligence. In unbearably overcrowded regions conditions for freedom and democracy will become impossible, almost unthinkable, as the primary needs of the overwhelming majority will never be fully satisfied. History has amply demonstrated that when the economic life of a nation becomes precarious, the central government assumes additional responsibilities. In similar circumstances in the future the central authority will be required to work out elaborate plans for dealing with critical conditions created by overpopulation and will impose many restrictions upon the daily activities of individuals. As the crisis deepens, the central government will be compelled to tighten its grip to preserve public order and its own authority. More and more power will be concentrated in the hands of governments.

The nature of power is such that even those who have not sought it, but have had it forced upon them, tend to acquire a taste for more. In countries where human numbers reach a crisis point, increasing economic insecurity and social unrest will lead to more control by authoritarian governments which will give way to oppression and corruption; 'Given this fact, the probability of over-population leading through unrest to dictatorship becomes a virtual certainty' (Huxley, 1983).

This situation in hard-pressed regions will affect the established democracies of Europe, Japan, Australia and North America – the West. It is more than likely that the dictatorial governments in the impoverished world will be hostile to the West and could reduce, or even stop, the flow of raw materials that the West needs for its survival. Due to lack of critical materials industrialized systems established in the West will break down and their highly developed technology, which until now has allowed them to sustain a high population in a small terri-tory, will no longer be effective. When this happens, the powers forced by unfavourable conditions upon Western governments may come to be used in the spirit of dictatorship.

Huxley asserts that in free and tolerant societies most of us know that pursuit of good ends does not justify the employment of bad means. But what about those situations in which good means have end-results that turn out to be bad? In par-ticular, advances in modern medicine, improved sanitation and compassionate social attitudes are allowing many children born with hereditary defects to reach maturity and multiply their kind. In the past, children with even slight hereditary defects rarely survived. Today, thanks to modern medicine, the lives of illness-prone individuals with terminal diseases are prolonged. In spite of 'improved' treatment, the physical health of the general population will show no improvement and may even deteriorate along with the fall in the average level of intelligence.

Huxley draws attention to research which suggests that such a decline has already taken place and is continuing along its downward trend as our best stock is outbred by stock that is inferior to it in every respect. With declining physical health and IQ levels how long can we maintain our traditions of liberty, toler-ance and sense of justice?

> To help the unfortunate is obviously good. But wholesale transmission to our descendants of the results of unfavourable mutations and progressive contamination of the genetic pool from which the members of our species will have to draw are no less bad.

Falling intelligence levels can only aid the transition from democracy to dicta-torship. Once totalitarian regimes are established in the West governments will try very hard to meet the basic needs of its subjects who, terrified by the prob-lems created by overpopulation, are likely to demand beefburgers and television before liberty and justice. The life for the thinking and caring minority will be hellish and they may turn to sedative drugs, the use of which will be encouraged by the authoritarian governments to keep them docile.

Huxley's predictions about expanding dictatorships and declining health and intelligence levels due to increasing numbers of inferior bodies may not convince everybody. His views on dictatorship were expressed at a time when communism appeared to be invincible and expansive. Today, communism is dead in its cradle and losing strength elsewhere. Despite painful transition problems, the masses who lived under communist dictatorships for long years are now making determined efforts to establish freedom, which is bound to have an effect on other forms of dictatorship elsewhere. The prospects for global democracy look better today than at any time in human history. On the falling intelligence levels it may be erroneous to imply that only healthy bodies can command higher intelligence. There are many examples of people of high intelligence and talent who have less than perfect bodies and Huxley himself was one. Likewise, there are many examples of successful individuals in all fields of life who came from extremely deprived social backgrounds.

1.2   RICARDIAN STAGNATION

Another pessimist in the history of economic thought on natural resources was David Ricardo, a contemporary of Malthus, who predicted a steady state of equilibrium which, on the whole, was far from enviable. He is also well known for his theories on international trade, the labour theory of value and rent.

The development of his classical theory of rent, in which Malthus played an important role, emerged from the so-called Corn Law Controversy which came about as a result of the Napoleonic Wars. The embargo on British ports during these wars kept foreign grain out of England and forced British farmers to increase their production of grain to feed the population. As a result, between 1790 and 1810 British corn prices rose by 18% per annum, on average. Land rents also increased, which pleased the landlords. In 1815 the Corn Law effectively prohibited the importation of foreign grain and was, in fact, one of the earliest examples of agricultural protectionism affecting economic growth and income distribution in the country.

Ricardo was in agreement with Malthus on the basic principles of population and rent (Ricardo, 1817). The reason for the increased price of grain was the law of diminishing returns. Ricardo argued that the price of the produce is determined by profits, wages and rent. In a situation where an increase in production takes place on old lands the incremental rent will be zero, or near zero, leaving wages and profits as the sole determinants of price. The rent on old lands would increase only when new territories were opened for agriculture. In his words

> When, in the progress of society, land of the second degree of fertility is taken into cultivation, rent immediately commences on that of the first quality, and the amount of that rent will depend on the difference in the quality of these two portions of land.
>
> Sraffa and Dobb (1951–55)

On the wages front Ricardo suggested that the Corn Law allowed a rise in wages but a fall in profits and thus less capital accumulation took place which slowed down the economic growth. In Ricardo's model profits are the engine of economic growth and wages the engine of population expansion. If wages rose above subsistence, the number of heads coming into the world would go up and eventually wages would fall back again to the level of subsistence. If they fell below subsistence, population would fall due to malnutrition and wages would rise again to subsistence level.

It is important to note that in the Ricardian model there is a conflict between wage and profit levels. When wages are bid well above subsistence, profits will be squeezed to a minimum, then capital accumulation will, temporarily, cease. The next thing is that the population will continue to grow, due to above-subsistence wages, forcing earnings of workers back to the subsistence level. When this happens, rising profits will increase capital accumulation and this in turn will initiate growth, and so on. Eventually, this process will come to a rest at a point where profits, due to diminishing returns, can no longer be increased. At this point there is no capital accumulation, no growth, wages are at subsistence and the economy is at the stationary state.

At the centre of the Ricardian model there is the notion that economic growth must eventually peter out due to scarcity of natural resources, which, at the time, were land and its food-production capacity. Envisage that the entire world is a giant farm of fixed size on which capital and labour (population) are used as

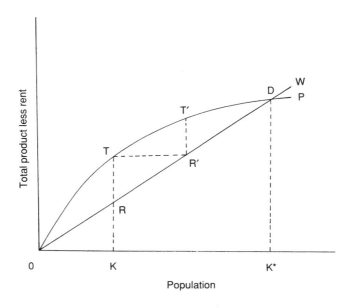

Figure 1.1   Ricardian model of economic growth and stagnation.

inputs to produce food, say grain. In Figure 1.1 population is measured along the horizontal axis and total product less rent along the vertical axis. We exclude the landowners because the size of land and its ownership is fixed. The output curve is OP and the subsistence curve is OW. When population is OK, the subsistence wage bill is KR and profits TR. The slope of OW is, of course, the wage rate, which is measured as KR/OK, i.e. total wage bill divided by number of workers. The working of the model is as follows: the profits will initiate growth, then the wages will go up above subsistence and thus encourage population to grow which will bid down the wage to subsistence, R'. At this point the level of output and population are higher than R, the starting point. Note that the existence of profit R'T' makes the system move again until we get to the stationary state D. Here wages are of subsistence and population and output levels are constant, all due to natural resource scarcity.

1.3   J.S. MILL

The influence of Malthus and Ricardo on J.S. Mill was strong. In his celebrated book, *Principles of Political Economy* (Mill, 1862), the idea of continued advancement which was popular among many of his contemporaries, was refuted. Mill pointed out that growth in nature was not an endless process; every growth, including the economic one, must eventually come to a lasting equilibrium. According to Mill a period of prolonged growth did take place especially in the eighteenth and nineteenth centuries because of humanity's fierce struggle for material advancement which essentially is not sustainable. It is undesirable as well as futile.

Only foolish people will want to live in a world crowded by human beings and their material possessions. Solitude is an essential ingredient of meditation and well-being. There is no point in contemplating a world where every square metre of land is brought into cultivation, every flowery waste of natural pasture is ploughed up, every wild plant and animal species exterminated as humanity's rival for food, and every hedgerow or superfluous tree rooted out. It will be the ultimate waste if the earth loses that great portion of its grandeur and pleasantness so that we can support a large population on it. 'I sincerely hope, for the sake of posterity, that they will be content to be stationary, long before necessity compels them to it' (Mill, 1862).

It is interesting to observe that what Mill feared has already occurred in some parts of the world. There are large parts of Europe and Asia where thousands upon thousands of square kilometres are divided between intensive farms, housing and industrial estates, cities and transport networks. For a large number of people who live in these crowded regions solitude has now become an expensive commodity.

Mill's ideas on human welfare, which are more than 100 years old, are still a source of inspiration for many modern conservationists who believe that economic growth will neither solve our problems nor improve future well-being.

They are also a great challenge for some welfare economists who tend to believe that the social well-being will be maximized when the greatest consumption capacity for the greatest number of individuals is attained. In the tradition of Mill, economic growth is necessary only for the developing countries. In the developed countries, however, the real issue is of income distribution, not its growth. He was not impressed with the kind of progress – growth and capital accumulation before anything else – that was fashionable in his time as well as now. Mill believed that the struggle to progress in terms of material goods in which people wage together and against one another is neither natural nor desirable for mankind.

Several attempts have been made to document the fact that despite constantly increasing gross national product (GNP) levels in many countries human welfare is not keeping abreast; in some cases it is actually falling (Douthwaite, 1992). As early as 1972 Nordhaus and Tobin (1972) made a pioneering effort to construct a meaningful index to measure human well-being, the measure of economic welfare (MEW), as opposed to GNP. The latter is a measure of production which many now believe cannot be used to measure human well-being. MEW excludes from GNP items of regrettable necessities such as national defence, police services, travel to work, sewage disposal, etc. Also expenditure on health and education are deducted on the grounds that they are investments in human capital. Disamenities endured by people are also deducted. Leisure time, which people enjoy, and housework, which is a productive activity, are added on annual figures. Nordaus and Tobin have noted that in the United States between 1928 and 1965 as per capita net national product grew by 1.7%, compound per capita MEW grew by 1.1%.

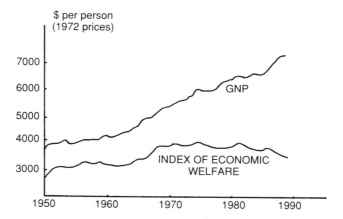

Figure 1.2   Growth of GNP and index of economic welfare in the USA. (Source: Daly and Cobb, 1989.)

More recently, Daly and Cobb (1989) constructed an index of economic welfare in which allowances were made for the adverse effects of pollution and destruction of natural resources. This index shows that economic welfare in the United States increased gradually from 1950 until 1969 and remained stationary for 10 years. Since 1980 it has been falling noticeably, even when GNP per head rose sharply (Figure 1.2).

1.4 W.S. JEVONS

Unlike Malthus and Ricardo, who lived in a largely agricultural society, Jevons lived at a time when rapid industrialization was well under way in Britain. The issue he looked at was the depletion of coal which, at the time, was the major source of energy. In his view coal was the most important constraint on Britain's economic development. Rapid industrialization was already depleting the rich and easily accessible reserves and forcing miners to extract from less easily accessible stocks.

In his book, *The Coal Question: An Inquiry Concerning the Progress of the Nation and the Probable Exhaustion of our Coal Mines*, Jevons (1865) emphasized that coal was of central significance in Britain's economic supremacy: 'coal alone can command in sufficient abundance either the iron or the steam; and coal, therefore, commands this age – the Age of Coal'.

A few years before the publication of *The Coal Question* it was argued that the possibility of exhaustion of British coal mines in the face of industrial development was very strong (Church, 1986). Indeed, during that time the coal output in Britain was increasing steadily (Table 1.2). Jevons made this an ominous issue by arguing that the hasty depletion of our most valuable seams was taking

Table 1.2 Annual estimates of British coal production, 1850–69

Year	Tons (million)	Year	Tons (million)
1850	62.5	1860	89.2
1851	65.2	1861	91.1
1852	68.3	1862	95.7
1853	71.1	1863	99.1
1854	75.1	1864	102.1
1855	76.4	1865	104.9
1856	79.0	1866	106.4
1857	81.9	1867	108.2
1858	80.3	1868	110.9
1859	82.8	1869	115.5

Source: Church (1986).

place everywhere and soon coal reserves would be exhausted or costs increased to a point where industry would no longer operate. 'I must point out the painful fact that a state of growth will before long render our consumption of coal comparable with the total supply. In the increasing depth and difficulty of mining we shall meet that vague but inevitable boundary that will stop our progress' (Jevons, 1865).

From the middle of the nineteenth century onwards there was a noticeable, but somewhat cyclical, increase in the price of coal in Britain (Table 1.3). During that time there was a substantial difference in price not only between kinds of coal, but between and within different regions. The price indices used in Table 1.3 consist of a number of separate series to give a general picture.

Some believe that Jevons had an exaggerating personality and was quite excited by the idea of the exhaustion of resources (Hutchinson, 1953). Humorously, Keynes pointed out that Jevons held similar ideas about paper, as he hoarded large quantities of packing and writing paper, so much that 50 years after his death his children had still not used up the stock (Spiegel, 1952).

1.5   A.C. PIGOU AND OTHER AUTHORITARIANS

The decision to save and invest today also involves a redistribution of income between generations and Pigou was one of the earliest economists to write, although briefly, in this area. He argued strongly that individuals distribute their resources between present, near future and remote future on the basis of wholly irrational preferences. When we have a choice between two satisfactions, we do not necessarily choose the larger of the two but often devote ourselves to obtaining a small one now in preference to a much larger one some years later. This is

Table 1.3   Index of pithead coal price estimates, 1850–69

Year	Price (1900 = 100)	Year	Price (1900 = 100)
1850	38.5	1860	46.9
1851	37.7	1861	48.5
1852	35.4	1862	45.4
1853	43.1	1863	44.6
1854	49.2	1864	48.5
1855	49.2	1865	53.8
1856	48.5	1866	59.2
1857	46.9	1867	57.7
1858	43.3	1868	54.6
1859	44.6	1869	50.0

Source: Church (1986).

because our telescopic faculty is defective and thus we see future events, needs, pains and pleasures at a diminishing scale. In this way we not only cause injury to our own welfare but also hurt the generations yet to be born (Pigou, 1929).

The harm to future generations is much greater than that to present individuals. As human life is limited, the fruits of saving which accrue after a considerable interval will possibly not be enjoyed by the person who makes the sacrifice. Consequently, efforts directed towards the more distant future will be much less than the efforts directed towards relatively near future. In the latter, the people involved will probably be the immediate successors whose interests may be regarded as nearly equivalent. In the former, the person involved will be somebody quite remote in blood or in time for whom each individual scarcely cares at all. The consequences of such selfishness will be less saving, the halting of creation of new capital, and the using up of existing gifts of nature to such a degree that future advantages are sacrificed for smaller present ones.

Pigou gives a number of examples to illustrate his point: fishing operations conducted with disregard to breeding seasons, thus threatening certain species of fish with extinction; extensive farming operations that exhaust the fertility of the soil; the depletion of non-renewable resources which may be abundant now but are likely to become scarce in the future, for trivial purposes. This sort of waste is illustrated when enormous quantities of coal are employed in high-speed vessels in order marginally to shorten the time of a journey that is already short. We may cut an hour off the time of our sailing to New York at a cost of preventing, perhaps, one of our descendants from making the passage at all.

In order to safeguard the well-being of future generations, Pigou suggests a number of policies:

1. Taxation which differentiates against saving should be abolished. Even without this kind of tax there will be too little saving, with it there will be much too little.
2. Governments should defend exhaustible resources by legislation, if necessary, from rash and reckless exploitation.
3. There should be incentives for investment, particularly into areas such as forestry, where the return will only begin to appear after a lapse of many years.

Obviously, Pigou's answer to the resource-depletion problem is highly authoritarian. Dobb (1946, 1954, 1960) seems to be in agreement with Pigou as he questions the rationality of individuals not only on matters regarding natural resources but in other aspects of economic behaviour as well. In Dobb's view, individuals are notoriously irrational in deciding about their future needs and could not make adequate provisions for themselves or for others if they were left in isolation in the marketplace. A centrally planned economy in place of the market-oriented one would provide a better alternative for present generations as well as for the future members of society. In effect, decisions made by the planning authority are full of implicit judgements about what consumers really want.

Holzman (1958) argues that individuals' savings versus consumption decisions, which will affect future generations, revealed in the marketplace are likely to be in conflict with decisions taken by the state. He observes many countries' explicit attempts to grow faster to increase per capita income. If the rate of saving and economic growth dictated by the free market were satisfactory, why should governments be obsessed by rapid economic development? By doing so, do governments damage consumers' sovereignty? Holzman contends that if governments enforced some saving on the community this would not seriously damage consumers' sovereignty. Accepting the short-sighted orientation of individuals, he points out that looking back on our past decisions many of us regret that these were not perfect. Most people, having lived through a period already, would be more than willing to trade part of the standard of living of the past for a higher standard in the present. However, the question is: how far should the state expand the level of saving beyond the point which is revealed by the market? According to Holzman, it is the level which the planners feel they can squeeze out of households. This does not violate consumers' sovereignty by as much as is generally assumed.

Despite a genuine belief by the socialist school that in a centrally planned economic system present as well as future individuals will be well catered for, it is difficult to find hard evidence for this. It is now clear that many of the East European countries where communist governments ruled for decades are environmentally the least agreeable parts of Europe, and back in 1972 Meadows *et al.* reported that the former Soviet Union was a major depletor of the world's natural resources, consuming 24% of the world's iron, 13% of copper and lead, 12% of oil and aluminium and 11% of zinc and had less than 5% of the world's population.

Unimpressed by the Pigovian school Hirshleifer *et al.* (1960) argue that even if the government wished to act as a trustee for the unborn there are a variety of instruments available other than legislation and direct state intervention. The manipulation of market rate of interest is a handy tool. If the rate is excessively high, undermining long-term investment projects in particular, the government can take a number of steps to ratify the situation. One is to let the private sector use the market rate but in government operations decisions can be taken on a lower figure. This already takes place in the United States as an accidental consequence of fiscal legislation rather than a deliberate policy. Many federal projects are discounted at rates between 2.5 and 4%, whereas private sector ones are discounted at 6–10%.

The problem with having two different sets of discount rates for public and private sector projects is one of inefficiency. Using a low discount rate for communal projects will expand the public sector towards less productive areas. Given the aggregate amount of present saving, less will be provided for future generations if low return projects are undertaken. As McKean (1958) puts it, we can do better for posterity by choosing the most profitable investment projects at all times.

1.5.1 Environmental abuse in the formerly communist countries

The 1990s began with the disintegration of communist regimes in Eastern Europe, with one country after another abandoning central planning in favour of free market and pluralist democracy. The collapse of these communist systems has led to scandalous revelations about the conduct of the former communist governments. The abuse of human rights and the environment in those countries was common knowledge, but what has startled the world is the extent of incompetence and the callousness of the former rulers in dealing with their people, the economy and the environment. Article 18 of the 1977 Soviet Constitution, now defunct, stated clearly that all necessary steps would be taken in the country to protect and make rational use of the land and its mineral and water resources, purity of air and seas for the well-being of future as well as present generations. Nevertheless, the regime left behind an appalling catalogue of environmental blunders, of which the reversible ones will take decades to sort out.

Pearce and Turner (1990) contend that one important reason for this environmental failure was that the Soviet territory was very fragile and that planning errors quickly broke its back. Only about 10% of the country was suitable for lucrative agriculture and only 1% of this 10% received annual rainfall of 70 cm or more. The bulk of the Soviet agriculture was conducted on arid or semi-arid lands supported by massive irrigation projects which were fundamentally flawed. As for the heavy industry, planners decided to locate industrial plants in and around population centres which required large quantities of mineral and energy deposits extracted and transported from distant regions. Concentration of population and industrial activities in such centres resulted in excessive water, land and air pollution. However, the fragility and the vastness of the former Soviet territory cannot be blamed entirely for the environmental devastation; the situation was equally bad in small but fertile communist countries such as Bulgaria and Romania.

The Stalinist period of 1928–53 witnessed a process of heavy industrialization and collectivization which inflicted great pain and hardship on the Soviet people. However, there were many notable achievements, especially after the Second World War when the Soviet economy grew rapidly. The launch of Sputnik into orbit in 1957 and the flight of Yuri Gagarin, the first man in space, shortly after, marked the heights of the Soviet success. In 1948 Stalin launched his Plan for the Transformation of Nature with grandiose projects designed to conquer, 'scientifically', nature and ultimately the great enemy, capitalism.

The most ambitious land-based projects were river diversion schemes which involved taking water from the northward-flowing rivers and diverting it to the more arid but warmer regions of the south. In European Russia, the Omega, Sukhona, Pechora and Vychegda rivers were diverted and in Siberia the rivers Ob, Irtysh and Yenese were altered (Gerasimov and Gindin, 1977; L'Vovich, 1978). In Soviet Central Asia the Aral Sea Project involved diverting waters from the Siri Derya and Amu Derya rivers to irrigation networks to grow mainly cotton. Further to the north-east, large-scale wood-processing industries were established around Lake Baikal. These discharged effluent either directly into the lake or into rivers feeding into it. Furthermore, extensive logging operations around the lake increased silt deposited into the lake.

These water projects turned out to be ecological disasters on a monumental scale. The Aral Sea has lost more than half of its water since the establishment of the irrigation network and a salt desert has formed on the dry sea bed. Wind carries the salt and the dust long distances, affecting crops and human health in a wide area. Today camels walk and ships lie on the dead bottom of the Aral Sea, which was once the fourth largest inland sea in the world. What is left of the fishing fleet now docks 40 km from its original harbour. Some reports suggest that the changes in the rainfall pattern affect food supplies as far away as Afghanistan and Pakistan.

According to some, north-to-south river diversion schemes disrupted the ecology of the region. Concern has been expressed for the wider consequences of such schemes, particularly on the Arctic Sea ice, if the volume of freshwater input diminishes significantly. The global water system has natural mechanisms by which it regulates and cleanses itself of pollutants. Different bodies of water have different rates of cycles. Waters of enclosed seas, lakes and aquifers turn over very slowly. The Baltic Sea, for example, is replenished only once every 80 years. As a result, pollutants dumped into the Baltic Sea by the communist regimes will remain a problem almost throughout the twenty-first century. Some rivers in the former Soviet Union are so polluted that occasionally they catch fire.

Perhaps one of the most worrying legacies of the former Soviet regime is nuclear pollution. The White Sea is already contaminated by nuclear waste. In 1991 millions of dead starfish were washed up onto many miles of the shore and the culprit was identified as radioactive military waste (Gore, 1992). In 1992 there were alarming reports in the Western media that Soviets had dumped large quantities of highly toxic nuclear wastes in unsuitable containers in and around Barents Sea.

The recent signing of the historic 'Start II' Nuclear Disarmament Treaty by George Bush and Boris Yeltsin in January 1993 was an important step forward towards the reduction of the world's nuclear arsenal. However, dismantling nuclear arms is an expensive business, especially for the Russians who are going through a very difficult economic transition period. According to the Start II Treaty, American and Russian nuclear weapons will be reduced by a further two-thirds within ten years, which will make a very substantial addition to the nuclear waste inventory, the management of which is a formidable problem (Chapter 10). The nuclear accident at Chernobyl in 1986 is a constant reminder of what could happen in other parts of the former Soviet bloc which house many unsafe nuclear power generating units. The Chernobyl accident contaminated a wide area and has affected and will continue to affect the health and safety of many individuals for years to come.

For more than 40 years after the Second World War the Erzgebirge Mountains in former East Germany provided the Soviet Union with its most important source of uranium. In 1992 German researchers announced that they had registered at least 5500 cases of lung cancer amongst former uranium miners working for Wismut, the joint Soviet–German company that operated the mines. Trade union officials now argue that between 300 000 and 350 000 people living in the Erzgebirge area have been exposed to excessive levels of radiation. The German Society for Work-Related Medicine recently described the Wismut operation as the world's third worst radiation disaster after Hiroshima and Chernobyl.

Recently it has been revealed by Russia's former chief nuclear safety inspector, Vladimir Kuznetsow, that the city of Moscow has 50 secret nuclear reactors and a huge nuclear waste dump (Campbell, 1993). Moscow is the only capital city in the world which runs many nuclear reactors and has a waste storage facility located next to a military airfield. Accidents in one of its essentially unsafe reactors or a plane crash on the nuclear waste dumping site could have catastrophic consequences not only for the city's 9 million residents, but also to tens of millions more who live in the region. These nuclear reactors, which were constructed mainly for research purposes, have no protective covers and staffing levels and safety precautions are inadequate. Furthermore, the attitude of staff at the reactor sites is dangerously lax. Alarm systems do not work, some safety equipment has been deliberately tampered with to make them insensitive because alarm systems were going off all the time. Mr Kuznetsow, who criticized the safety of Russia's nuclear institutions, was forced to resign from his job in January 1993. He predicted that sooner rather than later an accident will happen with frightening consequences.

There are many other environmental problems in the former communist countries that are gradually emerging. For example, after the collapse of communism in Romania we learned that in the black town of Copsa Mica the trees and grass are so dark that it gives an outsider the impression that someone has deliberately painted the whole area in black. The health of the people in the region has been extremely bad and yet they have little or no opportunity to settle elsewhere. In former Czechoslovakia the air in some industrialized areas has been extremely polluted; in fact it was so bad that in the past people who settled there were paid a bonus, called by inhabitants the burial bonus. In some areas of Poland children are regularly taken into underground mines to have a break from the build-up of pollution in the air (Gore, 1992). The Vistula river, which runs into the Baltic Sea near Gdansk is so full of poisonous and corrosive substances that its water cannot even be used to cool factory machinery.

1.6   THE UNITED STATES PRESIDENTIAL MATERIAL COMMISSION

Towards the middle of this century concern was growing, mainly in the United States, on matters regarding environmental quality and resource depletion. At the same time many realized that a fast-growing economy would become increasingly dependent on the importation of oil and other raw materials.

One of the earliest national studies of resource and environmental problems was carried out by the President's Material Policy Commission (1952), known as Policy Commission. Their report, *Resources for Freedom, Foundation for Growth and Scarcity*, revealed that in the United States alone the consumption of fuel and other minerals since the beginning of the First World War had been greater than the total world consumption of all the past centuries put together. The report concluded that natural resources are vital to the United States economy and urged the government to make plans for future needs. The recent history of oil makes it clear that the recommendation was timely, but it is less clear that the United States government has played a helpful role since that time. However, the President's Material Policy Commission helped to increase the interest in resource and environmental matters and a number of economists began to carry out extensive research in this area, some of which is outlined below.

1.7 SCARCITY AND HISTORIC PRICE TRENDS FOR NATURAL RESOURCES

The best-known study in this field is that by Barnett and Morse (1963) who tested the implications of resource scarcity on extraction costs and prices of natural resource-based commodities over an 87-year period between 1870 and 1957. This study was later extended for another 13 years up to 1970 (Barnett, 1979). In these studies, price trends for fossil fuels, minerals, forestry, fishing and agriculture in the United States were analysed as the nation progressed steadily from an underdeveloped stage to an advanced economy. During that period there was enormous pressure on natural resource-based commodities.

It should be mentioned at this stage that there are a number of methods available to test scarcity: (1) the unit cost of extraction in real terms; (2) the real price of natural resource-based commodities; and (3) the resource rent. The rationale for the first method is that if, due to scarcity, extraction is forced towards lower and less easily accessible deposits, more and more inputs will have to be used in the extractive sector and this will push the unit cost up. In the second method the market price of commodities is taken as a measure for scarcity which also captures the costs. This method is simple and can be very useful in a historic study. One problem with this criterion is that the results are very sensitive to the choice of price deflator. Some economists (e.g. Hartwick and Olewiler, 1986) argue that method (3) is the best approach as it captures the effects of technological change and substitution possibilities. The first two methods are likely to understate scarcity when substitution among factor inputs is possible, because prices and costs rise less rapidly than the resource rent.

The study by Barnett and Morse (1963) was based mostly upon the first method in which they defined the unit cost as

$$\text{Cost per unit} = \frac{\alpha L + \beta K}{Q},$$

where L is labour, K is capital that is used to extract resources, α and β are weights used to aggregate these inputs and Q is the output of the extractive sector. In their research all resources except forestry showed a decline in real costs. For the entire extractive sector the unit cost fell by about 1% per annum between 1870 and 1920 and nearly 3% per annum from 1920 onwards. A similar situation was observed in agriculture. Their conclusion was that all natural resource-based commodities except timber were becoming less scarce.

This increased availability was attributed to three factors: the discovery of new mineral deposits; the advance of technology in exploration, extraction, processing and production; and the substitution of more abundant low-grade resources for scarce high-grade ones. Table 1.4 shows the situation in sectors where costs fell.

In addition to unit costs Barnett and Morse also looked at real prices. Their view was that the price trend should follow the cost trend apart from changes in the degree of monopoly, taxes, subsidies and transport costs. Broadly speaking the test on prices confirmed the study on costs with some minor exceptions.

The work of Barnett and Morse has some shortcomings. First, the formula used to measure the unit cost lumps capital and labour together. Aggregation of capital alone is a substantial problem itself. Second, the unit cost index which emerges in the end is essentially a backward-looking indicator. Costs identified in this way do not incorporate expectations about the future. It is true that most people make their decisions on the basis of past observations and beliefs, but sometimes they turn out to be wrong. Third, the relationship between extraction cost and technical progress is quite a problem. As physical depletion comes near, it may well be that unit costs rise as deposits become harder to find, but it may also be the case that the effort to find new deposits will result in technological changes that may reduce exploration cost. In other words, no clear signal can emerge from their measure about future costs and prices.

Another notable work in the field of scarcity and prices is that by Potter and Christy (1962). While working for Resources for the Future Inc., an American non-profit making research establishment, they analysed the consumption and price trends for many natural resource-based commodities in the United States for a period comparable to that of Barnett and Morse. Their findings also

Table 1.4   Unit cost in natural resource-based sectors

Sectors	1870–1900	1919	1955
Agriculture	132	114	61
Minerals	210	164	47
Fishing	200	100	18
Total extractive	134	122	60

Source: Barnett and Morse (1963).

Table 1.5 The long-term price trend of lumber in the United States between 1870 and 1957, index 1959 = 100

Years	1870	1880	1890	1900	1910	1920	1930	1940	1950	1957
Price of lumber	20	23	27	30	34	51	48	61	103	95

Source: Potter and Christy (1962).

confirmed that all commodities except lumber were becoming less scarce. Table 1.5 gives the long-term price trend of lumber, which shows a strong upward trend nearly five times higher in 1957 than in 1870, an average annual increase of 1.9% in real terms. Potter and Christy also noted that the rise in the price of lumber may be understated because of the shift of the industry from the eastern United States to the West Coast, which placed the output further away from the consumer market.

In the United Kingdom research similar to that mentioned above confirmed that the price of timber had been increasing steadily. According to Hiley (1967) the annual price increase for imported timber was 1.5% in real terms between 1863/63 and 1963/64.

Ten years after the studies by Potter and Christy, and Barnett and Morse, Nordhaus (1973) published the details of his research on resource scarcity between 1900 and 1970. This study estimated the resource scarcity on the basis of the price of extracted material. The strength of his analysis is that (a) it is easy to compute, (b) it is a forward-looking measure because expectations about future supplies and costs will be reflected in the market price of the resource. It was mentioned above that one major problem in using prices is the choice of deflator. Is it to be GNP deflator, consumer price index, or something else? Nordhaus deflates the prices at the refined level by a manufacturing hourly wage

Table 1.6 The relative prices of some commodities to labour, index 1900 = 100

Commodity	1900	1920	1940	1950	1960	1970
Coal	459	451	189	208	111	100
Petroleum	1034	726	198	213	135	100
Aluminium	3150	859	287	166	134	100
Copper	785	226	121	99	82	100
Iron	620	287	144	112	120	100
Lead	788	388	204	228	114	100
Zinc	794	400	272	256	125	100

Source: Nordhaus (1973).

rate. Table 1.6 shows the results of his findings for a selective group of commodities. This analysis clearly shows that all these commodities were becoming less scarce.

Nordhaus and Tobin (1972) searched for specific functions consistent with historic values of factor shares in the national income. The production function employed consists of three factors: labour, capital and natural resources. One of their conclusions is that natural resources are not likely to become an increasingly severe drag on economic growth.

The findings of Barnett and Morse, Potter and Christy, Nordhaus, and Nordhaus and Tobin are quite heart-warming in the sense that there may be no need to worry too much about resource scarcity for many years to come. Will the factors they emphasize, such as technological advances, substitution and new discoveries continue to affect resource supply favourably? We do not know. Who can predict technical change or the potential for continued substitution possibilities?

1.8   THE UNITED STATES BUREAU OF MINES

Not everybody shared the optimistic findings discussed above that natural resource-based commodities were becoming less scarce. Some academics as well as government officials have felt that stock estimates and the number of years that they will last should be calculated with some reliable accuracy. One body which embarked on this task was the United States Bureau of Mines. At this stage it is worth mentioning that there are some fundamental as well as practical problems in estimating natural resource reserves. Stock estimates are normally established by the discoveries resulting from exploration. Sometimes the information is precise with a high degree of certainty, but at other times the information is subject to wide margins of error.

If information were costless, it would be desirable to have all possible estimates on existing resource stocks. However, exploration is very expensive and the information it yields is treated by agencies as a scarce input. From an economic viewpoint it would not be sensible to obtain complete information as it would most certainly involve a huge cost to the agency. Globally, it may not pay to eliminate all uncertainties of what the earth contains because a large chunk of stocks will not be exploited for many years. Furthermore, new methods of exploration such as satellite scanning of the earth's crust may make the estimation process much cheaper in the future when these methods are refined and well understood. Therefore, at any point in time there must be a sensible or optimum programme of exploration for mining companies as well as for individual nations. This is an important issue for countries that invest large sums in exploration with the hope of finding something valuable.

The terminology relating to stocks is quite simple because of the lead taken by the United States Bureau of Mines. In defining terms two aspects of stocks are recognized: (1) the extent of geological knowledge, and (2) the economic feasibility of recovery. Geologists are mostly concerned with exploration activities that increase the accuracy of our knowledge of stocks, whereas mining engineers concern themselves with improvements in recovery technology in order to bring costs down. Economists bring these two factors together in their analysis to determine the economic viability of exploitation.

Table 1.7 shows stock estimates for some important mineral deposits by the United States Bureau of Mines. Figures in column 3, the static index, are obtained by dividing the known global reserves by the annual consumption levels in 1970. This does not take account of growing consumption; the annual demand is assumed to remain constant at 1970 levels. Of the items listed in Table 1.7 coal is the most abundant commodity, with stocks which will last for about 2300 years. Column 4 shows the annual average growth rate of consumption and the last column gives the length of time stocks will last, given this rate of growth. For example, with a 4.1% annual growth in coal consumption the life of the known stock is reduced to 111 years – a substantial drop from 2300 years.

Table 1.7 Stock estimates for some elective resources and number of years that they will last with constant and growing demand

Resource	Known global reserves	Number of years resources will last with constant demand (static index)	Projected average annual rate of growth of consumption (%)	Number of years resource will last with growing demand (dynamic index)
Coal	5 x 10^{12} tons	2300	4.1	111
Natural gas	1.14 x 10^{15} ft^3	38	4.7	22
Petroleum	455 x 10^9 b.bls[a]	31	3.9	20
Aluminium	1.17 x 10^9 tons	100	6.4	31
Copper	308 x 10^6 tons	36	4.6	21
Iron	10 x 10^{10} tons	240	1.8	93
Lead	91 x 10^6 tons	26	2.0	21
Manganese	8 x 10^8 tons	97	2.9	46
Mercury	3.34 x 10^6 flasks	13	2.6	13
Nickel	147 x 10^9 lbs	150	3.4	53
Tin	4.3 x 10^6 lg.tons[b]	17	1.1	15
Tungsten	2.9 x 10^9 lbs	40	2.5	28
Zinc	123 x 10^6 tons	23	2.9	18

Source: Bureau of Mines (1970).
[a] Billion barrels.
[b] 1 long ton = 2240 lbs.

The formula used to obtain the dynamic index (the number of years a resource will last with growing consumption) is:

$$\text{Dynamic index} = \frac{\ln\left[(g \times s) + 1\right]}{g}$$

where g is the annual growth rate of consumption (i.e. 4.1% for coal) and s is the static index, i.e. number of years that the resource will last with constant demand at 1970 level (i.e. 2300 years for coal).

Example
On the basis of a 4.6% annual growth rate calculate the dynamic index for copper deposits.

$$\text{Dynamic index} = \frac{\ln\left[(0.046 \times 36) + 1\right]}{0.046}$$

$$= \ln(2.656)/0.046$$

$$= \frac{0.997}{0.046} = 21$$

1.9   SPECULATIVE ESTIMATES

Some resource economists (e.g. Rajaraman, 1976; McKelvey, 1972) argue that, in a stock's life, a calculation estimate based upon the known global reserves is too conservative, even inaccurate. A realistic estimate must include not only the known stocks but a 'guesstimate' about the quantities which are not yet decisively proven. In other words, the figures presented in Table 1.7 grossly underestimate the reality as they are based only on the proven stocks. Calculations of total reserves are normally made based on sample drill holes on the resource base. However, a sample can give a very biased picture of the total. There are numerous examples of mines started up on the basis of sample estimates that proved to be gross underestimates of the actual deposit. Furthermore, most total reserves are limited by current profitability, which takes into account the present state of technology, extraction cost, prices and political factors. Some scientists argue that we will never physically run out of mineral resources because the limits of materials in the ground are far beyond the likely economic limits of their utilization (Zwartendyk, 1972). Many metal deposits exist in almost all depths of the globe. At what point should we stop counting?

Hartwick and Olewiler (1986) report on a number of suggested cut-off points, but no system has been agreed upon. One measure is to count minerals that can be extracted without crossing an energy barrier, that is minerals are counted as long as extraction does not use up excessive amounts of energy (where 'excessive' is not defined). This problem is not acute in the case of fluids, such as petroleum, because there is a maximum depth in the earth at which they are found.

The stand taken by the 'absolutist' scientists can be described as a physical measure of the maximum stock of any mineral. But the question here is when will the potential stocks, which may be very large in quantity, ever become actual? Surely the economic measure of a mineral stock must be between the two extremes, i.e. proven and ultimate stocks. It is also worth remembering that over the last 50 years or so many new mineral and fossil fuel deposits have been discovered all over the world which enhanced the 'proven' statistics. Although geologists may have a notion that more deposits should exist they still have to be discovered if a sensible estimate of the mineral and fossil fuel stocks is to be made.

One simple criterion is the inferred estimates which include speculative, or hypothetical, quantities that may exist in unexplored areas of the world. Suppose that one-half of the African continent was explored for oil and x billion barrels were discovered. The advocates of this method believe that in the remaining half there should be another x billion barrels, making the total $2x$. That is to say that unexplored areas of the world contain just as much minerals and fossil fuels as the explored regions. Table 1.8 shows results based on inferred reserves for four deposits and compares them with the stock lives based on the proven reserves.

1.10   SPACESHIP EARTH

The father of the 'spaceship earth' idea, Kenneth Boulding, argues that anyone who believes in exponential growth that can go on forever in a finite world is either mad or an economist. He recommends with some urgency that the time has come to move from a throughput economy to the notion of spaceship earth (Boulding, 1966, 1970). Conventional economic thinking envisages a throughput system in which economic activity moves from extraction of natural resources to the rubbish dump and the ultimate physical product turns out to be the waste. Sooner, rather than later, this process is going to come to an end.

Table 1.8   Stock estimates on the basis of proven and inferred resources

	Stock estimates (million tons)		Average annual growth in demand (%)	Static index (years)		Dynamic index (years)	
Resource	Known	Inferred		Known	Inferred	Known	Inferred
Copper	370	1700	3.7	54	215	29	60
Lead	144	1854	2.2	38	489	27	112
Tin	4.7	41	1.0	18	158	16	95
Zinc	123	5606	2.9	23	1038	18	118

Source: Rajaraman (1976).

Up until now this theory led the economist to behave as if the earth were flat, a great plain, where there is always some new space to move to. However, the photographs of the earth from space showed that our world is really a small spaceship and there is nowhere else to go; we can only move inside the spaceship. The change from great plains psychology to spaceship is essential in economics; we must transfer our system from an endless throughput system to a sustainable state.

Introductory lectures in economics contain an economic system, the circular flow of economic activity, in which natural resource and waste problems are mentioned only in passing. In this it is largely assumed that wastes that result from production and consumption activities will be recycled by nature to be returned to the land, the natural and indestructible powers of the soil. Furthermore, introductory circular flow models assume either unlimited natural resources or an infinitely flexible and adjusting ecological system. This type of introduction to economics may create the impression in the minds of young students that they operate on a great economic plain with boundless opportunities for growth. It is a little like the early settlers in America arriving from Europe on the east coast and moving to the west for further and further settlements in the belief that America is infinitely accommodating. Eventually belief in the endless plain gets dented when settlers arrive on the shores of the Pacific Ocean. Boulding's message may not have to be taken to be that we have already arrived at the Pacific coastline, but rather that we can feel the air change as we move westwards towards the ocean.

Figure 1.3 adds natural resource scarcity and waste removal into the traditional circular flow of economic activity revolving around households and firms. Households supply factors of production to the firms where both extract from the natural resource sector. Likewise, both consumption and production create waste. In a throughput economy wastes accumulate and natural resources decline continuously as the economic activity is maintained. As the levels of economic activity and population keep on growing, which were highly conspicuous at the time when Boulding formed his ideas, both scarcity and waste problems will get worse. In a spaceship context this situation cannot continue as wastes must be properly managed and the natural resource base must be protected. In this respect there is a link, recycling, between the waste and the natural resource sectors. Our long-term survival depends upon the management of this link.

It has to be pointed out that the spaceship earth concept somewhat exaggerates problems. First, given sufficient time most wastes degrade, i.e. they are effectively recycled by nature. Of course, this does not apply to non-degradable material such as arsenic and some nuclear wastes. If the rate of increase in degradable waste is higher than the absorbing capacity of nature this could damage the natural resource base. For example, a high concentration of ultimately degradable waste, say sewage, can damage fish stocks in seas and lakes. Second, the diagram implies that production leads either to consumption or

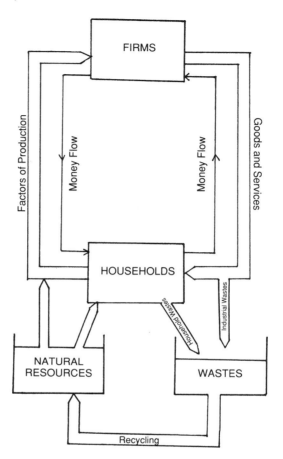

Figure 1.3   Circular flow of economic activity in spaceship earth.

waste whereas an important outcome of the economic system is capital. That is, we convert the iron ore in the ground into a durable hammer, steel girder, railway truck, bridge or whatever, which remains intact for a long period of time.

1.11   THE CLUB OF ROME

In April 1968, a group called the Club of Rome consisting of 30 individuals from ten countries – economists, natural scientists, mathematicians, businessmen, educators, etc. – gathered in Rome under the auspices of Dr Aurelio Peccei, who was one of the top managers of the Fiat and Olivetti companies, to discuss problems facing humanity, present and future.

The issues they aimed to discuss were very broad, and included population growth, unemployment, poverty, pollution, urban congestion, alienation of youth, inflation, rejection of traditional values and loss of faith in institutions. They viewed all these as contemporary human problems which occur to some degree in all societies, advanced as well as developing. These problems contain technical, social, economic and political dimensions which interact. In their view modern humanity, despite its knowledge and skills, fails to understand the origins of its problems and thus is unable to find an effective response. This failure occurs largely because it examines a single problem in isolation.

Phase one of the Club's project took definite shape in 1970 at meetings in Switzerland and at the Massachusetts Institute of Technology (MIT), where J. Forrester presented a global model containing most of the problems mentioned above. Later, with financial support provided by the Volkswagen Foundation, an international team examined five basic factors: population, natural resources, agriculture, industrial development and pollution. In 1972 the team published their report in a book *The Limits to Growth* which made front page news in many respectable newspapers around the world (Meadows *et al.*, 1972). This text attempts to illustrate that economic growth, with or without a growing population, is not only of questionable benefit but also potentially harmful and even disastrous.

Their basic argument is that there must be limits to exponentially growing economic activity, population and pollution simply because the world has finite arable land, energy resources, mineral deposits and pollution-carrying capacity. The global computer models constructed by the Club contain three groups of variables. First, absolute levels which relate to population, capital, non-renewable resources, land (divided into industry, agriculture and services) and population (divided into various age groups). The second group comprises changes in the levels which are normally measured in terms of growth rates, and the third group comprises auxiliary variables such as industrial production, food availability, effect of pollution on life expectancy, food production and pollution absorption time. The interactions between these three groups of variables are calculated by mathematical correlations. For a further explanation see Hueting (1980).

All the computer models contain eight explicit variables: population levels, industrial output, pollution, non-renewable resource stocks, services, per capita food availability, rates of birth and death. Among these the non-renewable resource levels always have a negative growth rate, i.e. they deplete all the time. The other levels may have positive or negative growth rates depending on a number of events taking place in world systems. There are two distinct phases, history and future, and the former describes the trend in the eight variables between 1900 and 1970.

The Club of Rome runs 14 models under various assumptions. In their first model, called the standard run (Figure 1.4), the proven stocks of natural resources are taken to be the major constraint on economic expansion. It is assumed that there will be no change in the established behaviour pattern of these variables, i.e. the past events will continue as usual. The horizontal axis shows a time scale between 1900 and 2100. Between 1900 and 1970 all curves

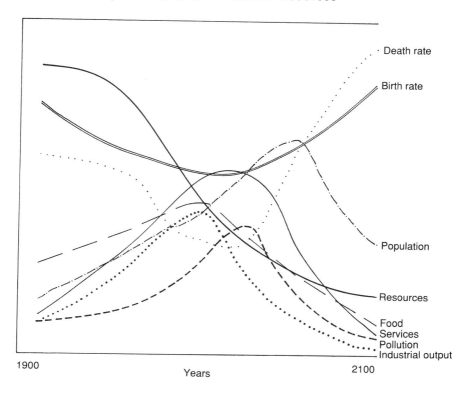

Figure 1.4 World model – standard run.

show the historic trends. Population rises from 1.6 billion in 1900 to 3.5 billion in 1970. Likewise, industrial output, food availability and services rise, resource levels decline and so on.

The chaos begins in the early part of the twenty-first century when the food curve falls below the population variable. Due to lack of non-renewable resources it is not possible to maintain the ageing agricultural machinery, nor is it possible to sustain the production of fertilizers. Malnutrition sets in, the death rate rises, the population begins to shrink and finally pollution recedes due to much diminished industrial activity. It is pointed out in the report that the exact timing of these events is not very important given the great aggregation and many uncertainties in the model. It is significant, however, that growth is stopped well before the mid twenty-first century. The Club claims that every doubtful case it was tried to make the most optimistic estimate of unknown quantities, and also ignored discontinuous events such as wars or epidemics, which might act to bring an end to growth even sooner than the model would indicate.

In the next model all the assumptions of the standard run are maintained, but the natural resource stocks are doubled, which only postpones the disaster. The reason for the end is pollution which halts the growth of food supplies as it is not possible to sustain agriculture in a contaminated world. In a series of models various other assumptions such as unlimited resources, unlimited resources plus pollution control, unlimited resources plus pollution and birth control, were used. Unfortunately, none of these prevent the collapse.

The Club demonstrates that stability will only be reached after all the key variables are kept under control. The particular recommendations are that: population must be stabilized by setting birth and death rates equal; effective antipollution measures must be implemented; production of industrial commodities must be reduced in order to avoid a non-renewable resource famine; efforts must be increased to expand the supply of services such as education and health; and great emphasis must be placed on food production by diverting capital to agriculture.

1.11.1   A criticism of *The Limits to Growth*

There are many serious problems with the analyses carried out by the Club of Rome and the conclusions drawn from them. All computer models treat the world as a single entity without geographical subdivision. The fact of the matter is that our world is a very heterogeneous place. A population-led collapse may occur in the hard-pressed and extremely crowded regions of the world, such as the Indian subcontinent, i.e. India, Pakistan and Bangladesh, but it need not happen worldwide nor as suddenly as the Club's models suggest. Population in most parts of Europe is now stable and great efforts are being made to minimize its growth in China. Most areas where population is growing, e.g. Africa, North and South America, are still sparsely populated regions of the world.

In *The Limits to Growth* no distinction was drawn between various types of destructible resources – they were all assumed to disappear over time. The only non-renewable resources, in the real sense, are fossil fuels and a few other materials such as potash and phosphate. All metal deposits are recyclable. An effective recycling policy would mean that, at least theoretically, the world would never run out of metal stocks. Furthermore, in most models 'proven' estimates were used which, as mentioned above, is a very conservative or even an inaccurate measure for scarcity. When 'speculative' estimates, as opposed to 'proven' figures, are taken into account, this modifies the results in most of the Club's models.

A number of world models fail because of widespread pollution, which stems from growing industrial activity. There are examples of countries such as Switzerland, Austria and Singapore, where industrial development has taken place with minimum environmental problems. It could be argued that if pollution becomes a real threat many countries will take steps to curb it. There are some signs that this is already happening. Green parties in many parts of Europe are making people and governments aware of environmental problems.

Another major problem in *The Limits of Growth* is that it ignores the possibility of a technological breakthrough in either energy or manufacturing/ agricultural sectors. History tells us that human development depends on innovations. The discovery of electricity and the invention of combustion engines, telecommunication methods, computers and micro chips initiated growth and changed our lives. A new technology may minimize the reliance on fossil fuels for energy by opening the way to exploit resources such as tidal waves, solar power and wind which are in endless supply. In the past, charcoal was replaced successively by coke, oil and now, to a limited extent, by nuclear energy. Technology may also help to minimize the depletion rates for metal deposits. For example, using thinner steel in manufacturing will certainly prolong the life of iron ore.

Last, but not least, *The Limits to Growth* ignores the role of the price mechanism in moderating, and even solving problems created by shortages. What will scarcity do to prices? This question did not seem to worry the creators of the world models. In effect, the price mechanism will have two powerful effects. First, high prices will encourage conservation and the use of substitutes, and secondly, high prices will encourage research and the resulting technological progress will take us away from commodities which are becoming scarce.

To give an historic example, in ninth-century Europe, whale oil became widely used in place of wood for light because of its clean-burning properties. Whaling expanded quite considerably to meet the increased demand. Increased pressure on the whale population in traditional hunting areas forced them to migrate towards the Arctic, which made capture more costly. Improvements in navigation and shipbuilding helped to overcome these difficulties to a large extent. The Indian Ocean became another hunting area for the industry. However, by the eighteenth century, the world's whale population was badly depleted and the industry confined most of its activities to the north-eastern United States. Due to scarcity, the price of whale oil increased by more than 400% between 1820 and 1860 and as a result consumers looked for substitutes. Howe (1979) points out that gas lighting became widespread in many European cities and drilling for oil intensified as a source of another substitute. By 1863 there were about 300 experimental refineries in the United States producing kerosene. Five years later kerosene had almost replaced whale oil and in 1870 its price struck an historic low.

A more recent example of the impact of high prices on demand is crude oil. Until the first oil shock of 1973, demand for oil in many industrialized countries was growing by about 10% per annum. The 1973 price increase, which was more than threefold, reduced the demand by more than 10% in many countries within one year of its occurrence.

However, despite all its shortcomings *The Limits to Growth* is one of the most remarkable documents published in the field of natural resources and the environment in recent years. In many ways it is influenced, substantially, by the views of early writers such as Malthus, Ricardo and Jevons. Global pollution is a recent phenomenon which does not appear in the works of classical writers but it is highly conspicuous in the Club of Rome's report. The most positive aspect of *The Limits to Growth* is that it intensified debates on issues regarding resource depletion and environmental degradation which are still continuing to this day.

The second study by the Club of Rome (Mesarovic and Pestel, 1974) was a considerable advancement on the first report on two grounds. First, in this study the world was divided into ten homogeneous regions. Second, the mathematical relationships between aggregate growth and ancillary variables were much more sophisticated than in the first report. The study concluded that a regional collapse in the food-deficit and crowded regions was more likely than a global one. It also reported on the dangers of nuclear proliferation and the widening gap between rich and poor countries. Unfortunately, unlike *The Limits to Growth*, the report by Mesarovic and Pestel failed to evoke a forceful public response.

At their 1982 meeting in Philadelphia the Club of Rome's consensus appeared to reject the notion of physical limits to growth. Instead, they argued that more attention should be given to the direction of growth, especially towards the boosting of the service sector and the problems confronting the poor countries of the world.

1.12 SUSTAINABLE DEVELOPMENT DEBATE

Behind the concept of sustainable development there is the notion that the environment is critical natural capital, essential for both direct consumption, e.g. breathing of clean air, and maintenance of flow of production. Therefore, damage to the environment can be seen as running down capital which will reduce the quality and quantity of its recurrent services. That is, breathing contaminated air or using dirty water diminishes human well-being; contaminated soil reduces agricultural yields; depleted fish stocks contract employment and incomes in the fishing industry as well as reducing consumption opportunities for fish eaters. Economists, along with others, have been concerned for some time to identify a level of environmental use that is consistent with preserving natural capital.

Some writers, such as Smith (1993), maintain that sustainability is a recent concept which originated at the 1980 World Conservation Strategy of International Union for the Conservation of Natural Resources. This body contended that sustainability is a strategic concept involving the lasting utilization of natural resources, the preservation of genetic diversity and ecosystem maintenance. Three years later, the United Nations established the World Commission on Environment and Development to formulate a global agenda for change. The Commission, headed by the former Norwegian Prime Minister, Harlem Brundtland, published its final report, *Our Common Future*, in 1987. This document, also known as the Brundtland Report (1987), defined sustainable development as 'development that meets the needs of the present without compromising the ability of future generations to meet their own needs'.

Pearce *et al.* (1990) describe how the publication of the Brundtland Report made sustainable development a popular concept among a wide range of disciplines and paved the way to many other definitions. I should like to point out that although it may have become a fashionable expression in some circles only recently, sustainable development is not a new concept. For example, in forestry the sustainable management concept has been known to forest biologists, engineers and economists since the work of Martin Faustmann in 1849. This work, which will be described in Chapter 3, suggests that a plantation forest should be maintained indefinitely by replanting after every harvest. In the management of fisheries, sustainability has been a buzzword amongst fishery economists since the work of Gordon (1954), Scott (1955) and Scheafer (1957) (Chapter 2), and in agriculture, writings on sustainable farming can be traced back to the eighteenth century (Eliot, 1760; also see Young, 1804). Mainstream economics is full of assertions linked to sustainability. For example, Hicks (1968), a quarter of a century ago, argued that we ought to define a person's income as the maximum amount he can spend during the week and still expect to be as well off at the end of the week. In this Hicks was clearly referring to sustainable consumption as opposed to reckless activity such as selling off the family silver for a week's consumption.

As pointed out by Winpenny (1991), it seems that a satisfactory definition of sustainable development has become the Holy Grail of environmental economics; Pezzey (1989), for example, suggests 60 definitions, whereas Pearce *et al.* (1989) put forward about 30. Many definitions of sustainable development are quite similar to the one by Hicks. According to Pearce (1991) 'sustainable development is readily interpretable as non-declining human welfare over time – that is, a development path that makes people better off today, makes people tomorrow have a lower "standard of living" is not sustainable'.

Perhaps the most relevant definitions of sustainable development are those that specify intergenerational aspects like the one suggested by the Brundtland Report which has already been mentioned. Some other examples are:

- 'According to the sustainability principle all resources should be used in a manner which respects the needs of future generations' (Tietenberg, 1992).
- 'Sustainable development is that which leaves our total patrimony, including natural environmental assets, intact over a particular period. We should bequeath to future generations the same capital, embodying opportunities for potential welfare, that we currently enjoy' (Winpenny, 1991).
- 'Sustainable development is a development strategy that manages all assets, natural resources, and human resources, as well as financial and physical assets, for increasing long-term wealth and well-being. Sustainable development as a goal rejects policies and practices that support current living standards by depleting the productive base, including natural resources and leaves future generations with poorer prospects and greater risk than our own' (Repetto, 1986).

- [Sustainability in the purest sense involves] 'embracing ethical norms pertaining to the survival of living matter, to the rights of future generations and to institutions responsible for ensuring that such rights are fully taken into account in policies and actions' (O'Riordan, 1988).

These definitions are interesting in the sense that they bring ethics and moral judgements back into economics. Economic analysis based upon classical utilitarianism largely denies that current generations have moral obligations to future individuals (Turner, 1991). In utilitarianism a rule is deemed to be desirable if it maximizes total utility even if there are some losers in the process. Especially an act utilitarian, who is the most inflexible believer of classical utilitarianism, would be bound to accept as right the action of framing and consequently hanging an innocent person if his/her death would prevent, say, a social disturbance which could take the lives of a number of individuals (Shradder-Frechette, 1991). In other words, in act utilitarianism dignity, the liberty and life of a single person, or a number of persons, is expendable if this is in the interests of a greater number in the community. A rule utilitarian, on the other hand, allows rules of equity into utilitarian doctrines, and would be unlikely to accept such a situation. A good number of utilitarian economists tend to subscribe to rule utilitarianism. If an economic investment decision is expected to enhance the well-being of a large number in the community but would make a few worse off, then the most widely accepted economic view would be that the gainers, at least in theory, should pay compensation to the losers.

When intergenerational environmental decisions are made at project level, for example, plantation of a forest, halting of slash-and-burn projects in the tropics, expansion of nuclear power which will create highly toxic and long-lived nuclear wastes, etc., normally, cost–benefit analysis is carried out to measure their economic worth to society. Some economists do not appear to be too keen to consider moral issues at this level as they see cost–benefit analyses as mere preference counters. For example, Pearce (1983) argues that

> As a procedure for aggregating the preferences of our set of individuals, we establish something of a fundamental importance at the outset: CBA [cost–benefit analysis] makes no claim to produce morally correct decisions. What CBA produces and what is morally correct may coincide if, and only if, we adopt a further rule, namely, that some aggregated set of preferences of individuals is the morally correct way of making decisions.
>
> (Pearce, 1988, p. 3).

Elsewhere, Pearce, with others, defines sustainable development as a vector of desirable social objectives such as an increase in real income per capita; an improvement in health and nutrition; educational achievement; access to resources; a fairer distribution of income; and increase in basic freedoms. Then he states that 'the elements to be included in the vector are open to ethical debate' (Pearce *et al.*, 1990, p. 3). In other words, sustainable development is open to ethical debate, whereas in cost–benefit analysis where actual decisions are made at micro level, ethical questions are swept under the carpet. Turner (1991) demonstrates that cost–benefit analysis is an integral part of sustainable resource management. Clearly, contradictions and confusions exist.

Perhaps with the help of rational and open debate inconsistencies will be eliminated in due course in debates on sustainable development. Alas, not everybody is optimistic about the future prospects. For example, Brown *et al.* (1987) argue that the sustainable development is already beginning to become a transcendent term. Others, such as Shearman (1990), point out that instead of focusing on the precise definition we need to be more concerned with the implication for any given context to which sustainability is applied. The concept is used as a modifier in development, growth, ecosystems, etc. and it is more important to understand the meaning of the term within the context in which it is employed. Most economists relate growth to increase in wealth of the nation but this process could cause suffering to the majority (Daly, 1990; Douthwaite, 1992). As Smith (1993) asserts, growth like anything else can cost more than it is worth at the margin. In particular, present patterns of resource distribution in a growing world economy can create more hardship for the masses by speeding up the process of environmental impoverishment.

Frank (1966) contends that the 'development of underdevelopment' has been underway for quite a while, and is a process bound to hurt future generations most. According to Clark (1986) humanity is entering an era of extremely serious and complex environmental problems. Relative to earlier ones, emerging environmental problems are characterized by profound scientific ignorance, enormous blunders in decision-making and institutional failures. Some of these problems, such as acid rain, the greenhouse effect, ozone depletion, tropical forest destruction and some nuclear contamination are irreversible. Many of these problems are linked, making individual *ad hoc* problem-solving insufficient (Smith, 1993). They are a function of social and political factors which render purely technical solutions inadequate.

One requirement in achieving sustainable development is that future generations should be compensated for damage caused by our present-day activities. According to Pearce *et al.* (1989) this is best secured by leaving the next generation a stock of capital no less than the stock we have now, which would enable them to achieve the same level of human welfare as us. The stock of capital includes both artificial and natural assets. If, for example, a part of the natural capital such as tropical forest runs down for agricultural expansion, then proceeds from this activity should be reinvested to build up some other form of capital (see also Pearce *et al.*, 1990). Since some natural resources, such as fossil fuel deposits, tropical forests, etc. are non-renewable, any positive rate of use by current generations reduces the stock available to the next generation. When running down of these assets is accompanied by building up of other forms of capital we would be maintaining a 'constant overall stock'.

This is not a convincing line of thought. First, it is doubtful whether increased agricultural space could be an acceptable compensation for the loss of tropical forests and the extinction of species. Second, even if the 'stock of capital' includes the technological know-how to maintain/enhance the levels of production, loss of species could undermine the very process of its creation. Third, the advocates of constant capital rule contend that what is at issue is not the physical stock of capital but its value. To assess this, environmental assets must be valued in the same way as humanmade assets. However, it is extremely difficult if not impossible to value the entire stock of environmental attributes at national level, let alone globally. Nevertheless, it is certain that with a growing population on the one hand and dwindling natural resources on the other hand, future generations will be worse off in terms of 'per capita physical'. It may be true that due to scarcity their prices will be higher in the future and in this way their overall value could be maintained, but this may not be a great comfort to future generations who may have to rely on physical availability for their survival. One further point is that it has been argued that since future generations are not here with us, we cannot know their values and preferences when we make intergenerational decisions. This is a defeatest if not dishonest argument to justify any reckless activity. As will be demonstrated in the final chapter, it is possible in Rawlsian framework to ascertain basic preferences for all generations in a fair and reliable manner.

Instead of conventional GNP figures, the use of some form of welfare index such as the measurement of economic welfare (MEW) suggested by Nordhaus and Tobin (1972), or the variant of it proposed by Daly and Cobb (1989) (pp. 10–11), would be useful in debates on sustainable development. Indeed, conventional national income statistics do not capture the adverse effects of pollution, noise and congestion on human well-being. Furthermore, growth measured by changes in GNP fails to allow for the depletion of natural capital (comparable to the running down of conventional capital such as buildings, machinery, tools, etc.). If a nation's fish stocks, fossil fuel deposits, natural forests, farm fertility, etc. are constantly eroding as the economy 'grows', resulting GNP figures will give entirely misleading signals to policy-makers. A nation cannot sustain the level of her economic expansion with dwindling natural capital. When we allow for these factors the resulting development figures could look very different from conventional ones.

Recently, some resource economists such as Repetto *et al.* (1989) have modified Indonesia's national income statistics by allowing for the depletion of her natural capital of oil deposits, forest stocks, hill farming and the like. This had the effect of lowering the national income by 17% in 1984. Furthermore, the real rate of economic growth turned out to be 4% per annum rather than 7% GNP based on analysis for 1971–84. There is indeed a commonsense case for modification of the way in which national income accounts are calculated but a number of fundamental questions would have to be considered first. Should we completely overturn the existing system of national accounts worldwide or should we keep them as they are and provide supplements? Should we consider the sale of metal and fossil fuel deposits as income or asset sales? Finally, where do we draw the line between defensive and other forms of expenditure? The former includes items such as the police force, national defence, motorway shielding against vehicle noise, human costs arising from stress created by modern industrial life, alcohol and drug abuse, etc.

1.13   ON THE BRIGHTEST SIDE

Maddox (1972) and Beckermann(1974), unimpressed by *The Limits to Growth*, argue that the future is uncertain and cannot be predicted by looking at past events. In other words, past and future are not samples drawn from the same population of events. This, indeed, is a sound criticism which hits at the heart of the Club of Rome's first study in which history was used to predict the future. Furthermore, it can be argued that economic growth creates technical and financial resources for remedying most problems. In other words, the problem may become the solution.

If *The Limits to Growth* is a super-pessimistic document then Herman Kahn's *The Next 200 Years* in which the idea of physical limits to growth is rejected must be described as a super-optimistic publication (Kahn, 1976). His view is that although humanity has problems in the short run, the long-term future is glowing with promise. The world is just passing through the point of most rapid population growth which will stop within the coming two centuries at around the 15 billion mark, three times the present level. The major reason for this is growing affluence and education levels which will bring the birth rate down.

Kahn argues that, even with today's level of agricultural technology it is possible to feed a global population of 15 billion. Yields of rice in India, for example, could be greatly expanded with known methods of water control, fertilization and cultivation. There are untapped food resources such as krill which is abundant in the South Atlantic. Discoveries of miracle grains and pesticides are all genuine possibilities. The only danger to food production would be a chain of weather disasters taking place in several important food-producing countries.

As for the raw materials, Kahn believes that there is no early or even remote danger of depletion of resources apart from the fossil fuels. Soon we shall see the end of the petroleum age. Alternatives to oil and gas do exist and one or several of them will be developed. Coal and solar energy, for example, could become the principal energy source in the United States of America.

Regional pollution is not a huge problem as it is possible to eliminate it at a cost of not more than 2% of the GNP of the rich countries. As for global pollution, such as the greenhouse effect and acid rain, Kahn acknowledges that problems do exist, but points out that the scientific view on these issues is neither convincing nor conclusive. The need for research and monitoring is essential and when we fully understand problems then solutions will be implemented.

All in all, Khan concludes that environmental and resource problems are manageable and there is no need for a defeatist attitude. At present the world is simply going through a transition period which is painful, but the future is something to be envied not a monster to be feared.

In challenging a rather pessimistic document, the Global 2000 Report to the President, Simon (1984) argues that if present trends continue, the world in 2000 will be less crowded (though more populated), less polluted, more ecologically stable and less vulnerable to resource–supply disruption than the one we live in now. Furthermore, people will be richer in general, and stresses involving resources and the environment will be less than now.

The main evidence for such optimism is as follows:

1. Life expectancy has been rising fast throughout the world which is a sign of scientific, demographic and economic success.
2. The birth rate in developing countries has been falling during the last two decades.
3. Food supply has been increasing steadily for many decades.
4. Fish catches, after a decline, have resumed the upward trend.
5. Deforestation is troubling only in some parts of the world such as the tropics. In the rest of the world there is no alarming trend.
6. The climate does not show signs of unusual and threatening changes.
7. Mineral resources, including oil, are becoming cheaper.
8. Threats of air and water pollution have been exaggerated.

Simon does not suggest that all will be rosy in the future but he maintains that in general trends are improving rather than deteriorating.

1.14 PESSIMISTS AND OPTIMISTS

Looking at the history of economic thought on environment and resources it is difficult to identify a definite trend of ideas. Issues raised by Malthus and Ricardo seem to persist, in some modified form, in the writings of a number of contemporary economists. Others, however, maintain that almost 200 years after publication, it can be said that the prophecies of Malthus and Ricardo did not materialize. On the contrary, life in many parts of the contemporary world is infinitely better than it was 200 years ago. If one was to act on the basis of historic evidence then it may be possible to show that the human situation is likely to improve, with some ups and downs, in the future rather than deteriorate. The major objection to this argument would, of course, be that past is past and it cannot be used to predict future events. Resource and environmental problems are genuine and they will not go away unless humanity acts sensibly and without delay.

Most economists, past as well as present, can be put into two distinct groups – the pessimists and the optimists – and the conflict between the two will no doubt continue indefinitely. It is possible to discover common grounds within each school. Take the pessimists first. They tend to emphasize the central role of population growth in aggravating resource and environmental crises. This problem frightens even some optimists but the degree of fear among the former is overwhelming. Second, the pessimist school put a great deal of emphasis on the finiteness of world's resources. Given this fact, economic expansion cannot be sustained indefinitely and sooner or later it is bound to come to a halt. When the process of slow down bites, it may generate conflicts between nations competing for critical commodities and this may endanger world peace. Third, pessimists are always suspicious about advancing technology, pointing out that modern technology is imposing incalculable risks on humanity at every stage of its development. Thalidomide, DDT and accidents in the nuclear industry are some examples for this. Finally, pessimists blame the materialistic and expansionist ethics which are inherent in capitalist as well as in communist systems. In the former, the emphasis is on conspicuous and wasteful consumption to sustain the profit motive and hence economic growth. In the latter, conspicuous production is actively encouraged at every level.

As for the optimists, they point out that many stock estimates used by pessimists in their analysis are too conservative and even inaccurate. Proven stocks constitute only a small part of the world's mineral and fossil fuel wealth. Furthermore, optimists stress the importance of substitutes in stock use. Most natural resource-based commodities have substitutes; if fossil fuel runs out, then nuclear or even solar energy may take its place and so on. Second, technology is not a villain but a saviour in many ways. Extraction from less easily accessible deposits becomes a possibility with technological advancement. North Sea oil is a good example. Technology also helps to economize on existing stocks. Using thinner but equally durable steel in the manufacturing sector will make the resource base last longer. Advancing technology can also improve agriculture. New irrigation methods, cloud seeding, discovery of miracle pesticide and grain are all possibilities. And finally, the role of the price mechanism is always emphasized in optimistic arguments. If the supply of a particular raw material contracts its price will increase, curbing consumption in related sectors. A new equilibrium price that corresponds to the altered demand and supply conditions will be quickly reached. Furthermore, high prices will encourage research to discover cheaper substitutes and technological developments will take place in the right direction.

1.15   RELIGION, ETHICS AND THE ENVIRONMENT

According to some, the root of the resource and environmental crisis lies in the teaching of Western religions, in particular Jewish/Christian theology (White, 1967). The Old Testament states that 'Then God said, let us make man in our image, after our likeness; and let them have dominion over the fish of the sea, and over the birds of the air, and over the cattle, and over all the earth, and over every creeping thing that creeps upon the earth' (Gen. 1: 26). This gives a powerful message that humanity is given dominion over all forms of life and given permission to exploit them for its own ends. God made humans the dominant species on earth and everything was created for them to use.

On the population problem most religions do not favour birth control. In particular, the Roman Catholic Church has been criticized as being a major contributor to the population problem, especially in devout Catholic countries. The objections of the Catholic clergy to any form of birth control except the so-called rhythm method in reducing birth rates in developing countries has been branded by some as irresponsible.

Islam, which is a close cousin of Christianity, also comes under attack from time to time for its fatalistic outlook to life and its belief that all is predetermined and all is the will of God. If a person, or a nation, is possessor of a non-renewable resource, say oil, this is God's gift to the owner who has the right to exploit it to the full. If, on the other hand, a person or nation is not endowed with much natural resources then this must be its faith, and must be accepted as God's making.

Such criticism is often unfair. It is true that in Judaeo-Christian and Muslim theologies humanity is the central point in life on earth and is given many gifts of nature and has the right to use them. However, right without responsibility does not make a person a good follower in any faith. Reckless depletion of nature's gifts or misuse of the environment harms others; I know of no established religion that condones injury to others.

As for the population problem, the anti-abortion and anti-contraceptive views of the Roman Catholic Church are well known. However, many other faiths also frown upon these acts although they may not make their position known as strongly. High birth rates are not only a feature of the developing Catholic countries; they are also prevalent in Muslim, Buddhist and Hindu nations.

White (1967) contends that scientific development cannot get us out of the environmental crisis; we have to find a new faith or rethink our existing ones. Sylvan and Bennett (1988) argue that in Aristotelian thought the end of all things is seeking well-being by following the natural order in an active and rational manner. When the ideas of Aristotle came into Christian Europe during the twelfth century the initial reaction of the Church was negative. But in later years some Christian scholars, such as Thomas Aquinas, saw the possibility and desirability of incorporating Aristotelian ideas into Christian beliefs (Roover, 1970). In the synthesis of Christian theology and Aristotelian philosophy the society is an integrated system, like the universe, where God, nature and humanity each has its own place. The virtuous life requires a societal structure in which there is a mutual exchange of functions within a hierarchy in which priest, farmer, worker, artisan, merchant, civil servant perform their tasks in accordance with the laws of God and nature, and all are properly rewarded for their contributions.

The influence of Judaism, Christianity and Islam on the Western way of thinking around ecology, or other things for that matter, cannot be denied. Other influential views are those of Aristotle, which has already been mentioned, and those of the philosophers Hobbes and Locke who saw the state of nature as one of disorder. When there is chaos then order by political and economic means must be imposed by humanity, the central player on earth. By contrast Eastern religions take a fundamentally different view. Taoism, for example, rejects the Western model of the relationship between humans and nature. In Taoism humans do not have dominion over the earth. Furthermore, nature is not in disorder but in order. Other differences between Western and Eastern religions are that while the former see the world as value-neutral except for the human species, the latter believe that values are built into the environment, and are an integral part of the way of things.

While the Western view is one of dominance and stewardship of humanity over the environment, Taoism offers a passive, let-it-be approach to our relations with nature. Like Buddhism, Tao upholds the notion 'Do nothing and from unforced order greater order results'. In Taoism undominated things are naturally self-governing and self-creating. Furthermore, nature is not a mere instrument for other ends, i.e. a resource, but something of great value in itself. In other

Table 1.9 Dominant paradigm of modernism contrasted with deep ecology and Taoism

Western view	Deep ecology (DE)	Taoism
Domination over nature	Harmony with nature	Elaboration of DE
Nature is just a resource, an input	Nature is valued for itself	Much as for DE
Human supremacy	Biocentric egalitarianism	Wide impartiality
Ample resources/ substitutes	Supplies limited	Supplies ample
Economic growth is an important objective	Non-material goods knowledge/self-realization	Following Taoism
Consumerism	Doing with enough/ recycling	Doing with enough/ no recycling
Competitive lifestyle	Co-operative lifestyle	Much as DE
Centralized/urban/ national focus	Decentralized/bio-regional/neighbourhood focus	Much as DE
Hierarchical power structure	Non-hierarchical/ grassroots democracy	Hierarchy without power structure
High technology	Appropriate technology	Limited technology

Source: Sylvan and Bennett (1988).

words, nature is something to be cherished, to be left alone to take its own course, and it is not to be interfered with or destroyed by policies created by humans. In Eastern thought the dominant view is reversed; value for humans is achieved above all by identification with nature.

Sylvan and Bennett (1988) point out that although Taoism appears to be a deeply environmental philosophy, it seems plain that it did not face up to such environmental problems as overpopulation and incremental resource degradation. The scriptures say little about deforestation and soil erosion, which have been historic problems in China. In contrast, Plato was concerned about deforestation in Greece. Furthermore, Taoism does not offer much on contemporary problems such as extinction of species, animal welfare, urban degradation, global pollution, etc. though some of the problems are hardly new.

Many modern scholars are beginning to search for environmental ethics, some of which are secular. In the concept of deep ecology, Devall and Sessions (1985) give equal rights and protection to all species. In the Gaia hypothesis, Lovelock (1979) contends that living and non-living parts of the earth interact; the former constantly modifying the latter and therefore, living and non-living is one entity. Humanity, according to the Gaia hypothesis, wants an environment that can sustain itself without constant human intervention. There are various other ethical positions (for a survey see Collard et al., 1988). Table 1.9 compares three different philosophies: Western, Taoism and deep ecology.

2
Economics and policies in fisheries

Due to the open access problem fishermen are not wealthy, despite the fact that the fishery resources of the sea are the richest and the most indestructible known to man. By and large, the only fisherman who becomes rich is one who makes a lucky catch or one who participates in a fishery that is put under a form of social control that turns the open resource into property rights.

H.S. Gordon

2.1 PROPERTY RIGHTS

Establishing property rights on resources is crucial for their rational and efficient utilization. A resource which is owned by nobody is open to misuse. If no one owns trees, then no one will have wealth incentive to preserve them until they are mature and ready for felling. In a situation like this a 'first come first served' principle will prevail and the trees will be cut down prematurely. Whoever does it first will take the wood. Likewise, as long as no-one owns lakes and rivers then nobody will be motivated to use them in their most valuable ways. Some may dump rubbish in an unowned lake since the lost value of polluted water is not thrust upon them with sufficient rights or self-serving incentive to control pollution.

There is a consensus among economists that setting up private or public ownership on previously unowned resources will create an opportunity for their rational use. If all resources were owned by private or public bodies and property rights were fully enforceable in courts of law, then everybody would think twice before engaging in a 'use as you please' principle. For example, the owner of a lake polluted by, say, a chemical company could obtain damages, get an injunction against continuing pollution or charge the firm for the privilege of using the lake as a dump for its waste. These added costs for the chemical company would shift the costs of pollution from the owner of the lake to the firm itself.

However, economists seem to be divided on the issue of which type of ownership, private or public, is more desirable. At the beginning of the nineteenth century Arthur Young (1804) argued that the magic of private property turns sand into gold. In his travels throughout the British Isles he observed the benefits of a change from communal to private farming. Under the old system, large open fields had been farmed in common by local groups. The enclosure movement resulted in individual farms enclosed by a fence or stone wall. Only then did each farmer capture all the benefits from the hard work he put into his land. He was also free to try out new farming methods whereas before he had been compelled to follow the traditional communal farming technology. The result was a great improvement in productivity.

Many economists believe that an entrepreneur should be allowed to retain more than a small fraction of the profit he/she makes, which is a reward for skills, risk taking and mental strain created by business life, otherwise they may decide that the game is not worth the cadle. In the absence of private property rights innovation, capital accumulation and risk taking may decline. In the end, the community may end up with less wealth than it might otherwise have had under the system of private entrepreneurship and property rights. There is no point in abolishing private property rights in favour of a greater equality which will create disincentive effects on production. It may be better to have 10% of £1000 than 20% of £300. Private ownership can take many forms such as single ownership, partnership and corporations. The owners should have unrestricted rights to buy, sell or exchange their property.

At the other end of the spectrum economists who belong to the socialist school argue that public ownership of productive property, such as land and capital, is more desirable than private ownership. They believe that capitalist ideals lead the owners of property to produce results that impoverish workers. A socialist economy substitutes ownership by society as a whole, or by workers as a group, for private ownership. The fundamental idea is that an individual participates in the means of production because of his/her membership in a group rather than because of his/her legal ties to the property itself. In other words, the individual is a member of the group and the group owns the means of production. Here again, the forms of ownership may vary widely, ranging from ownership of entire industries by central government, to ownership of public utilities, such as water systems by local governments, to ownership of a factory by the workers who work there. Social ownership of productive resources distinguishes socialism most clearly from capitalism.

Today, socialists blame capitalism for most of our environmental problems such as water and air pollution, noise and urban congestion. The advocates of private property, on the other hand, argue that these problems are created because of lack of well-defined, transferable and marketable property rights in certain areas of economic activity. In other words, the failure to apply capitalism fully is the main source of environmental and resource problems. However, at this stage I am reluctant to enter into a controversy regarding the superiority of capitalism over socialism or socialism over capitalism. Instead, I shall try to demonstrate that the absence of property rights, private or public, will create a gross inefficiency and waste.

What exactly are property rights? The term property rights refers to the entire range of rules, regulations, customs and laws that define rights over appropriation, use and transfer of goods and services. One of the main problems in economics is the allocation of scarce resources between competing needs and the market mechanism is an efficient way of achieving this. However, the operation of markets depends on the existence of well-defined and easily transferable property rights. In effect, when individuals trade in the marketplace they actually trade rights to goods and services. The more clearly these rights are defined and the easier they are to transfer the more efficient the market will work. For example, when I buy a house I am actually buying the right to live in it, to let it or to sell it. A theatre ticket bought by an individual gives him/her the right to see the play at a particular time, or if he/she wishes they can transfer the right to somebody else.

We buy goods and services all the time and by doing so we are actually buying the rights to use them as we please under certain conditions. In other words, there are always legal restrictions attached to the rights to use property. For example, I cannot disturb others while living in my new house nor can I turn it into a place of unlawful activity. Likewise, the ticket holder cannot disturb others while watching the play in the theatre.

When the government regulates the market by various means such as price ceilings, subsidies and taxation it actually modifies individuals' property rights. For example, by imposing a price ceiling on goods, the government is redefining the rights which sellers have with regard to property they sell. Tariffs, subsidies, direct and indirect taxes all have similar effects. A customs duty is a price which must be paid by importers so that they can acquire the right to bring the commodity into the country. A family income supplement enhances the rights of the recipient to buy goods and services. Conversely, an income tax limits the taxpayers' entitlement for goods and services by reducing their disposable income.

2.2 COMMON ACCESS

There are cases in which property rights cannot be clearly defined on the resource base. Take, for example, a situation in which three individuals can drill from their own land into an oil deposit. In this case there would be great haste to deplete the stock, the rule for each individual being 'extract as fast and as much as you can, if you do not others will beat you to it'. Consequently the resource will be depleted very quickly. Multiple access was indeed a genuine problem in the early days of oil extraction in some parts of the world.

Open grazing lands and forests are another area of difficulty. When shepherds take their herds to an open pasture they tend to allow their animals to overgraze it. The grass which is eaten by one person's sheep cannot be eaten by another person's animal and in this way each imposes a cost on the other. Eventually the land may be stripped of its grass, which will damage all shepherds.

In the above cases the open access problem is not entirely hopeless. Take the oil example. One person can negotiate successfully the purchase of land from others and becomes the only oil producer. Alternatively owners may form a joint company to extract oil from the reserve in question. Another solution would be for the government to decide to nationalize the resource by means of a compulsory purchase, in which case the ownership will be transferred from private individuals to the public sector. Each case solves the problem of multiple access successfully.[1] As for open grazing lands and forests, again the problem can be remedied by establishing property rights on the resource base. When a farmer acquires property rights on previously open land by putting a fence around it, he would be able to exclude others from his territory.

However, there are situations in which the problem of open access is extremely difficult to solve. As its name suggests, open access resources are the ones open to many. Any individual who has the necessary skill and equipment can dip into these resources. Fishing is one of the most vulnerable natural resources to open access. In the open sea it is extremely difficult to establish property rights on the resource base. Even nationalization would fail to solve the problem for a number of reasons. First, in an ocean even if each coastal state had some exclusive zone with, say, a nationalized fishing industry, this would still leave a very large part of the open sea as common property. In 1977 there was a rush to proclaim 200-mile exclusive fishing limits by many coastal states. Even with this, rich fishing grounds in the Antarctic and in many other parts of the oceans have remained outside any nation's territory. Second, fish stocks travel from one country's exclusive zone into another country's territory, thus complicating the issue even further. The problem is no easier to solve in closed seas such as the Mediterranean or the Black Sea. It is not politically feasible to divide these seas into exclusive fishing zones between the coastal nations. For example, the Mediterranean is shared by over 20 sovereign nations. If this sea was to be divided completely into exclusive fishing zones by these nations, a practice which is highly unlikely, outsiders who have always fished in the area would strongly oppose it. Countries such as Portugal, Russia, Romania and Bulgaria do not have a coastal strip in the Mediterranean but have always fished in the area.

---

1. Despite the fact that bringing crude oil production under a single management (unitization), is the most complete solution to the common pool problem, complete field-wide unitization is not widespread in the United States. Libecap (1989) reports that in 1947 only 12% of some 3000 oil fields in the United States were fully unitized and as late as 1975 only 38% of Oklahoma production and 20% of Texas production came from field-wide units. To promote rationalization, most states have adopted compulsory unitization rules, where a majority of leaseholders on a deposit were able to force a unit. However, a majority agreement proved to be difficult until towards the end of the life of the deposit because of conflict over the distribution of revenues and costs. That is to say that by the time an agreement was reached the oil field was almost exhausted.

The ideal situation would be to establish an international authority by the co-operation of interested nations to regulate the fishing activity. This is not easy to achieve where historic rivalries exist, for example, in the Mediterranean between Greece and Turkey, Greece and Albania, Israel and Egypt. Despite all the apparent benefits, a fruitful co-operation becomes very difficult in this situation.

For all these reasons, large parts of the world's seas and oceans are exploited under the conditions of unrestricted access and as a result fish stocks are badly depleted. Once a resource is open to users without restrictions, reckless exploitation becomes inevitable. Not many fishermen in isolation would give proper consideration to breeding seasons. Not many fishermen would return a fish to the sea to grow to a larger size because they know that it will end up in somebody else's net. In the open sea fishery there is always a scramble to catch more fish, and in this situation it is the fishermen themselves who lose most.

2.3   A COMPARATIVE STATIC ECONOMIC THEORY OF FISHERY

The economics of fishing is a young and developing subject. The earliest articles to give rigorous treatment to fishing were published by Gordon (1954) and Scott (1955). Given the open access problem, governments have often been called on to regulate fishing activity. The crucial question is, of course, what should be the basis for fishery regulation? In Figure 2.1 the stock of fish in a fishery is meas-

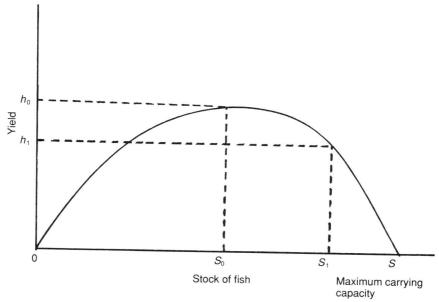

Figure 2.1   Fishery yield curve.

ured along the horizontal axis. In the absence of fishing activity, at any given stock level there will be a certain number of births and deaths and the difference between these will be the yield or the growth rate of the fishery which is measured along the vertical axis. The shape of the yield curve is damped exponentially. As the stock rises the death rate will rise relative to the birth rate due to increased competition for food and increased vulnerability to predators. Hence as stock rises, yield rises more slowly and, indeed, beyond S_0 it actually starts to fall. At S the population is stabilized; this level can be called the maximum carrying capacity. At this point the birth rate equals death rate. If it is assumed that the aquatic environment in which the fish lives and grows is not changing and in and out migration is negligible, then the size of the stock will be fixed at the maximum carrying capacity, *ad infinitum*.

Now let us introduce a new predator, fishermen. When the stock level is S_0 the fishermen's catch (measured along the vertical axis) would be h_0, the surplus of births over deaths. Of course, at this point along the curve, deaths plus harvest equals births, and the stock of fish will be maintained at S_0 At point S_1, which corresponds to harvest level h_1, again deaths including the harvest will exactly equal births, and the stock of fish will stay at S_1. Therefore, the sigmoid curve tells us that for each stock size a certain level of fish could be caught without disturbing the stock, and this situation could persist for ever.

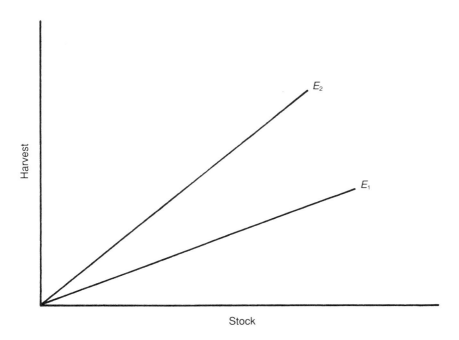

Figure 2.2 Fishery harvest levels.

Our task is to identify a sustained harvest which will give the best result. For this let us consider first the technology of fishing. In order to catch fish we need manpower, boats, gear and petrol, together called fishing effort. *Ceteris paribus*, we would expect that the more effort put in, the more fish would be caught, under the constraints imposed by the yield curve. In fishing there is one other input which affects the level of catch, that is the stock of fish. For any given level of activity, the more fish there are in the fishery, the more will be harvested (Figure 2.2). When the fishing effort is E_1 the harvest level rises along with the stock of fish. An increase in level of activity from E_1 to, say, E_2, shifts the relationship upwards, so for any given level of stock, a larger harvest results.

Figure 2.3 brings together the yield and harvest curves. If the effort level is E_1, then the steady-state stock and harvest levels will be S_1 and h_1, respectively. When the effort rises to E_2, then the steady-state stock and harvest levels shift upwards. This is repeated for large numbers of effort levels, and plotting each level of effort against the corresponding steady-state level of harvest, converts the yield/harvest curve into the yield/harvest/effort curve (Figure 2.4).

Let us finally introduce prices into our fishery model. Assume that the price of fish is given and constant in our model. This is another way of saying that our fishery is only a small part of the fish market and levels of catches here would not significantly affect the price of fish in the market. If the harvest level is multiplied by the constant price of fish then the harvest curve becomes the total rev-

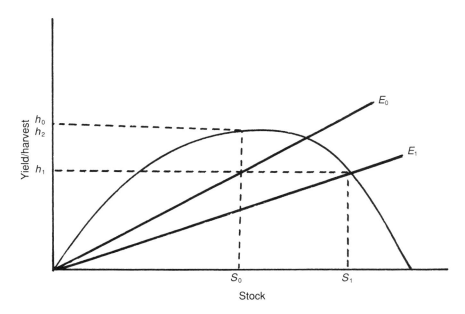

Figure 2.3   Steady-state harvest.

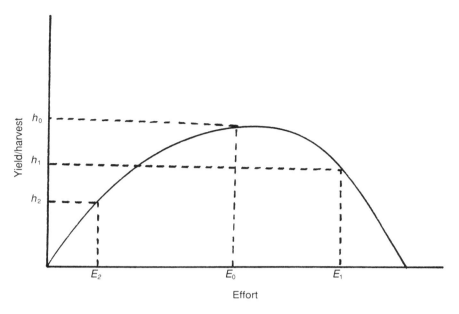

Figure 2.4   Yield/harvest/effort curve.

enue curve, i.e. $TR = (P)(h)$, where TR is total revenue, P is the price of fish, which is externally determined, and h is the level of harvest or catch.

Turning our attention to the cost of harvesting or catching fish let us assume, for the sake of simplicity, that the total cost curve is linear as shown in Figure 2.5. The cost curves include the wages of fishermen, the cost of gear, rental charge for the use of boats, fuel and reward for risk taking by the entrepreneur, which is normal profit.

2.3.1 The maximum sustainable yield

Our simple fishery model enables us to identify the optimal level of fishing activity. What is that level? The answer would depend on whether you are a marine biologist or an economist. Some marine biologists try to identify a level of fishing activity which brings the maximum catch. This point corresponds to the turning point of the harvest curve, which is now transferred into total revenue curve in Figure 2.5 by using a constant price for fish. This criterion is called the maximum sustainable yield (MSY).

Note that the MSY identifies itself in Figures 2.1, 2.3 and 2.4. It is the turning point of the curves. This gives the maximum sustained yield, harvest and revenue. What is wrong with this? The answer simply is, quite a lot. The maximum sustainable yield gives no consideration whatsoever to the cost of catching fish.

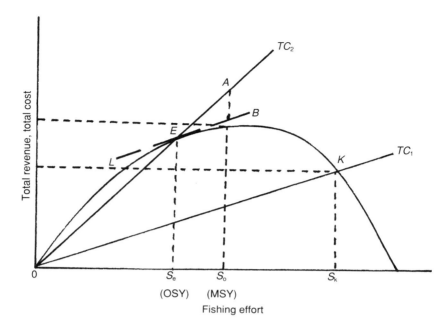

Figure 2.5   A comparative static model of fishery.

Suppose that the relevant cost curve is TC_2. Due to extremely difficult conditions – such as adverse currents, winds and cold – the cost of catching fish in this particular fishery is quite high. In this high cost situation, the cost of maintaining the level of activity at MSY is much higher than the corresponding revenue. In other words, at S_0 the total cost in the fishery is AS_0 and the total revenue BS_0. The difference between the two is net loss, AB, which comes about as a result of maintaining the level of activity at MSY. Also despite a substantial change in cost there is no change in the maximum sustainable yield as it is fixed at S_0.

2.3.2   The optimum sustainable yield

In Figure 2.5 the gap between total cost and total revenue is the economic rent which is a gift of nature to the fishermen. The economic criterion, the optimum sustainable yield (OSY) which is also called the maximum economic yield (MEY) or optimum economic yield (OEY) aims to identify the level of fishing activity which yields maximum economic rent. It corresponds to the point where the distance between total revenue and total cost curves is widest. In order to find this point the total cost curve (TC_1) is pushed towards the total revenue curve in a parallel fashion until it is tangent to it, point E. The optimal sustainable yield is in accordance with the objectives of all the other sectors in the economic system where consumers aim to maximize their utility and producers their profits. In the fishery the aim is to maximize the economic rent.

As far as economists are concerned, fishing effort on either side of the OSY is suboptimal. At OSY the slope of the total revenue curve is the same as that of the cost curve, i.e. where marginal revenue is equal to marginal cost. An increase in effort beyond OSY is undesirable because for each additional unit of effort the additional revenue generated will be smaller than the additional cost incurred. That is to say that an expansion of fishing beyond OSY will involve a marginal loss. Another way of seeing this is that any attempt to increase the harvest further must reduce the stock of fish and so imply the need for more effort and higher cost, and this is not profitable.

Restricting the fishing effort at OSY should not be confused with the restrictive practices carried out in monopoly where the monopolist, by lowering the level of output, gains excessive profit at the expense of consumers. In this analysis it is assumed that the price of fish is exogenously determined by the supply and demand conditions prevailing in the wider market, not by fishermen operating in this fishery. This is because our fishery is a small one and thus variations of output in it will have no effect on the market price. Furthermore, envisage a situation in which the fishing effort is expanded to S_k. The same revenue can be achieved at L at a much lower cost. So unlike the situation in a monopoly, expanding the level of activity would not necessarily yield a much better result for customers.

It is most important to note that the economically optimal level of fishing is below the point which is suggested by some marine biologists. In other words, here it is the economist who is arguing for a reduction in the level of fishing activity to maintain a larger stock of fish.

If the fishery is owned by a single firm, the management will want to run it in a way that maximizes the steady-state profits at OSY. The owner, whoever it is, will also take necessary steps to protect his property from poachers. The type of ownership, whether by public or private sector, is not an issue as both should be able to implement rational management practices which will ensure that overfishing does not take place.

Note that if, say, due to high cost of fuel, the total cost curve shifts upwards this will widen the gap between the OSY and the MSY. Conversely, a decline in the cost conditions will bring these criteria closer.

2.3.3 Open access in fisheries

Now suppose that anyone is free to enter the fishery and catch fish. As in the sole ownership case, it is reasonable to suppose that individual operators are interested in making maximum profits. Since there is no restriction on the number of boats that enter the fishery, we need to ask what will determine whether or not a boat will enter the industry. Of course, a boat will compare the price of fish with the average cost of catching fish and will enter as long as price at least covers average cost. In other words, the existence of economic rent, which is the area between total revenue and total cost curves in Figure 2.5, will attract more and more fishermen into the industry and the level of activity will expand. For example, with cost curve TC_1 the boats will enter until price equals average cost, or total revenue equals total cost at S_k.

At point S_k the economic rent evaporates. Beyond S_k fishermen operate at a loss and the economic logic dictates that they should withdraw from the fishery and operate elsewhere in the 'economic universe' where higher rewards are available. It is possible to think of a number of situations in which fishing effort could expand well beyond S_k. First, let us assume that entry stopped at S_k. Then due to a large increase in the cost of fuel the total cost curve shifted upwards, creating a gap between total cost and total revenue (Figure 2.6, net loss 1). Alternatively we may envisage that there is no change in the cost curve but for some reason the price of fish fell, thus shrinking the total revenue curve inwards. In either case the fishing effort should be reduced to, at least, where total revenue equals total cost.

This reduction may not happen in reality for a number of reasons. First, fishermen establish themselves in tightly knit communities and are most reluctant to move to other jobs. For many of these communities fishing is not only a job but a way of life and a cherished tradition. Furthermore, skills which fisher-

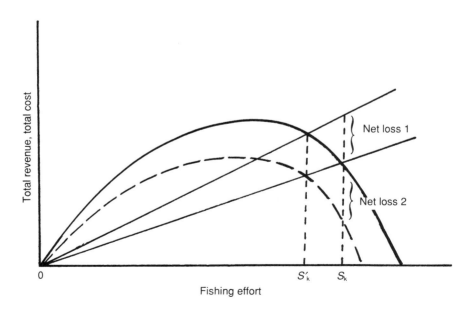

Figure 2.6   Adverse changes in fishery.

men inherited and acquired may not be readily used elsewhere in the economy. In the event of a withdrawal, fishermen may have to go through extensive job training schemes before they fit into new work. Kinships and friendships established in fishing cannot be maintained outside the industry and therefore a resistance develops among fishermen against withdrawal.

Second, vessels and fishing gear committed to fishing represent sunk capital which cannot be recovered by withdrawal. Fishing gear, for instance, is of no use in any other industry but fishing. Fishing boats may be converted to carry tourists for sightseeing in and around picturesque fishing villages. Some fishermen may branch out into this sector but it has to be said that the tourist season is short and the trade is unreliable, especially in the British Isles.

Third, it has been argued that fishermen are natural optimists and gamblers: they always dream about a big catch or an extraordinarily good fishing season which may moderate their financial problems. In effect, this happens very seldom in the fishing industry and fishermen always hope that the next season will be the best.

For all these reasons labour and capital get trapped in over-expanded fisheries and, to a certain extent, this explains why many fishing communities are poor in most parts of the world.

2.4   A DYNAMIC ECONOMIC THEORY OF FISHERY

It would be interesting and revealing to look at the fishery problem from a different angle, especially one containing time. The dynamic theory of fisheries was developed in the 1970s by Copes (1972), Clark and Munro (1975), Clark (1976) and Clark et al. (1979), all of which lean on the earlier work of Schaefer (1957). In the following analysis it is assumed that the aquatic environment in which the fish live and grow is not changing. The water has a fixed carrying capacity which determines the maximum size of the fish stock. The size of the stock – the biomass – is measured in terms of weight rather than number of fish. The biomass grows as a result of new fish entering the fishery as well as infant fish growing in size.

Figure 2.7, which is normally referred to as a Schaefer curve, illustrates the situation in the fishery. The size of the biomass is measured along the horizontal axis and its growth over time, dQ/dt, along the vertical axis, where t denotes time and Q the biomass. Growth rate, which is a function of stock size, is slow in the early stages because of a relatively small number of fish in the fishery. It speeds up to a maximum level, the turning point of the curve, then moderates until the final point K, the maximum carrying capacity, is reached which is also the saturation level.

More formally:

$$\frac{dQ}{dt} = f(Q),$$

(2.1)

where the rate of growth depends on the size of the biomass which is eventually halted by the environmental limitation at K. Equation (2.1) can be specified as:

$$\frac{dQ}{dt} = rQ\left(1 - \frac{Q}{K}\right),$$

(2.2)

where r is the intrinsic rate of growth. In the absence of fishing activity, the maximum carrying capacity can also be referred to the natural equilibrium population level, which is at its highest point.

Now let us introduce a fisherman, a predator, into the model and modify the growth function as:

$$\frac{dQ}{dt} = rQ\left(1 - \frac{Q}{K}\right) - h(t)$$

(2.3)

where $h(t)$ is the harvest function.

Equation (2.3) simply says that the change in the fish stock over a small interval of time will be given by the difference between the biological growth func-

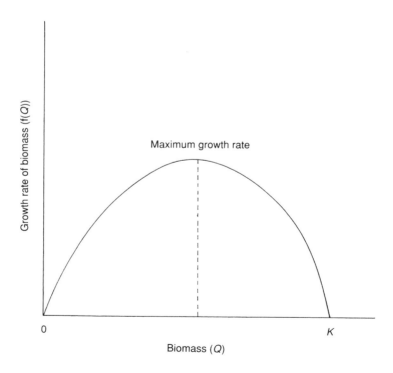

Figure 2.7 The Schaefer curve.

tion and the level of fishing activity in that time. Equation (2.3) can be solved for an equilibrium in which a certain level of fish stock can be maintained.

There are two factors which determine the level of $h(t)$: the size of the biomass, which is determined by nature, and the level of fishing activity, determined by man. If $h(t)$ is turned into a production function with these two inputs, then we have:

$$h(t) = h(Q, E), \tag{2.4}$$

where E is the fishing effort measured in terms of labour and capital, i.e. fishing fleet, including crew and all the necessary gear. Let us now specify this function in a mathematically convenient form as:

$$h(t) = \alpha Q \cdot E, \tag{2.5}$$

in which the elasticity parameters for both inputs are unity and, of course, they are constant; α is an efficiency parameter which is also constant.

What we need now are cost and revenue functions. Let us employ a simple cost function in which total cost depends on the level of fishing effort:

$$TC = \beta E, \tag{2.6}$$

where β is a constant. As in the static theory, explored above, here too we employ a constant price for fish, P. The aim is to find the market value of the fish stock, i.e. to transfer the sustained growth function (equation (2.2)) into a total revenue function. This could be done by multiplying equation (2.2) by the price of fish. Again, for the sake of simplicity assume that $P = 1$, then the growth curve (sustainable) will also measure the sustainable total revenue (STR), i.e.

$$STR = rQ - \frac{rQ^2}{K}. \tag{2.7}$$

Figure 2.8 shows the situation in the sustained yield fishery. The biomass level is measured along the horizontal axis, sustainable revenue and cost of catching fish are given along the vertical axis. The sustainable total revenue function, STR, is extracted from the Schaefer function (Figure 2.7, equations (2.1) and (2.2)). Note that the maximum sustainable yield, MSY, can easily be identified at the turning point of the sustainable total revenue curve. In the absence of harvesting cost and discounting the MSY (Q_0) will be the most desirable equilibrium for the fishery and a sustained harvest level of h_0 will achieve this.

In our model fishing is costly and thus we have to illustrate it in Figure 2.8. But before that let us carry out the following operation. First, substitute equation (2.5) in equation (2.3)

$$\frac{dQ}{dt} = \left(rQ - \frac{rQ^2}{K} \right) - \alpha QE, \tag{2.8}$$

which must be equal to zero when the harvest is sustained.

Then

$$\alpha QE = rQ\left(1 - \frac{Q}{K}\right).$$

Solving for E, the level of fishing which will yield a sustained equilibrium, gives:

$$E = \frac{r}{\alpha}\left(1 - \frac{Q}{K}\right). \tag{2.9}$$

This is substituted in equation (2.6)

$$TC = \frac{\beta r}{\alpha}\left(1 - \frac{Q}{K}\right) \tag{2.10}$$

or

$$TC = \frac{\beta r}{\alpha} - \frac{(\beta r)Q}{\alpha K} \tag{2.11}$$

to give a constant cost function as β, r, α and K are all constants. The cost varies only with the size of the biomass which can be increased or decreased by the fishing operation.

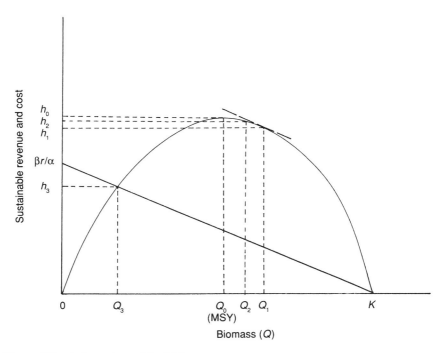

Figure 2.8   A dynamic model of fishery.

When $Q = K$, equation (2.11) will be zero. This means that when the size of the biomass, Q, equals the maximum carrying capacity, K, there is no fishing and the cost is zero, which gives us point K along the horizontal axis. When $Q = 0$, equation (2.11) will be $\beta r/\alpha$. This means that the entire fish in the fishery is caught at a cost $\beta r/\alpha$, the stock size is zero, and the fish is extinct. By connecting K and $\beta r/\alpha$ we get a linear cost curve. Note that total cost declines as the size of the biomass expands; it increases as the fish in the fishery dwindles. In other words the cost is an inverse function of the biomass level.

At biomass level Q_1 the difference between the sustainable total revenue and the total cost is at its widest point as the slope of TC is equal to the slope of STR. In the absence of discounting, Q_1 is the optimum sustainable level of equilibrium. Once again notice that the optimum sustainable yield is different from the maximum sustainable yield.

The absence of discounting, which has been the case so far, implies that the fishermen are indifferent between acquiring their income now or at a later date. In other words, £1 profit from fishing is valued the same whether it is earned this year or next. However, with a positive discount rate, the owner of the fish stock faces an intertemporal trade-off. A harvest next year would simply mean less than an equal amount of harvest today. The owner may not be willing to operate with a large fish stock and small catch today in anticipation of rising future harvests. The existence of a positive discount rate as compared to a zero discount rate implies larger harvests this year with a smaller biomass level. The higher the discount rate, the more impatient the individual is for harvest now.

To find the optimum level of fishing over time first we have to remember that the fish stock, biomass, is a capital asset. Of course the central problem in capital theory is to determine the optimal stock of capital over time as a function of time. The capital stock can be adjusted by investment or disinvestment. In fishery economics investment would occur when the harvest level is less than the sustained biomass level,

$$h(t) < f(Q),$$

then the stock will grow over time. Disinvestment will take place when the harvest level is more than the sustained yield,

$$h(t) > f(Q),$$

and consequently the stock will decline over time.

Now in deciding to disinvest or invest in a particular fishery the owner will want to compare the marginal yield on fishing with the alternative investment opportunities elsewhere in the economic system. It must be intuitively obvious that the optimum equilibrium biomass level is the one when the marginal yield on fishing equals the marginal yield on other investments. At this point it is also worth mentioning that risk varies from one sector of the economy to the next. In the fishery the biomass is actually a living thing and thus the risk would differ from an investment in, say, bicycle manufacturing. For the sake of simplicity let us argue that all investment projects are either riskless or carry the same level of risk in the economy, and there is a unique interest rate, s, which measures the rate of return.

The marginal yield in fishing is related to the marginal change in sustainable profit, the difference between total cost and the sustainable total revenue which is brought about by an incremental change in the biomass level. If we increase Q we shall be able to capture a stream of future benefits through time. In Figure 2.8, for a biomass level below Q_1 there will be a positive reward for investment and the incremental rent will go up when we move towards Q_1.

Let us refer to the marginal profit as dSTR/dQ. To express the marginal yield as an internal rate of return dSTR/dQ is divided by the cost of the incremental biomass level which is also a capital stock. The cost of making this investment is the marginal current net benefits forgone by not harvesting the biomass now, i.e. the market value or price of incremental fish minus the cost of harvest. Since we assumed that the price is 1 then the cost is

$$1 - c(Q),$$

where $c(Q)$ is the cost of catching additional fish. Then the marginal yield on the stock will be

$$\frac{\text{dSTR} / \text{d}Q}{1 - c(Q)},\qquad(2.12)$$

which is the internal rate of return in the fishery. The optimum level of fishing over time is that where the internal rate equals the rate of interest, s:

$$\frac{\text{dSTR} / \text{d}Q_2}{1 - c(Q_2)} = s.\qquad(2.13)$$

Q_2 is the optimum sustained biomass level which will satisfy equation (2.13). Q_2 can be at a point of maximum sustainable profit level, Q_1, only under a restrictive condition where the interest rate is zero. In other words, at $Q_2 = Q_1$ there is a situation in which yields on alternative capital assets equal zero. But when the interest rate is positive it is likely to be anywhere between Q_1 and Q_3, as shown in Figure 2.8. Could the optimal level be at Q_3? Yes it could, but only if the interest rate, s, is infinitely large. This also means that the owner of the fishery ignores the future completely and focuses on present harvest levels only. The conclusion must therefore be that the higher rate of interest the nearer the equilibrium to Q_3 – no rent situation. Conversely, the lower the interest rate the closer the equilibrium to Q_1.

2.4.1 Some further points

It is clear that the time optimum fishing level is a function of interest rate, s. It is also obvious that the economically optimum level will only be obtained if the fishery is properly managed by the owner. What if nobody owns the fishery? In other words, what if we have unrestricted access, i.e. the fishery is a common property resource? In this case the fishing effort will expand up to Q_3 where the sustained rent is zero. This is because the individual fisherman acting alone has no incentive to invest in or conserve the resource. He will realize that the fish he refrains from harvesting will be captured by another fisherman. It is important to emphasize that undesirable results of an open access fishery arise not because of the naïvity or irrationality of individual operators but because of a market failure.

Q_3 is a stable equilibrium, as h_3 is taken at each point in time. But this is a kind of knife edge equilibrium. If, given the constant level of harvest, the size of biomass falls below Q_3, the fish in the fishery will become extinct. For example, a toxic chemical spill may wipe out a good part of the fish population or abnormally warm water temperatures may interfere with their ability to spawn. In such cases, as long as the harvest level is maintained, the fish stock will disappear completely. This is because to the left of Q_3 the natural growth rate of the biomass lies everywhere below h_3, the sustained harvest rate. In each period, less and less fish will enter the fishery because existing ones are captured before they get a chance to reproduce.

When the fishery is properly managed under single ownership the time optimum sustainable yield will depend on whether the stocks are owned privately or publicly. One reason for that is the alleged difference between the private and social interest rate. There is a vast literature which argues that the private interest rate, for various reasons, is likely to be well above the public sector discount rate.

One reason for possible divergence between social and private interest rates is intertemporal altruism. Some economists (e.g. Baumol, 1952; Sen, 1961; Marglin, 1962, 1963a) have argued that due to the absence of any collective mechanism in the marketplace, individual rates cannot be used as a base for the social discount rate since each one acts separately. In isolation, individuals may be more self-regarding than when they see themselves as part of the community. Therefore, governments will be justified in determining the social discount rate paternalistically, which will be lower than the individual rates. This is discussed further in Chapter 11.

Another reason for a gap between private and social interest rates is market imperfections. The private rate of return on capital may be higher than the social one due to market imperfections such as monopoly or oligopoly. Furthermore, when a private rate of return is calculated on capital the external effects, such as pollution, noise and congestion, which may be created by private investment projects, may not come into the picture. In other words, the private rate of return may be higher than the social one not because of an efficient operation but as a result of restrictive practices and the exclusion of externalities, which are social costs. Many argue that the stock market's view of profit may not be a proper measure for social profitability.

These are, indeed, all legitimate points on the issue of divergence between private and social interest rates. However, some research shows that, for the USA and Canada, this social interest rate may coincide with post-tax rates that consumers have historically been able to earn in the stock market (Dorfman, 1975; Kula, 1984a).

2.5   THE SITUATION IN WORLD FISHERY

In the 1950s and 1960s fishing was a booming activity throughout the world. Between 1956 and 1965 the world fishery output increased by 50%. In European Community waters catches doubled between 1958 and 1968, and this put tremendous pressure on many species of fish (European Documentation, 1985). For example, in the 1950s, the average annual catch of adult herring in the North Sea was around 0.6 million tonnes. This increased to 1.7 million tonnes in the 1960s. Ten years later catches had fallen to 0.5 million tonnes and by 1977 spawning stocks were reduced to a critical 150 000 tonnes. In the North Atlantic as a whole, herring catches dropped from 3.3 million tonnes in 1964 to 1.6 million tonnes in 1974.

A similar fate befell cod. In 1964 annual catches by French, German and British boats amounted to 0.7 million tonnes in the North Atlantic. Ten years later the figure had dropped to 0.2 million tonnes. The collapse of cod and herring put further pressure on mackerel. The total catch of mackerel in the North Atlantic was about 0.2 million tonnes in 1964. By 1974 1 million tonnes were being taken. Similarly, in the United States as species such as halibut declined, pressure on others increased.

Fish forms an important part of the human diet in many countries as shown by Table 2.1, and as the world population grows, increasing pressure will, no doubt, be put on fish stocks (Lawson, 1984). Already, there has been a huge increase in the world catch in recent years. In 1960 the total catches were slightly over 40 million tonnes but this grew to over 100 million tonnes in 1987 (Table 2.2). The Food and Agricultural Organization (FAO) divides the world fisheries into 27 fishing areas. The largest is the Eastern Central Pacific with about 16% of the total area, and the smallest is the Mediterranean and Black Sea with O.8% of the total area. The expansion of fisheries in most of these areas has been noticeable in both developed and developing countries and the rapid increase in activity has begun to threaten species throughout the world.

According to the FAO, in 1950 only 27% of the world's catch was by developing countries. This figure increased to 46% in 1977 and it may go to 58% by the year 2000. In Peru, for example, in 1960 the total catch of anchoveto amounted to 2.9 million tonnes. Ten years later the figure was 12.3 million tonnes, before it slumped to 0.5 million in 1978 due to overfishing. Many early fishery management regimes did not explicity support incentives to manage and conserve the resources and only a few areas of the sea were monitored.

Although in most countries the commercial fishing industry does not represent a large share of GNP, it can be extremely important for some regions. For example, in the United States the fishing industry's contribution is about 0.3% of the total GNP (Anderson, 1982), but in the state of Alaska income generated in fishing accounts for about 13% of the total state income. In Canada, 14% of the state income in Newfoundland comes from commercial fishing. In the European Community the importance of the fishing industry varies from country to country. For example, in Portugal income from fishing constitutes 1.6% of the GNP, whereas in Germany it is 0.02%, and about 0.1% in the United Kingdom. These figures somewhat underestimate the importance of fishing in national economies. Recreational fishing is an important industry in most developed countries, generating millions of pounds of income (O'Neill, 1990; O'Neill and Davis, 1991). In some regions, the revenues from recreational fishing even exceed those from commercial fishing (Hartwick and Olewiler, 1986).

Table 2.1 Per capita intake of fish protein as a proportion of total protein consumed, 1966–69 average

Country	Fish flesh (kg/year)	Fish protein (g/day)	Animal protein (g/day)	Fish protein as % of animal protein
Japan	64.1	15.8	28.2	56
Portugal	56.5	13.9	31.7	44
Denmark	44.5	11.0	60.2	18
Norway	38.6	9.5	50.4	19
Korea	34.4	8.5	11.5	74
Jamaica	25.2	6.2	18.7	33

Source: European Documentation (1985).

Table 2.2 Catches in world fisheries (1000 tonnes live weight)

Country/ region	1981	1983	1985	1987
Japan	11 463	12 088	12 211	12 587
USSR	9 713	9 854	10 593	11 155
EC	6 726	6 717	6 779	6 591
Norway	2 544	2 813	2 087	1 892
World	85 393	85 736	97 289	103 113

Source: Eurostat 1990, Agriculture Annual Statistics.

2.6 THE LAW OF THE SEA CONFERENCES

In the past most countries had 3-, 6- or 12-mile territorial limits which brought only a very small part of the world's marine stocks within the jurisdiction of single states. In the 1950s and 1960s some governments sought to increase their control over the waters off their coasts, arguing that this was the best way to protect fish stocks and combat the open access situation. For example, in 1952, Chile, Peru and Ecuador extended their fishing territories. Around the same time in Europe the Icelandic government was arguing for a special jurisdiction over fish stocks around a wide area of Iceland. In 1952, the fishery limits had been extended to 4 miles, in 1958 to 12 miles and in 1972 to 50 miles by Iceland.

Nations such as Britain, France, Germany and Denmark who had always fished in waters around Iceland found themselves denied access to some of the world's richest fishing grounds. The Icelandic government's argument was that fishing has always been the most important activity for the nation so there should be a special dispensation for Iceland. Indeed, the country exports about 90% of its catch and is heavily dependent on fishing. However, the unilateral declaration of new exclusive fishing territories raised the tempers between Britain and Iceland and the so-called 'cod war' began in the 1970s.

In 1958 the first United Nations Law of the Sea Conference was convened with a view to explore commercial and legal matters regarding the use of seas. No consensus was reached among the participants on the extent of the territorial fishery limits. The second conference took place two years later again without a common agreement. After the second meeting the pressure was building up by some nations to expand their territorial limits. By 1974, 38 states had expanded their limits beyond 12 miles. Iceland was the only European country to do so, together with India and many African, South American and South-East Asian countries. The leading group in this action was the Organization of African Unity (OAU) which with the solidarity of some South American states recognized the right of each coastal state to establish an exclusive economic zone up to 200 miles.

At the third conference on the Law of the Sea in 1977 a document called the Informal Composite Negotiating Text (ICNT) was produced with the recommendation that coastal ocean states should have exclusive economic zones of 200 miles where appropriate (UN, 1977). In view of the fact that fish stocks move back and forth across the 200-mile zones, the text also recommended that nations who have a common interest in the optimal use of fish stocks should co-operate with one another. Following this conference many nations, including Australia, China, Canada and the United States of America, unilaterally declared 200-mile fishing limits.

By December 1982, when the Convention on the Law of the Sea was signed, most fishing nations had extended their exclusive economic zones to 200 miles. Article 56 of the Convention establishes the foundation for the rights of the coastal state within its 200-mile zone. The coastal state has 'sovereign rights for the purpose of exploring and exploiting, conserving and managing the natural resources, whether living or non-living of the waters superjacent to the seabed and of the seabed and its subsoil'. Specific rights of navigation and overflight and laying of cables and pipelines are preserved in Article 58, but no right to the living resources of the exclusive economic zone is granted to third states.

Article 61 of the Convention states that the coastal states shall determine the allowable catch in their exclusive zones. They should also ensure by proper conservation and management measures that the maintenance of the stocks in their exclusive zone is not endangered by overexploitation. As appropriate, the coastal state and competent international organizations, whether regional or global, should co-operate to this end. Measures should also be designed to maintain or restore populations of harvested species at levels which can produce the maximum sustainable yield, as qualified by relevant environmental and economic factors, including the economic needs of coastal fishing communities and the special requirements of developing states, taking into account fishing patterns, the interdependence of stocks and any general recommended international standards. In taking such measures the coastal state should take into consideration the effects on species associated with or dependent upon harvested species with a view to maintaining or restoring populations of such species above levels at which their reproduction may become seriously threatened. Available scientific information and economic statistics should be contributed and exchanged on a regular basis through competent international organizations and with participation by all states concerned.

These provisions refer to biological, political, economic and sociological requirements when determining the allowable catch to ensure the maximum sustainable yield in accordance with Article 61. The allowable catch should also be designed to maintain or restore populations of harvested species at levels which can produce the maximum sustainable yield, as qualified by relevant environmental and economic factors, including the economic needs of coastal fishing communities and the special requirements of developing states. Defining allowable catch has an important bearing on the international sharing of fish resources within the exclusive economic zones. Article 62 states that the coastal state should determine its capacity to harvest the living resources of the exclusive economic zone. Where the coastal state does not have the capacity to harvest the entire allowable catch, it should give other states access to the surplus.

Determination of allowable catch could be quite a complex process. The problem becomes even more complex when stocks that are common to several exclusive economic zones are involved. In such cases Article 63 invites coastal states to agree upon the measures necessary to co-ordinate and ensure the conservation and development of such stocks. The Convention also specifies that nationals of other states operating in an exclusive economic zone should comply with the conservation measures and with other terms and conditions established in the regulations of the coastal state. Article 73 mentions actions that can be taken by

64   Economics and policies in fisheries

coastal states to enforce its laws and regulations. They include the boarding of boats, inspection, detention of boats and arrest of crew members who should eventually be released on receipt of suitable bond and security.

Before the signing of the United Nations Convention on the Law of the Sea few states had fixed total allowable catch quotas. Today, many countries establish their allowable catch in line with national management schemes. There are, however, some countries which determine total allowable catch only for stocks which they consider to be threatened.

2.7   FISHING IN THE WATERS OF THE EUROPEAN COMMUNITY

The European Community has been monitoring these developments closely. Following the gradual establishment of 200-mile fishing limits by Iceland and Norway in the 1970s the Council of Ministers decided to take up the matter formally. In June 1977 at the European Council of Head of States meeting in Brussels it was declared that the Community was determined to protect the legitimate rights of its fishermen. The following month the Council of Ministers announced their intention to introduce a 200-mile limit. In September, the Community also decided to extend the principle of 12-mile coastal bands with access only on the basis of historic rights to all member states until a common fishery policy was worked out in 1982. It was also agreed to introduce a system of quotas to limit the catch inside the 200-mile zones. Special dispensations beyond the 12-mile coastal band were given to fishermen in the Republic of Ireland and the United Kingdom as they are island states. The main objective was to protect fertile fishing grounds in the North Sea and around Ireland from East European fishermen.

In 1983, the ten member states of the European Community agreed to a common fishery policy (CFP) with the aims of protecting fish stocks, promoting their efficient use and improving the livelihood of those employed in the fishing industry. The CFP replaced the gentleman's agreement that had existed previously between the member states. Some of the provisions of the CFP are:

1. rational measures for managing resources;
2. fair distribution of catches;
3. paying special attention to the Community's traditional fishing regions;
4. effective controls on fishing conditions;
5. providing financial help to implement the CFP;
6. establishing long-term agreements with countries outside the European Community.

These policies have binding legal force in all member states and are enforceable through the European Court of Justice and national courts.

The European Community has long realized that a rational management of the fish stocks cannot be done by individual countries. The adult fish caught in one country may have been spawned and matured in the waters of another. For example, the UK has a high proportion of the Community's mature fish in its waters, but is heavily dependent on conservation in other countries' territories to ensure that stocks are maintained. Cod is a migrant fish that moves into British waters as it matures. Therefore it is necessary to maintain a discipline not only in Community waters but also round them in order to manage fish stocks on a rational basis.

The European Community believes that protecting the future of fish stocks is the best way to maintain a stable level of employment in the industry. Overfishing weakens the catch that boats can expect from a voyage. As the stocks contract, faced with the high cost of fuel, repayment on loans and the need to find wages for the crew, many boats will become uneconomic, forcing skippers to buy off their men. The Community is in the process of restructuring fishing with an emphasis on inshore fleet, fish farming, training men for rational and efficient fishing methods and improving the quality of its fleet. Over the years a number of regions, such as Hull and Grimsby, which depended heavily on cod from Icelandic and Scandinavian waters, suffered unemployment as access to these areas was gradually denied in the 1970s. Because of these problems the number of fishermen in the Community of nine declined from 1975 to 1980. The membership of Greece in 1981 pushed the figure up (Table 2.3).

It had been argued that for every job that exists at sea, another five are created on land (European Documentation, 1985). These on-shore jobs are mostly in boat building, repairing, gear making, fish processing, marketing and transportation. Increasingly sophisticated manufacturing techniques and the need to process many of the lesser-known species to make them attractive to the public have created many new jobs throughout the Community. In some countries such as Germany and the UK the business is concentrated in the hands of multinational companies. In Denmark, on the other hand, the firms involved are small family-run businesses. The processing industry is particularly important for the fishing industry and thus it is eligible for aid to develop plant or the necessary infrastructure within the framework of the CFP.

Before the introduction of the CFP there were numerous aid packages in the Community. At one time the Community was examining as many as 20 national aid schemes that governments had introduced in the absence of the CFP. In some cases aids were straight payments to producer organizations, in others they were a fuel subsidy. Nowadays the European Community does not favour such schemes because they lead to no lasting improvement in the sector. The Community tends to favour the following aid packages:

1. temporary or final laying up of fishing boats;
2. research for the discovery of new fishing grounds, development of existing ones and the exploration of previously unfavoured species of fish;
3. improvement of fishing vessels;
4. improvement in processing and marketing structures;
5. manpower training; and
6. promotion of fishery products.

Table 2.3   Numbers of fishermen in the EC

Country	Year			
	1970	1975	1980	1982
Belgium	1 264	1 072	894	865
Denmark	15 457	15 316	14 700	14 500
France	35 799	32 172	22 019	20 177
Germany	6 669	5 767	5 133	5 229
Greece	50 000	47 000	46 500	–
Ireland	5 862	6 482	8 824	8 975
Italy	62 075	65 000	34 000	–
Netherlands	5 514	4 619	3 842	4 206
Portugal	35 309	30 562	35 579	–
Spain	69 059	71 810	109 258	106 584
UK	21 651	22 970	23 289	23 358

Source: European Documentation (1985).

Between 1983 and 1985 the Community allocated 156 million European Currency Units (ECU) to modernize the fishing fleet and encourage fish farming. Some specific measures in this package were:

1. 11 million ECU for restructuring, modernizing and developing the fishing fleet and improving storage capacity;
2. 34 million ECU to aid the processing installations;
3. 2 million ECU to construct an artificial reef;
4. 44 million ECU as aid for the temporary laying up of vessels over 18 m in length, provided they were commissioned after 1 January 1958;
5. 32 million ECU for scrapping some vessels;
6. 11 million ECU to aid exploratory vessels to discover new grounds and species;
7. 7 million ECU to explore fishing possibilities outside European Community waters.

2.7.1 Third country agreements

Agreements with third countries, which take a number of forms, are all negotiated by the European Community and adapted by the Council of Ministers after consultation with the European Parliament. Some, such as those with Norway and Sweden, are based on the principle of reciprocal fishing arrangements, where fish is traded for fish. Another category of agreement is based on access to surplus stocks, one current example of which is with the United

States. In accordance with the rules of the Law of the Sea Convention the United States grants the European Community a fishing allowance every year from its own surplus stocks.

A third category of arrangement is that the European Community pays financial compensation for the right to fish in some countries' waters. The funds for these come from shipowners as well as from the Community budget. The rationale for this type of arrangement is to restore or maintain fishing rights for a Community fleet, especially for French, Italian and Greek boats. Some recent examples are agreements with the West African states, the Seychelles, the People's Republic of Guinea, Equatorial Guinea and Senegal. A final category of agreement involves trade facilities. An example of this is the agreements with Canada, where Community boats can fish in Canadian waters, and in return, the Community opens tariff quotas at reduced rates for certain fish originating from the North Atlantic fish stocks.

These agreements have no specific content. They merely establish a framework within which the parties will deal with all fisheries problems for a number of years. They contain general regulations covering access, licence fees, compensations, scientific co-operators and procedures to settle disputes. In the agreements with developing countries the Community aims to ensure a secure supply of fish for which it gives technical and financial assistance in return. The agreements with the developing countries are expected to grow quite substantially in the future.

Until now the Community has had no fishery agreements with Eastern European countries, but has been involved in lengthy negotiations since the introduction of the 200-mile zone. A system of licensing was introduced for the Soviet Union, Germany and Poland to enable the European Community to enforce some quotas but this is now under revision following the changes in the political map.

The European Community is a full member of many international organizations, such as the North Atlantic Fisheries Organization, North East Atlantic Fisheries Convention, North Atlantic Salmon Convention, Convention for the Conservation of Antarctic Marine Living Resources, Convention on Fishing Conservation of the Living Resources in the Baltic Sea and Belts, International Convention for the Southeast Atlantic Fisheries, International Convention for the Conservation of Atlantic Tuna, Food and Agricultural Organization and the International Whaling Commission.

2.8   SOME ASPECTS OF FISHING IN THE UNITED STATES WATERS

Until 1976 there was no act regulating the fishing beyond 3 miles outside the United States coastlines. In 1976, the Fisheries Conservation and Management Act (FCMA) established the right of the federal government to manage the fish stocks between 3 and 200 miles off the entire United States shores. Within the 3 miles fisheries are regulated by the state governments.

The major objective of the FCMA is management and conservation measures to prevent overfishing while achieving sustainable optimal yields. In the interpretation of the optimal yield the maximum sustainable yield was modified by taking account of economic, social and ecological factors. Also it was emphasized that in the management of the fisheries costs should be minimized. For example, a number of researchers (e.g. Crutchfield and Pontecorvo, 1969) argued that the United States and Canada could maintain the same catch of Pacific salmon at an annual cost of $50 million less than the costs that existed at that time. Christy (1973) revealed that in many other countries the situation was even worse. At that time in Peru, which has one of the largest Pacific catches in the world, the fishery was so over-capitalized with excess vessels that the same catch could be maintained at savings amounting to $50 million per year.

There are a number of management techniques used by the United States fishery authorities. These are: (1) regulations relating to the fishery itself, such as closed areas and seasons and quotas on catches; (2) regulations relating to the kind of effort such as trawling techniques and mesh sizes; and (3) regulations concerning the level of fishing effort such as licensing of fishing boats. The overall management of a fishery would probably require a mixture of all three techniques although conditions in a particular location may make one method dominant.

Establishing fishery departments to manage fish stocks is costly. One of the early steps must be to construct and implement policies. Then the fishery must be monitored to ensure that the regulations are being followed. Young (1981) estimated that in the Pacific Northwest region alone the annual cost of enforcement of the regulations was about $20 million in 1978. Day-to-day regulation involves daily publication of a fleet disposition report. During the high season there can be over 300 foreign boats in the region. The patrol boat has the task of identifying each boat. Licence permits, equipment, restrictions on species, areas and seasons, and national quotas must all be checked for boats operating in the fishery.

Following the Fisheries Conservation and Management Act the federal government established eight regional fisheries management councils. They are: New England, Mid-Atlantic, South Atlantic, Gulf, Caribbean, Pacific, North Pacific and Western Pacific. The major objective of these councils is to determine the optimal harvest in their areas. If a particular species is important for more than one council then they are expected to manage the species jointly. Councils range in size from 8 to 16 individuals, half of whom are appointed by state authorities and the other half consisting of interested members of the public, mostly from representatives of the fishing industry. Each council also employs professional staff including marine biologists, lawyers and economists.

The point of emphasis may vary in policy-making from council to council depending on the aspirations and political inclinations of the constituent members. In some councils economic efficiency in fishery may be considered to be a prime target whereas in others distributional issues may gain prominence. Fairness can indeed be a crucial objective necessary to thwart any lawsuits by injured parties as a result of the implemented management plan.

Even in a particular council the management plan can change quite drastically from one period to the next in view of changing circumstances. For example, it was reported by Hartwick and Olewiler (1986) that in the New England Council, the management plan initiated in 1977 for cod, haddock and lowtail flounder consisted of a quota on each species. The plan did not restrict entry. With the annual quota, the harvest was taken early in the year and fishermen were furious that their activity was curtailed for the rest of the year. The plan in the following year was modified to allow for two harvests, but this was not a long-term situation. Quarterly quotas were tried in later years, but these did not work satisfactorily. Then quotas based on vessel size were tried, by which the emphasis was shifted to fishing effort. However, most boats incurred high harvesting costs because they frequently had to rush to port before they filled their holds. Owners of small boats felt particularly pinched because they were frequently affected by adverse weather conditions and missed filling their quota. The quota was revised again, this time based on the size of the crew; this resulted in large crews appearing on each vessel. Additional regulations were then laid down for each type of boat. In the end the whole system was so complex and unmanageable that it was abandoned. From 1981 onwards, the fisheries have been regulated mainly by restrictions on mesh size and closed areas to protect the young fish.

With regard to international trade in fish and fish products, the United States pursues reasonably liberal and fair trade policies. There are no import quotas except those on marine mammals, but fish cannot be landed from a foreign flag vessel. When ground fish is imported from Canada there is just over 5% countervailing duty which is in line with the GATT principles. Imports of fish products are routinely checked for compliance with health regulations by the Food and Drug Administration. There are also strict labelling and packaging requirements for imported as well as domestically produced fish products. If imported fish and fish products are linked to operations which could diminish the effectiveness of an international fishery conservation programme the authorities are empowered to prohibit the entry of such products.

As for exports, there are no monopolies or export restrictions on fishery products except on marine mammals. The United States authorities do not make direct payment to firms on export performance. However, the Government Corporation Control Act allows the Commodity Credit Corporation to support the price of fish through loans, purchases and by other means. The corporation can also act in the development of foreign markets for fish and fish products.

The United States is the largest provider of fishing possibilities in her territorial waters to foreign fishermen. For example, in 1984 the total allowable level of foreign fishing amounted to 1.9 million tonnes. There are 12 governing international agreements in force, and each one is only a framework not containing any commitment on the part of the United States to allocate quotas. All agreements concern access to portions of surplus determined for each fishery on the basis of optimum utilization. Allocation to each country is made on the basis of an evaluation of performance under a number of criteria established by the Magnuson Fishery and Conservation Act of 1976.

2.9   RIGHTS-BASED FISHERY MANAGEMENT

During the 1980s the focus of attention among fishery economists had changed towards the concept of the rights-based fishery. It is now well recognized that a fishery management scheme works better when authorities do not encourage fishermen to race for the allowable catch. In a rights-based fishery, regulators grant each participant a right to gain access to a geographic area at certain times, allow them to land and market certain species of fish and issue permits to employ vessel and gear. For example, a boat may be licensed to fish in territory X during July, and to land no more than 200 tonnes of haddock.

The main factor which encouraged fishery managers to think in these terms was that licence and entry limitations were not working satisfactorily. As explained above, early management objectives were aimed at stock conservation mostly in the area of the maximum sustainable yield. Since the Second World War, improved fishing technology, coupled with short seasons, resulted in over-capitalization and wasteful racing to catch fish in many world fisheries. It became obvious to some fishery managers that there had to be a better way of managing stocks.

Scott (1989) argues that conceptual origins of rights-based fishing are ancient, and that they existed, in various forms, in Anglo-Saxon, Japanese, Nordic and Aboriginal cultures. In the old English tradition the common law principle regarding individuals' rights to fish was a complex one and depended on many factors: whether the fish were swimming at large or captured, and when at large, were they to be found offshore, in tidal waters or in rivers and lakes? Although the development of private fishery rights began early in England, it was ended early, especially in the tidal and offshore fisheries. Private rights, however, did continue on inland fisheries but were based on property rights on land. When the private rights ended in open waters the concept of public rights of fishing took over. After that there were no essential changes in practice or in ideas regarding individual or national rights until very recently.

In the 1950s the 'horrors of commons' was the major focus of attention for economists who wrote on fishery management. Accepting common access as inevitable, most economists gradually turned their attention to improving the biological production function and to studying regulations, which went on until well into the 1970s. The regulatory literature expanded along the lines of licence and entry limitations in which the main questions were: how many licensed vessels? which types? at what initial price? what procedures? Nevertheless, in practice, regulations were not operating to everyone's satisfaction. Entry limitations and licence withdrawals resulted in resistance by fishermen, leading to costly litigation and extensive political lobbying. Regulators usually granted a licence to those vessels or owners already in the fishery and tried to reduce numbers by natural wastage. In order to speed up the process of reduction some governments introduced buy-back schemes which began to identify themselves as creating new characteristics of ancient rights-based fishing. Buy-back schemes implied an official recognition that a legitimate interest had been vested in the fishery by operators.

In the 1980s the concept of individual transferable quotas (ITQs) rapidly gained acceptance. ITQs displayed at least three of the six broad characteristics of property rights identified by Scott (1989):[2] duration, exclusivity and transferability. Duration enables the holder to save fish for harvest in a later year by conservation; exclusivity minimizes rush and unnecessary competition among fishermen to fill their quotas, assists co-ordination among those who hold similar rights and reduces competitive investment which leads to overcapitalization; and transferability allows fishermen to trade their rights.

The ITQ system was thought to bring distinct advantages over the old policies of restricted entry and regulation in which fishermen spent a good deal of their time out-racing rivals, out-witting regulators and disputing gear and other rules. Under the new quota system it was thought that the environment would change. Fishermen would stop racing, co-operate with enforcers, choose their own gear and set their own pace. Furthermore, the new system would tend to shift the enforcement function from the government to the holders of these rights.

Various criticisms have been levelled at rights-based systems. First, just as under the old system, quota owners can easily cheat and poach, and a quota system may require equally extensive monitoring and enforcement. This criticism assumes that fishermen will continue to behave as evasively as they do under the old regulatory system, although the strength of the quota system may be that it creates incentives for self-enforcement. Second, judging on their experience with past fishery reforms, fishermen tend to be sceptical about the longevity of the ITQ system. They may think that the new system is just another brainwave which will soon lose momentum and be replaced by yet another new policy. Furthermore, their experience in old fisheries where rules and quotas changed so many times will not encourage fishermen to accept their quotas. Instead they will try to increase their allowance by protest, lobbying and whatever other methods they may find useful. Third, there may even be resistance to the quota system from the bureaucrats who, as professionals, are concerned about their own job security, power of influence and self-esteem. They may find it hard to believe that their role as useful managers in the complex world of fishery can be replaced by the invisible hand of a rights-based system. Even if they are convinced about the usefulness of the new system they may begin to worry about their own careers which may be undermined by the new system. Therefore, scepticism by fishermen and bureaucrats alike may guide them to contribute to the failure of the rights-based system.

---

2. The other characteristics are quality of title, divisibility and flexibility.

Some fishery economists (e.g. Huppert, 1989; Libecap, 1989) are highly sceptical about the success of the transferable quota system. No doubt with the
assignment of exclusive private property rights some fishermen will be made
better off while others will become worse off as they are denied access to the
fishery. In theory, it may be possible to think of a compensatory system to pay
off the potential losers, but in practice questions such as the size of compensation, who should pay, who should receive and when may be difficult to deal
with.

Another crucial question is, how are the quotas to be allocated? In view of
their administrative convenience, if the authorities go for uniform quotas fishermen who were highly efficient and successful under the old regulatory system
will be hit hard and will strongly oppose the change in *status quo*. Previously
less successful fishermen will no doubt welcome the uniform quota regime.
Quotas based on historical catch reduce some of the problems associated with
uniform quotas, but raise the problem of how to measure and validate historical
catch claims. There will be long disputes over permanent ratification of shares
that are based on past catch records. Furthermore, in addition to opposition from
those fishermen who are likely to lose out with the ITQ, there are other groups
such as vessel and equipment manufacturers, retailers and even sports fishermen
who have a stake in the current regulatory regime. The outside groups may be
the most effective in lobbying politicians to prevent the introduction of the
rights-based system.

The opposition to a rights-based system is likely to vary from region to
region. There are areas where highly skilful or talented fishermen excel in the
open access conditions. Reorganization there will not only deprive such
operators but also injure other interested parties. Huppert (1989) gives the
example of the thriving fishing town of Kodiak, Alaska, where well-being
depends on open access competition in crab, halibut, salmon and ground
fisheries. Rationalization will not only injure the fishermen directly involved,
but those business communities such as processing plants, boat repairers and
many others who locate themselves close to fishing grounds. Huppert also
points out that Pacific halibut is perhaps the most promising candidate for
rights-based fishing in the United States, but the system of political decision-
making there frustrates all attempts to alter the existing system of open
access.

Retting (1989) believes that a right-based system developed in close consultation with all the interested parties is likely to be more successful than the
one designed by aloof government officials or model-building academics alone.
Furthermore, allocation of rights among people with common cultural and social
ties is likely to be more successful than programmes involving diverse groups.
When fisheries change over time, the success of a rights-based system will
depend very much on its flexibility. In particular, the system should be flexible
enough for changes resulting from new information and better understanding of
stock behaviour and evolving social conditions.

2.9.1   The ITQ system in New Zealand

The first comprehensive application of an individual transferable quota management system was introduced in New Zealand, a small country where fishery is regarded as an important national asset. The progress of this system is being watched with keen interest and great hopes by fishery economists and managers who have been recommending quantitative harvest rights for many years. No doubt the New Zealand system will be given the most intense scrutiny by professionals all over the world.

Since the introduction of the New Zealand Fisheries Act of 1908 the management of stocks has gone through a series of fundamental changes which were highly confusing to many involved in the industry. Between 1938 and 1963 the inshore fishery was managed under a system of strict gear and area controls coupled with restrictive entry, confined mainly to a depth of approximately 200 metres. In 1963 the inshore fishery was completely deregulated with the hope that the new situation would encourage investment for which capital grants and tax incentives were provided. Consequently domestic industry expanded rapidly and with the declaration of a 200-mile zone in 1978 the future looked very promising indeed. Before 1978 the stocks around New Zealand had been exploited by foreign fishing vessels, mainly from Russia, Korea and Japan, and the government was suddenly faced with developing a plan to manage resources in a very large and unfamiliar territory.

In April 1982, a limited quota management system was introduced for some deep water species which were relatively unexploited stocks. In 1983 a new Fisheries Act was passed which introduced the concept of fisheries management plans. For the first time in New Zealand fishing history recognition was given not only to biological objectives but also the concept of optimum sustainable yield to maximize the economic rent (Clark *et al.*, 1989). The government also introduced a rent-orientated management system for deep water fisheries and based up transferable quotas in the new 200-mile zone. This was the beginning of the rights-based system. In a 1986 amendment to the 1983 Act economic goals were more comprehensively recognized. The ITQ system became fully effective on 1 October 1986.

Initially the quotas were allocated for a period of ten years and for seven key species. The basis of allocations was prior investment by nine firms. In 1985 the government turned these quotas in perpetuity and included other inshore stocks. The fundamental point about New Zealand's ITQ is that it is a transferable property right allocated to fisheries in the form of a right of harvest to surplus production from stocks. The quotas are normally issued on the basis of historical catches. At the early stages of the ITQ programme quotas were reduced by a buy-back programme. For further details see Clark *et al.* (1989).

The main objectives of New Zealand's ITQ system are: (1) to achieve a level of catch which maximizes the benefit to the nation as a whole while ensuring a sustainable fishery; (2) to achieve the optimum number and configuration of fishermen, boats and gear to minimize the cost of taking any given catch; (3) to minimize the implementation and enforcement costs. Before the introduction of the ITQ system the enforcement method was a standard game warden approach, apprehending law breakers and discouraging all illegal behaviour, which was costly. The new role of the authorities is not so much policing but monitoring, following product flow from vessels to retailers. Enforcement activity now takes place on land rather than at sea, which is more cost effective and can be carried out by officials who are more auditors than game wardens.

3
Economics and policies in forestry

What do we plant when we plant the tree?
We plant the ship, which will cross the sea.
We plant the mast to carry the sails;
We plant the planks to withstand the gales –
The keel, the keelson, the beam, the knee;
We plant the ship when we plant the tree.

What do we plant when we plant the tree?
We plant the houses for you and me.
We plant the rafters, the shingles, the floors,
We plant the studding, the lath, the doors,
The beams, the siding, all parts that be;
We plant the house when we plant the tree.

What do we plant when we plant the tree?
A thousand things that we daily see;
We plant the spire that out-towers the crag,
We plant the staff for our country's flag,
We plant the shade, from the hot sun free;
We plant all these when we plant the tree.

H. Abbey

Like fish stocks, forests are a renewable resource. However, if this endowment
is misused and depleted badly, the process of its replenishment can be extremely
long and painful, a case which will be explained below with reference to the
British experience. In the past, forestry has been a contracting sector throughout
the industrialized world, mainly for three reasons. First, the growth of population increased the pressure on land to grow more food, which resulted in forest
clearance. The decline of forests in many regions became inevitable to sustain
the growing agricultural sector. Second, the use of wood as fuel and construction
material took its toll on forests, and such exploitation was most reckless in the
early industrialized countries such as Britain and Holland. Third, wars had a
devastating effect on forests, especially in Europe. The scars of these events
have been deepest in the British Isles.

Unfortunately, history seems to be repeating itself in the Third World. In their haste to catch up with the West and to feed their rapidly growing population, many developing countries are putting an overwhelming emphasis on agricultural and industrial expansion, without due regard to forestry. It is essential that the Third World countries should not repeat the mistakes that were made by some industrialized countries in the past when they recklessly depleted their timber deposits, from which deforestation they have not yet recovered.

Towards the end of the Middle Ages, the British Isles were clothed with thick forests. The southern parts were largely covered with broadleaf forests, whereas the northern parts contained a mixture of pine and silver birch. Today these islands are one of the least forested regions of Europe. There were four main reasons for forest destruction. First, most of the trees located on fertile soils had to be cleared to make room for agricultural expansion to support the growing population. Second, clearance also took place to provide fuel for the iron smelting industry in the sixteenth and seventeenth centuries. This was so extensive that at the end of the seventeenth century turf firing had to be introduced in many areas in place of wood. Around this period a traveller to Scotland wrote that 'a tree in here is rare as a horse in Venice' (Thompson, 1971). Third, the rise of Britain as a major industrial and naval power took its toll on forests. The industrialization process was based entirely on a single fuel – coal – and timber has always been a major input used as pit props in the coal mining industry. An expansion of this sector meant a decline of forests. Also, in the past timber was widely used in the shipbuilding industry. However, the building of the first iron ship for the Royal Navy in 1860 signified an end to the shipbuilding industry's insatiable demand for oak. Fourth, destruction also took place for military purposes, especially in Ireland. During the colonial struggle settlers cleared strategic locations of trees because they were providing cover for the local resistance groups. This was particularly extensive in Ulster.

Towards the end of the eighteenth century a revival of forestry took place in many parts of the British Isles. At the time the British aristocracy realized that woodlands were a desirable source of wealth and amenity which led to extensive plantation for commercial as well as ornamental reasons. Forests in private estates were retained for generation after generation for reasons of sentiment and prestige. Selling trees was usually an indication of the decline of the family. The decline of these private forests began around the middle of the nineteenth century, a time when many landlords, for various reasons, were in financial difficulties. Travelling mills moved into many estates to clear the land of trees. By the turn of the twentieth century the destruction was complete and the one-time forest-rich British Isles had become more or less treeless.

3.1   REGENERATION OF THE FORESTRY SECTOR – A LONG AND PAINFUL PROCESS

The year 1903 was an important one for forestry in the British Isles. In that year a forestry branch of the Department of Agriculture was established in Ireland for the purpose of training young men as practical foresters. The Department also bought land in various areas in order to establish forestry centres with a view to afforestation. Avondale in County Wicklow was the first forestry centre where a training school was established.

Progress was slow to start with but it gathered pace in later years and the Department set up many other similar centres elsewhere in Ireland. Unfortunately, the First World War put a stop to the expansion of forestry projects. During the war Britain suffered a severe shortage of timber. The German submarine campaign reduced imports to almost nothing and domestic stocks were nearly exhausted. At the time timber was a strategic commodity used extensively for pit props – the country's vital source of energy being coal. Many found this situation intolerable as problems co-ordinating timber supplies increased. Forest exploitation was carried out partly by the established trade and partly by the hastily created Timber Supplies Department. There were great logistical problems in shifting timber from the remote areas to places where it was needed. Transportation was carried out by horses and trains and in many areas there were severe shortages of horses as well as manpower. The administrative posts of the Timber Supplies Department were quickly filled by university-trained foresters, and many prisoners of war were sent to do the manual work. After the war the Prime Minister, Lloyd George, admitted that Britain was nearer to losing the war from lack of timber than from lack of food.

In 1917 the Acland Committee was appointed to examine the implications of insufficient supplies of home-grown timber. It recommended establishing some kind of forestry authority to remedy the timber deficiency. The committee also recommended that, in the long term towards the end of the twentieth century, 1 770 000 acres (716 000 ha) should be planted. On the basis of this recommendation, in 1919 the Forestry Commission was established. Some of the members of the Timber Supplies Department became the backbone of this Commission. As a first step it was hoped that, within the first decade, the Commission would create 200 000 acres (81 000 ha) of new plantations as well as restoring 50 000 acres (20 000 ha) of pre-existing forests. It was also agreed to encourage private planters by giving them state aid. Unfortunately, this early target figure of 250 000 acres (100 000 ha) was not achieved because a persuasive antiforestry lobby convinced the government that state forestry was not a profitable venture. When the government imposed an expenditure cut during the 1922–24 period the afforestation programme suffered especially badly.

Then came the Second World War which reminded the government of the reasons why the Forestry Commission was established in the first place. Bulky timber imports by sea proved to be an easy target for the German Navy, and the trees planted in the 1920s were not ready for felling. Nevertheless, the production of home-grown timber was increased to many times the level during peacetime and this took its toll mainly on the mature and remote forests. Although the availability of mechanical equipment speeded up the haulage operation, there were many logistical problems. The demand for softwoods was far greater than that for hardwoods, so the broad-leaved trees of the south suffered less severely than the coniferous forests of the north.

Having had two bitter experiences, after the war further and quite extensive plantations were created in many parts of Britain and forests in general have enjoyed a period of growth under the protective policies of post-war governments. Today the Forestry Commission is the largest landowner in Britain and manages well over 1.5 million ha of land. The Commission also gives aid to private foresters, and at present there are over 1.5 million ha of privately owned plantations. The duties of the Forestry Commission are: buying land, planting trees, harvesting and selling timber, establishing forest recreation centres for the benefit of the community, aiding private foresters by giving them advice and cash grants, doing research, and employing and training manpower to carry out all of these activities.

When the government of Northern Ireland was formed in 1921 it assumed responsibility for forestry in the province, and the new Department of Agriculture became the forestry authority. Today, the Department of Agriculture's Forest Service is in charge. This is a separate body from the Forestry Commission, but has similar powers and duties. In the Republic of Ireland the relevant government authority is the Forest Service of the Department of Energy. The Irish government's aim is to expand forestry projects as fast as possible, hopefully with the help of the European Community since Ireland, despite her very favourable climatic and soil conditions has, after Iceland, the fewest trees per square kilometre in Europe.

The rigour of forestry policy in the UK can be adjusted in line with changing domestic and world conditions regarding timber demand and supply. Economic circumstances may lead to future governments reconsidering their investment in forestry as a result of major shifts in the domestic market. As for global needs, it is worth mentioning that currently the world's natural forests are under great pressure and future supplies are bound to rely more on plantations. Therefore, in the future a timber famine is more likely than a glut. For these reasons the UK government intends to carry out five-yearly reviews to judge the progress of forestry programmes in the light of changing circumstances.

At present, forestry policy in the UK is governed by three factors. First, there is the compelling need, demonstrated by two world wars, to reverse the process of deforestation and create domestic resources of timber. It is now well known that dairy and beef sectors are overexpanded in the UK and some other northern countries in the European Community, creating surplus output, a source of embarrassment to governments. On the other hand, the UK and many other European Community countries are continuing to import large quantities of timber from outside. In effect, at present the European Community is only 50% self-sufficient in wood. The figure for the UK is 10–20%. The import bill for forest products in the UK amounted to about £6 billion in 1989. In many people's minds over-production of dairy and beef and under-production of timber are powerful signs of misallocation of land between the sectors in the country.

Second, there is a need to provide productive work in rural areas where structural unemployment prevails. Afforestation projects seem to offer a solution to the problem. The employment creation aspect of forestry, especially in the rural sector, was strongly emphasized during the creation of the Forestry Commission. Third, the government also believes that the private sector should be encouraged to participate in the expansion of forestry. In this respect it allocates public money in the form of cash aids and tax exemptions to private individuals or companies who are investing in forestry.

In spite of a 70-year effort by the authorities, forestry in the British Isles has never been as important as it is on the Continent. Only a little over 5% of the land is under trees in the whole of Ireland, North and South. In mainland Britain the figure is slightly higher, about 8% at the time of writing. Compare these figures with those for some other European countries: Holland 10%; Denmark 12%; Spain, Greece and Turkey over 15%; France, Italy and Belgium about 20%; Portugal, Germany and Norway over 30% and Finland a colossal 72%. On average it takes 50 years to grow coniferous timber in mainland Britain. The figure for the whole of Ireland is between 30 and 45 years, depending on the location. The gestation period for similar species in Germany and the Benelux countries is 80 years, and it is over 90 years in some parts of the Scandinavian peninsula.

3.1.1 Problems with discounting and consequently forestry becomes a special case

In a 1961 White Paper, *The Financial and Economic Obligations of Nationalized Industries* (Cmnd 1337), it was strongly argued that in the UK publicly owned industries should operate in a manner so as to earn a sufficient rate of return on their capital investments. In this respect, in 1967 in another White Paper, *Nationalized Industries: a Review of Economic and Financial Objectives* (Cmnd 3437), it was suggested that the discounted cash flows (DCF) method (also called by some the net present value criterion, NPV) together with an 8% test rate of discount should be used in evaluation. In August 1969 this rate was increased to 10%. In 1972 the rate was reviewed but not changed.

In 1967, some economists seemed to be satisfied with the choice of test rate of discount, and almost all were happy about the use of the discounted cash flows method in the appraisal of public sector projects. For example, Alfred (1968) first calculated a typical rate of return of 6.2% for the UK economy as a whole. After considering the income tax on profits he then increased his figure to 7.1% which was close to the government's first proposal of 8%. In view of the government's replacement of income and profit tax by a 42.5% corporation tax in April 1965, Alfred finally modified his estimate to 10.1%. This figure was also remarkably close to the increased test rate of discount.

However, some economists pointed out a number of problems with the choice, as well as the magnitude of the test rate of discount. First, the test rate of discount was based entirely on the opportunity cost rate argument. Indeed, there is a substantial body of economists who have always advocated that a public project would involve the sacrifice of some other projects. Thus the proper rate of discount must be the social opportunity cost rate. This rate is defined as the one that measures the value to society of the next best alternative investment project in which funds could have been employed. Generally, those next best alternatives are sought in the private sector and the objective behind the use of the social opportunity cost rate is to avoid displacing better investments in the private sector. If, for example, new investment projects in the private sector are earning a real rate of return of, say, 10%, the public sector projects should earn at least the same rate. However, this is only half the story. There is an equally powerful school of thought which has suggested the use of a different discount rate, the social time preference rate, also known as the consumption rate of interest. This school argues that the reason behind putting money in investment projects is to enhance future consumption capacity. In other words, an investment project involves a trade-off between present and future consumption. Therefore, what we need to do is to ascertain the net consumption stream of investment projects and then use the consumption rate of interest as a deflator. A much wider argument is given in Steiner (1959), Feldstein (1964, 1974), Arrow (1966), Arrow and Kurz (1970) and Kay (1972). The social time preference rate is defined as a rate that reflects the community's marginal weight on consumption at different points in time. The actual derivation of this rate for a number of countries is given in Kula (1984a, 1985, 1986a). The economic theory suggests that in the choice of a social rate of discount the two rates, i.e. the social opportunity cost rate and the social time preference rate, should play a joint role (Fisher, 1930; Eckstein, 1957, 1961; Feldstein, 1964, 1974; Marglin, 1963a). Nowhere in these White Papers was there a mention of the social time preference rate.

Second, in estimating the social opportunity cost rate as a basis for the test rate of discount, the government used extremely crude measures. The Treasury specified, in a memorandum submitted to the Select Committee on Nationalized Industries in 1968, how the figure was chosen (HCP 371-III appendices and index, appendix 7). It was the minimum return which would be regarded as acceptable on a new investment by a large private firm. Obviously this commercial rate of return on private capital was adjusted neither for market imperfections nor for externalities. In other words, the stock market's view of rate of return was taken to represent the social opportunity cost rate. Moreover, a crude average rate of return, rather than the marginal one, was taken in the choice of a figure for the test rate of discount.

Third, the funds that were used to finance the capital investments of nationalized industries were assumed to displace private investment rather than private consumption. In order to finance public projects, if the government borrowed the entire funds from the capital market, then the social opportunity cost rate would have been relevant provided that the private rate of return was adjusted for market imperfections and externalities. However, in most cases projects are financed by the tax revenue, a large proportion of which comes from the reduction in private consumption. The appropriate course of action in this case would be to use the consumption rate of interest rather than the social opportunity cost rate.

The strongest objection to the test rate of discount came from the pro-forestry lobby as forestry economists such as Price (1973, 1976) and Helliwell (1974, 1975) argued that the 10% test rate of discount had ended the hopes of an economic rationale for forestry investment in the UK. Indeed, some earlier studies on forestry in Britain had revealed that a very low discount rate was needed, of the order of 2%, so that a positive discounted cash flow figure could be obtained (Walker, 1958; Land Use Study Group, 1966; Hampson, 1972). Some economists (e.g. Thompson, 1971) argued that in most cases it is not possible to earn more than 3% compound from forestry in Britain without subsidies. However, he also maintained that in most cases this figure was higher than the return obtained in other countries of the temperate northern hemisphere.

In the discounted cash flows method with a discount rate as high as 10% the power of discounting practically wipes away the distant benefits that arise from felling in forestry. In order to salvage forestry projects, forestry economists have persistently argued for the use of lower rates of interest than the test rate of discount. They defended their arguments on the grounds that, unlike fabricated goods, the risk of land-based investments such as forestry becoming worthless in the distant future is extremely low. Since forestry is more or less a risk-free investment, then it should be discounted at a specially low discount rate.

All these criticisms of the test rate of discount turned out to be fruitful. In 1975 an interdepartmental committee of administrators and economists was set up to review the test rate of discount and to consider its relevance to a wide range of public sector ventures. These included the investments of public sector trading bodies and the nationalized industries' new capital projects, which take place in an environment of changing technology and market demand. The committee recognized the fact that in public sector investment appraisal the chosen rate of discount should reflect ideas on the social time preference rate as well as the social opportunity cost rate. The committee also stated that a justifiable figure for the social time preference rate would almost certainly be below 10%.

Additionally, there was no point in imitating the private sector unless the rate of return there correctly reflected society's view of profitability which was likely to be very different from the stock market's view. Also, it was hardly convincing to regard a public investment as diverting resources entirely from private investment rather than private consumption. In view of all these considerations the committee recommended a rate of 7% in real terms as striking a balance between the social profitability of capital and the social time preference rate.

After the committee finished its study in late 1976 there was a great deal of discussion during the run-up to the White Paper, *Nationalized Industries* (Cmnd 7131) in 1978. The appropriate rate of discount was considered in more detail with the departments directly concerned. The main change from the original recommendation was a reduction in the figure. The original 7% was the average of the 6–8% range. The later data showed that the appropriate profitability in the private sector was between 5 and 7%, and in the end 5% was chosen. This figure was called the required rate of return.

Even a 5% discount rate was too high to justify the UK's forestry programme on commercial grounds. Figures revealed by the Forestry Commission in 1977 showed an expected rate of return of about 2–2.5% on 60% of the Forestry Commission's acquisitions in Britain without subsidies (Forestry Commission, 1977a). These low figures were no surprise to foresters. After the publication of the 1967 White Paper, which came as a shock to foresters, many tried to inflate rate of return figures by putting up the following arguments. First, returns from forestry fail to take into account the increasing value of land acquired for afforestation. If this factor was considered, then it would no doubt improve the profitability. Second, cost–benefit analysis should not be based on current prices, which fail to take into account the historical fact that the price of timber rises. The future rather than the current price of timber should be used in any meaningful cost–benefit study. Third, recreational and environmental benefits from afforestation must also be incorporated into cost–benefit studies. Assumptions must be made about the expected visits to plantations by the general public as trees grow older and more attractive, about wildlife conservation, etc.

Going back a few years, in 1971 an interdepartmental government team was set up to carry out a cost–benefit analysis of British forestry. In this study a wide definition of profitability was taken as the criterion. The team considered the environmental, landscape and recreational benefits of forestry together with the value of the timber produced. Also the team tried to impute shadow prices, i.e. true prices based on the scarcity value of resources for the inputs as opposed to distorted market prices. Various sensitivity analyses were also carried out. In one, a 20% premium for import saving was levied on the home-grown timber. After all these painful and time-consuming calculations to improve the profitability, the maximum rate of return from forestry was increased to only 4%. The results of this study were published in 1972 (HMSO, 1972a).

Following this study, in 1972 the government revealed its forestry policy in a document in which it was stated that afforestation would continue in Britain as before (HMSO, 1972b). This document also indicated that this policy was guided by, but not based on, the conclusions of the cost–benefit study. With regard to the discount rate, the Forestry Commission was given a target real rate of return of 3% to be used as a discount rate in the discounted cash flows method in evaluation of forestry projects. The justification for such a low rate was the unquantifiable benefits of maintaining rural life and improving the beauty of the landscape, both of which are associated with forestry. In 1973 the Forestry Commission commissioned a study (Wolfe Report, 1973) to provide an argument to counter the 1972 cost–benefit analysis. In this study assumptions and omissions in the 1972 cost–benefit analysis were emphasized.

In the spring of 1989 the government revealed that since 1978 the rate of return in the private sector had risen to around 11%. In the light of this, the government decided to raise the required rate of return for nationalized industries and public sector trading bodies from 5 to 8% in real terms before tax. The discount rate to be used in the non-trading part of the public sector was to be based on the cost of capital for low-risk projects in the private sector. Under such conditions this indicated a rate not less than 6% in real terms. In the government's view these proposals would ensure that the appraisal of public projects would be no less demanding in the non-trading sector than in the trading sector, both public and private (*Hansard*, 1989). As for afforestation projects, the government did not change its discounting policy.

3.2   AFFORESTATION IN THE BRITISH ISLES

For the purposes of economic analysis the growth of trees is normally quantified in terms of volume, although it is possible to measure it by means of height, weight or dry matter. There are two important parameters in the measurement of volume: current annual increment and mean annual increment. The former represents the annual rate of increase in volume at any point in time and the latter the average rate of increase from planting to any point in time. Figure 3.1 illustrates the behaviour of these two rates. The current annual increment curve cuts the mean annual increment curve at the maximum point. For, *ceteris paribus*, if, in an even aged plantation, felling and replanting were repeatedly carried out at this point, then this maximum average rate of volume production should be

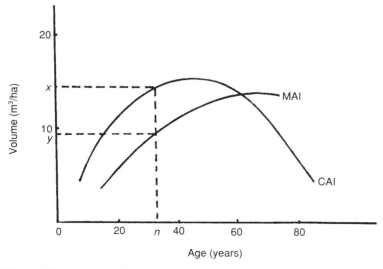

Figure 3.1   Patterns of volume increment, MAI, mean annual increment; CAI, current annual increment.

maintained for ever. Needless to say, differences in rates of growth occur within the same species on different sites. The growth figures also vary from species to species. In British conditions the annual growth can be as low as 4 m³/ha for some hardwoods and up to 30 m³/ha in the case of some softwoods.

Yield classes are created by splitting the growth range into steps of 2 m³/ha per annum, and normally even numbers only are used. For example, a stand of yield class 16 has a maximum mean annual increment of about 16 m³/ha, i.e. greater than 15 but less than 17.

There are a large number of species which can grow in the British climate. Softwood species that are well established in the British Isles include cedar, cypress, fir, hemlock, larch and spruce; hardwoods include ash, beech, birch, chestnut, elm, oak, poplar and sycamore. General yield class curves for oak and Sitka spruce are shown in Figures 3.2 and 3.3, respectively. The range of maximum mean annual increment is between 4 and 8 m³/ha for oak and between 6 and 24 m³/ha, or even higher, for Sitka spruce. Because of its high yield and relatively short gestation period, Sitka spruce is commonly planted in the British Isles by the official forestry authorities as well as by private individuals.

One of the advantages of planting, as compared with natural regeneration, is that it allows foresters to select the species that will achieve the aimed objectives. The choice of species is a crucial factor in the success of any afforestation programme. In addition to economic considerations the forestry authorities in the British Isles plant trees for reasons of water conservation, prevention of landslides, climatic improvement and providing protection from winds. In each

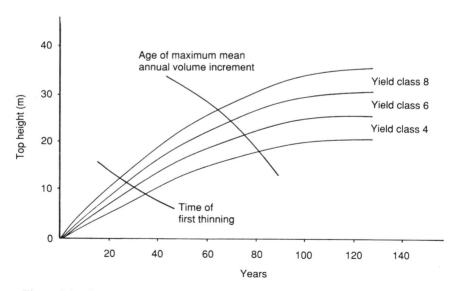

Figure 3.2   General yield class curves for oak. (Source: Forestry Commission, 1971.)

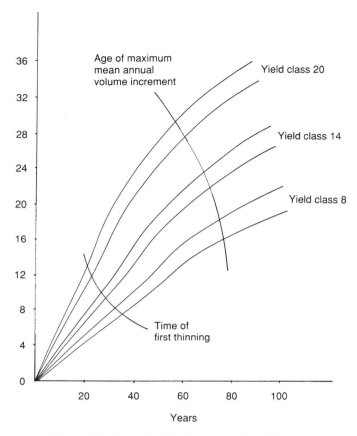

Figure 3.3   General yield class curves for Sitka spruce.

of these cases different species may be required. Foresters also take into consideration the influence of particular species on soil fertility. Species which are likely to impoverish the soil in a particular area are carefully avoided. The protection of land is as important a factor as the production of timber. Of course climate and soil types in a locality are the natural constraints on the choice of species. Where conditions for tree growth are favourable, there may be a wide range of choice and many different species can be considered. Where conditions are unfavourable, the choice may be limited to a few species only. The natural distribution of trees also follows this rule. At high altitudes and high latitudes most natural forests are composed of a single species or a mixture of very few species.

The climate of the west coast of Britain is closer to that of the west coast of North America than to that of continental Europe. Districts with high rainfall and a humid atmosphere have proved especially suitable for growing the American west coast conifers such as Sitka spruce, noble and grand fir, and therefore these species have been planted more frequently in the west. European species such as Norway spruce and some hardwood trees are more suited to the east coast of Britain. Chestnut and Corsican pine, both of which come from the Mediterranean region, are valuable species in southern Britain but are seldom seen in Scotland or northern England. Scots pine, on the other hand, yields greater volume in Scotland than anywhere else in the British Isles.

Sitka spruce now dominates the British scene and is likely to become even more widespread in the future because of the good results obtained in its plantation and management so far. It is native to British Columbia and to southern Alaska. It thrives even in subarctic conditions, growing right up to the timber line at the edges of glaciers. It grows very quickly once it reaches about 3 m. If allowed to, it can grow for several hundred years and in favourable conditions can attain a height of over 60 m.

Sitka spruce was introduced to Britain in 1831, but attracted little attention until after the end of the First World War. It is usually turf-planted on ground previously ploughed and often artificially drained. It tolerates poor soil but is intolerant of competition from heather. The most attractive aspect of this species is its high and quick yield of timber. The timber quality of Sitka spruce grown in the British Isles is not, however, as good as that of timber grown in North America, but it does make very good pulp. It is also used widely in the coal mining industry for pit props, and in house building mainly as interior construction material.

3.2.1 Development of a Sitka spruce plantation

It is important to bear in mind that in afforestation reliance entirely on a single species such as Sitka spruce is not a prudent decision. After all, trees are living things and like any other living thing they are vulnerable to disease, especially on non-native soil. An epidemic like the Dutch elm disease which occurred some years ago may inflict heavy damage. Just as a financial investor tries to diversify his/her portfolio, a prudent entrepreneur should plant more than one species. Indeed, Sitka spruce can be successfully planted with other types such as Norway spruce, lodgepole pine and Scots pine. In this section the basic procedure for establishing a Sitka spruce plantation will be explained briefly.

Ground preparation

Planting Sitka spruce on a good location is plain sailing and it can be carried out without much soil preparation. The young plants take root quickly and need little attention. Lush and fertile lowlands are ideal locations for many cultivers. Areas of heavy and wet clays require more work by way of ground preparation. Afforestation on blanket bog requires intensive draining before planting.

Planting

In most peaty locations the trees are planted by hand along the continuous peat ribbons which are turned out to one or both sides of the drain. After planting, a small amount, normally about 2 oz (110 g), of ground mineral or other phosphate is scattered around each plant. Planting-out age for Sitka spruce is normally 1–2 years.

Fencing

After young trees are planted the area must be fenced off against animals. There are various types of fencing to provide protection against sheep, rabbits, stock or deer. Although fencing is an efficient method of protecting the forest from animals, it is an expensive operation. Sometimes natural boundaries may form an effective barrier to animals and thus there may be no need for fencing. In some cases a fence may be obligatory because of some agreement in a lease or statute. Sometimes in a particular locality the damage which animals can inflict may be minimal. The types and methods of fencing are given in Forestry Commission (1972). The larger the area fenced, the lower the cost per hectare. The cost per hectare of fencing areas from 4 to 40 ha decreases considerably with increasing size. For areas over 40 ha the decrease is negligible.

Weeding

Weeds frequently compete with newly planted tree crops during the early years of a plantation's life. The main risk to the crop arises from smothering and excessive shading. Smothering also creates conditions in which fungi can flourish. Weeding on fertile soils can be an expensive operation as weeds are most persistent. Sitka spruce can force its way through lighter weeds and even bracken, which is sufficient to destroy most other trees. However, this species is more vulnerable to competition from heather than any other type of tree. A dense cover of heather causes nitrogen and phosphate deficiency, which is not easily corrected until the heather is controlled. For full details of weed control see Forestry Commission (1975).

Brashing

Brashing – knocking off the lower branches of the crop trees – takes place about 10 years after planting and well before the first thinning. The reason for brashing is to provide easy access into the forest so that observation paths can be created to monitor the progress of tree growth and so that thinning workers can operate in the plantation. Pushing through an unbrashed Sitka spruce plantation is an extremely difficult task. Sometimes trees are to be sold standing, and very few merchants would be prepared to make an offer for unbrashed wood as they would normally want to see the merchandise at close hand before buying.

Cleaning and pruning

Cleaning is the removal of weed growth between the trees. It can be done together with pruning. Weeds can be suppressed by covering the ground with tree branches from the pruning operation. Pruning is carried out to secure knot-free clear timber and is usually done on fast-growing conifers such as Sitka spruce. It is normally confined to the final crop of trees, i.e. the trees that will be kept until clear felling. Pruning is a way of investing in a tree.

Road construction

Forests cannot be managed without roads. Timber is a very bulky and heavy commodity and in most cases it is produced in places remote from main roads or railways. The roads that must be built inside the forest estate are more important for thinning than for clear felling. In the latter case it is possible to use heavy tackle and to cut gradually into the plantation to winch the logs over long distances to points where they can be loaded on lorries. For thinning, however, the operation must be carried out between very long rows of trees without causing damage to the crop, and for this the inner roads are essential. Since thinning has to be carried out at regular intervals, these roads will be in frequent use. In the early periods of forest management in the British Isles, forest roads were constructed when the plantations were newly established. However, the authorities soon realized that postponing road construction would improve the net present value figures because of discounting. Nowadays, it is a part of good forestry management to delay road building as long as possible, usually until about year 15, just before the first thinning operation gets under way. This also shortens the period for road maintenance. (For details of optimal road spacing see Forestry Commission, 1976.)

Maintenance

In this, the most relevant items are the maintenance of fences, drains and roads and ensuring good hygiene to prevent damage by pests. Apart from these items there are others which may be relevant in certain cases, such as beating-up and fire protection. Beating-up, also called supplying or beating, is the replacement of plants that succumbed in the initial planting. In many plantations the take is so good that no beating is necessary. In a minority of cases the area has to be entirely replanted. Sometimes beating-up is required only in patches.

In the British Isles, forest fires occur very rarely due to the wet and humid climate. Dry spells do not last long, even during high summer, and daily temperatures are seldom above 25°C. Unlike some southern European countries and the United States of America the fire-prevention cost in the British Isles is not very important in forestry. The damp climate is indeed a blessing for investors in forestry.

Thinning

Thinning is the cutting out of select trees from a plantation to gain a harvest and improve the growth and quality of the remainder. Economically, thinning can only be justified if the increased benefit from thinning outweighs the cost. There may be circumstances in which thinning may not yield benefits large enough to offset costs. In areas where road building is expensive, it may be worthwhile postponing the construction of roads until the clear felling. Roads are built early mainly because of thinning operations. Also the cost of maintenance following thinning may be high. For example, expenditure on drain cleaning after thinning may be quite high, which would make the thinning worthless. Alternatively, the revenue difference between the thinned and unthinned crop may be too small and it may pay off financially not to thin at all. The major benefits of thinning lie in providing earlier returns by encouraging the growth of the best trees so that the final yield is larger and higher in quality and also occurs earlier. Hiley (1967) gives a comprehensive discussion on various methods of thinning.

Felling

Some woods may contain unevenly aged trees, which may come about by design or by accident. In this case, felling is normally confined to small groups, which gives better results than complete clearance all at once. Clear felling indicates a complete clearance of the site at the intended period.

In forest management it is not necessary to stick rigidly to the intended period of rotation. Market conditions may change, favouring an earlier or later felling than was originally intended. Flexibility allows managers to take advantage of changing market conditions. A number of types of rotation are recognized by most foresters: technical rotation, rotation for maximum volume production, rotation for natural regeneration, and financial rotation. In the first, the felling takes place at the time when the plantation meets the specification of a given market, e.g. veneer logs, saw logs, pulpwood, etc. In the second, the objective is to determine a rotation which yields the maximum mean annual increment. Rotation for natural regeneration is determined by the age at which maximum viable seed is produced. Financial rotation is the one that yields a maximum discounted revenue.

3.3   BASIC PRINCIPLES OF COST–BENEFIT ANALYSIS FOR FORESTRY

The forestry authorities in the British Isles, i.e. the Forestry Commission in Great Britain and the Forest Services in Northern Ireland and the Republic of Ireland, pursue a similar set of objectives which broadly are:

1. to maximize the value and volume of domestically grown timber in a cost-effective manner;
2. to create jobs in the rural sector; and
3. to provide recreational, educational and conservation benefits and enhance the environment through visual amenity.

Afforestation projects must meet these basic criteria before approval can be given. As well as being technically sound, each acquisition must represent value for money in terms of the net benefits for society as a whole. To demonstrate that an afforestation proposal meets these criteria an evaluation by means of a cost–benefit analysis is normally carried out. Although there may be differences in emphasis, the basic procedure employed by each forestry authority is quite similar. It contains identification of costs and benefits: discounting; weighing up the uncertainties, e.g. future prices and felling age; and assessing environmental and distributional effects. Below, some of the general principles used by the Forest Service in Northern Ireland are outlined.

3.3.1 Valuation of costs and benefits

The concept of opportunity cost is the main principle in evaluation of costs and benefits for afforestation projects. The prices attached to the costs and benefits should reflect society's valuations of the final goods and resources involved which, in some cases, may differ from the prices actually paid. Indeed, most resources have alternative uses in the economy and the cost of using them is then the alternative use that is forgone. All costs and benefits should be included, no matter who accrues them. Both costs and benefits should be allocated in the years in which they occur. These costs and benefits should be expressed in real terms. With this approach there is usually no need to forecast future changes in the general price level. Price changes only need to be taken into account if there is a change in relative prices.

3.3.2 The cost of land

While the market price of land should normally reflect its opportunity cost, it may also reflect and include some element of the capitalized value of subsidies to agricultural production. Where applicable, an adjustment to exclude all subsidies from the market price of land should be made. The preferred approach is to take a percentage reduction in the market price of land to eliminate the effects of subsidies in identifying the real return to land. In Northern Ireland, most of the forestry acquisitions have generally been confined to less favoured areas and there has been a reduction of 55% in the market price for some years. The reduction for non-less favoured regions has been 45%.

3.3.3 Operational forestry costs

These costs include labour, machinery and materials. Cost estimates for these are based on average costs and it is assumed that these are equal to the marginal costs of additional projects. The opportunity costs of all three inputs have been assumed to equal actual costs.

In processing labour cost some additions are made on its market value. These are:

- 35% for sickness, leave and employers' National Insurance;
- 53% for unproductive time in view of bad weather, travel to and from work and subsistence;

- 33% Civil Service superannuation; and
- 91% supervisory and administrative addition.

This brings the labour conversion factor up to 212% of the cost of labour. These figures were for the 1985/86 financial year and they are revised when conditions change.

3.3.4 Benefits of forestry projects

The benefits, both quantifiable and unquantifiable, are: revenue from the sale of timber; recreation, educational and conservation benefits; employment creation and terminal values such as land and roads.

Revenue from the sale of timber depends very much on the future price of timber. The pricing of timber at some future date requires judgement on the price increase in real terms and the base price to which any such changes should be applied. The consensus among experts in this area is that timber prices will increase between 0% and 2% per annum and therefore real price increases of 1% and 2% should be used only as a sensitivity analysis. For discussions on the future price of timber see Potter and Christy (1962), Barnett and Morse (1963), Hiley (1967) and Kula (1988b).

In areas adjacent to existing forests it is thought unlikely that an additional investment will yield significant additional recreational or educational benefits. Therefore, these benefits are normally not considered in the economic evaluation of marginal projects. If, however, at some future date an appraisal is required for a forestry project located away from existing sites, particularly if near a large centre of population and therefore able to provide a net increase in recreational benefits, an estimate of these benefits will be calculated on an *ad hoc* basis.

As for the employment creation aspect, in order to be consistent with other investments no shadow values are applied to the cost of labour in forestry. Furthermore, the multiplier effect or the knock-on effect of investment expenditure on further income and employment is not normally taken into account as it is not unique to forestry. However, one of the aims of forestry investment is to alleviate rural employment. Therefore, if a project fails purely on financial grounds, there may be circumstances in which it could be justified on the grounds of relatively cost-effective job creation.

The terminal value of land and roads should be included in a cost–benefit study. These values are assumed to be equal to their initial value.

3.3.5 Discounting

As explained above (p. 82) the Forestry Commission in the past normally used a 3% target rate of return. In Northern Ireland, however, the Forest Service has used a 5% test rate of discount. It is common procedure to take the base date for discounting as the date of the initial investment or the date of the appraisal if that is different. This date would then be taken as year 0.

3.3.6 Distributional effects

A statement on distributional equity should form an important part of any cost–benefit analysis. To the extent that one of the aims of forestry is to redistribute income from the taxpayer to the unemployed through the job creation process, it may be that the distributional effects are self-evident. However, additional information may be necessary to reveal the extent to which the project provides employment for those already in part-time or seasonal employment. A knowledge of such effects might influence the decision as to whether or not forestry investment is a more effective means of reducing rural unemployment compared with other forms of job creation. It may also be worth while identifying potential losers from the scheme such as conservation, sporting and amenity groups.

3.3.7 Sensitivity analysis

It is desirable to carry out a sensitivity test to show the effects of possible variations on the assumptions made and their effects on the stream of costs and benefits. Assumptions to which the analysis is particularly sensitive should be noted.

3.4   A CASE STUDY IN NORTHERN IRELAND

In this section a cost–benefit analysis is carried out on a 100 ha plot on high ground which has a moderate slope and is located in the western part of Northern Ireland. The area is described as severely disadvantaged by the Department of Agriculture as its agricultural potential is poor. The species to be planted is Sitka spruce and the estimated yield class is 10. It is worth mentioning that in the past, the Forest Service focused its activities on the high grounds where the productivity of land is poor. Currently, efforts are being made to put afforestation projects on low regions where trees grow better. However, the plot chosen for this analysis is not unusual even in present circumstances.

The gestation period is assumed to be 45 years which includes a thinning regime. Table 3.1 shows the plantation costs in terms of 1986/87 prices. Although the Forest Service advice is to use the past 12 years average timber price as a base in revenue calculations, I am reluctant to do this for a number of reasons. First, if we have to take the 12-year average for the price of timber, then we should take the 12-year average in cost calculation. Indeed the cost of establishing a stand may fluctuate together with timber prices. Second, in view of the fact that as the long-term price of timber is increasing, taking the average of the past 12 years only will severely misrepresent the reality. This is because the past

Table 3.1   Cost details of afforestation, Northern Ireland, 1986/87 prices

Operation	Year	Total cost (£)
Draining	0	3 120
Ploughing	0	4 890
Planting	0	23 920
Fertilizer	0	9 050
Fencing	0	4 050
Maintenance		
Fertilizer	8, 16	14 600
Drains	15, 25, 35	1 980
Re-space	20	5 200
Rent	throughout	2 250

12 years contain two severe recessions, 1974–75 and 1981–84, when the price of timber fell quite sharply.

The rental value of this plot is about £5000, £50/ha per year. In view of the argument presented above (p. 90) a 55% reduction is made to the market value, thus reducing it to £2250 per annum for the whole area. Weeding is not carried out as the soil there does not encourage weed growth. Scrub cutting is not relevant as it is not typical in this area of Northern Ireland. Other operational assumptions are: no fence repairs, no beating-up, no brashing, no pruning and no new roads as the existing structure is assumed to be sufficient.

Table 3.2 shows the output details resulting from thinning and felling. Thinning starts in year 25 and continues at 5-yearly intervals. The output is divided into three groups: centimetre top diameter (cmtd) 7, which is good for

Table 3.2   Output details, Sitka spruce

Yield class 10	Thinning output (m³) in years				Felling, year 45
	25	30	35	40	
Cmtd 7, pulpwood[a]	2 900	3 400	3 300	3 200	17 500
Cmtd 18, palletwood	100	100	200	300	8 300
Cmtd 24, sawwood	–	–	–	–	1 300
Total output	3 000	3 500	3 500	3 500	27 100
Money benefits (£)	23 925	27 835	28 300	28 765	260 436

[a] Cmtd, centremetre top diameter.

pulp; cmtd 18 used as pallet and box wood; and cmtd 24 and over which is saw-quality timber. The 1986/87 roadside prices per cubic metre were: cmtd 7, £17.82; cmtd 18, £22.47; and cmtd 24, £25.45. From these about £10 must be deducted for felling and extraction costs. The net prices then become: cmtd 7, £7.82; cmtd 18, £12.47; and cmtd 24, £15.45. By using these prices the money benefits are obtained and are shown in the last row of Table 3.2.

A rate of return analysis is carried out by using the internal rate of return formula

$$\sum_{t=0}^{n} \frac{1}{(1 + x)^t} NB_t = 0,$$

where t is time (years), n is length of rotation (45 years), NB_t is net social benefit at time t, and x is internal rate of return.

The result is 2.5%. Note that this is with the assumption of constant prices.

3.5   PRIVATE SECTOR FORESTRY IN THE UK

3.5.1 Grants

Forestry authorities in the British Isles, i.e. the Forestry Commission in Great Britain and the Forest Services in Northern Ireland and the Republic of Ireland, have planting grant schemes for private investors. These grants are modified from time to time in view of changing economic circumstances. For example, inflation erodes the real value of the cash grant which reduces the incentive for private foresters. This may necessitate an increase in the nominal value of the grant.

In his annual budget speech on 15 March 1988 the Chancellor of the Exchequer announced new improved forestry planting grants for Britain and Northern Ireland. These grants apply to the establishment and restocking of broadleaved, conifer and mixed woodlands, whether by planting or by natural regeneration, and to the rehabilitation of neglected woodlands under 20 years of age. The objectives of the new grants were:

1. to divert land from agriculture and thereby assist the reduction of agricultural surpluses which have been creating waste;
2. to encourage domestic timber production and reduce the balance of payment deficit in the wood sector;
3. to provide real jobs in rural areas;
4. to enhance the landscape, to create new habitats and to provide for recreation and sporting uses in the longer term; and
5. to encourage the conservation and regeneration of existing forests and woodlands.

The new grants are paid in three instalments: 70% immediately after planting; 20% five years later; and the remaining 10% in year 10. Instalments under the

Table 3.3   Forestry planting grants in the UK as from April 1988

Areas of plantation (ha)	Conifers (£/ha)	Broadleaved trees (£/ha)
0.25–0.90	1005	1575
1.00–2.90	885	1375
3.00–9.90	795	1175
10.0 and over	615	975

old scheme were 80% when the approved area was satisfactorily planted and 20% five years later. Table 3.3 gives details of the 1988 grant scheme for conifers as well as broadleaved trees in the UK.

On receiving an application the forestry authorities first ensure that the land is suitable for forestry. They may then undertake consultation with the appropriate agriculture department, local authority or other authorities concerned with conservation, land use and amenity aspects of the proposal. An inspection of the land is normally carried out prior to planting, and a further inspection undertaken as soon as possible after the applicant has notified the forestry authority in charge that planting has been completed. The applicant is expected to maintain the plantation to the forestry authority's satisfaction.

Grants form a part of the investor's income and this must be incorporated into the financial analysis.

3.5.2 A case study for private sector profitability

In this section private rate of return analysis is carried out for an afforestation project located in County Tyrone, Northern Ireland. The soil in this location is wet and low lying which gives poor results in agriculture, but it is good for afforestation. The 30 ha plot is used for rough grazing during the summer months. The rental value of the land is £1000 per annum.

The species to be planted is Sitka spruce and the yield class is estimated to be 22 with a felling age of 30 years. Table 3.4 shows the cost details of plantation. Thinnings start in year 15 and continue at five-yearly intervals. The road building, which is a costly operation, takes place in year 14. Insurance payments increase as the stand matures and becomes progressively more vulnerable to wind blow. All costs are based on 1986/87 estimates.

Table 3.5 shows the output details which are divided into three groups: pulpwood, palletwood and sawwood. In 1986/87 the net prices (net of felling and extraction costs) in Northern Ireland were:

Pulpwood	£7.82
Palletwood	£12.47
Sawwood	£15.45.

Table 3.4   Afforestation costs, 30 ha, 1986/87 prices (£), Northern Ireland

Year	Plough	Drain	Plant	Fence	Fert.	Beating up	Road	Rent, insurane and maintenance	Total cost
0	4500	7000	10 000	1000	1000	–	–	1600	25 100
1					1500	3000	–	1400	5 900
2					500		–	1300	1 800
3							–	1200	1 200
4							–	1200	1 200
5							–	1200	1 200
6							–	1200	1 200
7							–	1200	1 200
8							–	1600	1 600
9							–	1200	1 200
10							–	1225	1 225
11							–	1250	1 250
12							–	1275	1 275
13							–	1300	1 300
14							25 000	1325	26 325
15							–	1350	1 350
16							–	1375	1 375
17							–	1400	1 400
18							–	1425	1 425
19							–	1450	1 450
20							–	1475	1 475
21							–	1500	1 500
22							–	1525	1 525
23							–	1550	1 550
24							–	1575	1 575
25							–	1600	1 600
26							–	1625	1 625
27							–	1650	1 650
28							–	1675	1 675
29							–	1900	1 900
30							–	2100	2 100

Using these prices the output is converted into money benefits and the results are shown in the last row of Table 3.5.

The planting grant due for the whole area (30 ha) expressed in terms of 1986/87 prices is £17 527 and the instalments are:

First payment, year 0	£12 269
Second payment, year 5	£3 505
Third payment, year 10	£1 753
	£17 527

Table 3.5   Output details, Sitka spruce

Yield class 22	Thinning output (m³) in years			Felling, year 30
	15	20	25	
Cmtd 7, pulpwood	750	2 220	2 040	4 680
Cmtd 18, palletwood	30	90	270	4 620
Cmtd 24, sawwood	–	–	–	1 860
Total output	780	2 310	2 310	11 160
Money benefits (£)	6 233	18 483	19 320	122 346

Table 3.6   Net benefit stream, 30 ha, 1986/87 prices (£), Northern Ireland

Year	Cost	Grant aid	Thinnings	Felling	Net benefit
0	25 100	12 269	–	–	–12 831
1	5 900	–	–	–	–5 900
2	1 800	–	–	–	–1 800
3	1 200	–	–	–	–1 200
4	1 200	–	–	–	–1 200
5	1 200	3 505	–	–	+2 305
6	1 200	–	–	–	–1 200
7	1 200	–	–	–	–1 200
8	1 600	–	–	–	–1 600
9	1 200	–	–	–	–1 200
10	1 225	1 753	–	–	+528
11	1 250	–	–	–	–1 250
12	1 275	–	–	–	–1 275
13	1 300	–	–	–	–1 300
14	26 325	–	–	–	–25 325
15	1 350	–	6 239	–	+4 889
16	1 375	–	–	–	–1 375
17	1 400	–	–	–	–1 400
18	1 425	–	–	–	–1 425
19	1 450	–	–	–	–1 450
20	1 475	–	18 483	–	+17 008
21	1 500	–	–	–	–1 500
22	1 525	–	–	–	–1 525
23	1 550	–	–	–	–1 550
24	1 575	–	–	–	–1 575
25	1 600	–	19 320	–	+17 720
26	1 625	–	–	–	–1 625
27	1 650	–	–	–	–1 650
28	1 675	–	–	–	–1 675
29	1 900	–	–	–	–1 900
30	2 100	–	–	122 946	+120 846

Table 3.6 shows the entire cash flow situation for this investment. The internal rate of return method is used to ascertain the commercial value of this investment, that is:

$$\frac{B_0 - C_0}{(1 + x)^0} + \frac{B_1 - C_1}{(1 + x)^1} + \frac{B_2 - C_2}{(1 + x)^2} + \ldots + \frac{B_n - C_n}{(1 + x)^n} = 0,$$

where t is time (years); n is the project's life (30 years in this case); B is cash benefit (subscripts refer to a particular year); C is cost (subscript refer to a particular year); and x is the internal rate of return. Using this formula on cash flow figures (Table 3.6) yields the following result:

Internal rate of return $(x) = 4.3\%$ in real terms;

for similar results see Convery (1988).

3.5.3 Taxation and private forestry in the UK

Until the beginning of the 1988/89 financial year where woodlands were managed on a commercial basis and with a view to the realization of profits, the occupier was assessable for income tax under Schedule B unless he had elected in respect of those woodlands for assessment under Schedule D. Schedule B assumed that the owner received an annual rent equal to one-third of the annual value of the land in its unimproved state, i.e. without trees growing on it. This assumed rent was of the order of £3/ha and the taxable income was thus £1. No tax was payable on the proceeds of any sale of timber or on receipt of planting grants. The occupier was able to elect to place his forest under Schedule D when all profits were treated as trading profits and taxed accordingly in the annual balance between income and expenditure.

Under Schedule D all the costs of forest establishment and maintenance could be offset against other sources of income. Also, grants were taxable under this schedule. This relief against other taxable income was particularly attractive to high-rate taxpayers and thus forestry management companies were successful in encouraging such people to invest in afforestation projects. In particular, famous and wealthy snooker players, musicians, television personalities, etc., were very active in buying forests in Scotland.

In his 1988/89 budget speech, the Chancellor of the Exchequer announced that the tax system which had been in operation for many years could not be justified as it enabled top tax payers to shelter other income from tax by setting it against expenditure on forestry, whereas the proceeds from any eventual sale were effectively tax free. He then stated that as from 15 March 1988 expenditure on commercial woodlands would no longer be allowed as a deduction for income tax and corporation tax, but equally, receipts from the sale of trees or felled timber would no longer be liable to tax. The effect of these changes, together with the increased grants, was to end an unacceptable form of tax shelter, to simplify the tax system and abolish the archaic Schedule B in its entirety, and to enable the government to secure its forestry objectives with proper regard for the environment.

The payment of inheritance tax can be deferred on timber until the trees are sold. Capital gains tax only applies to the land. For those occupiers who were committed to commercial forestry before 15 March 1988, the pre-budget tax and grant levels remain for five years, i.e. until 5 April 1993.

3.5.4 The rationale for planting grants and tax incentives

Why should the government single out forestry and make it a special case by giving planting grants and tax concessions to the private sector? There are a number of reasons for this. First, forestry is still a young sector in the UK and like most infant ventures it requires nursing. As explained above, the forestry culture which existed in the distant past disappeared along with the forests during the course of history. Although the forestry authorities in the British Isles were established more than 70 years ago, creation of a viable forestry sector proved to be a painfully slow process. The size of the forest estate in these islands is still minute compared with other European countries, and no doubt tax exemptions and planting grants are likely to stay in force for many years to come.

Second, it is now a well-publicized fact that some traditional sectors in agriculture such as meat, dairy and cereals are overexpanded in the northern countries of the European Community. On the other hand the Community is only about 50% self-sufficient in timber and there is a compelling need to transfer land from overexpanded agriculture to forestry.

Third, forestry provides employment opportunities for the population in rural areas where structural unemployment prevails. In effect, the employment creation aspect of forestry in the rural sector was strongly emphasized when forestry authorities were first established in the British Isles. Especially in areas where the land is not suitable for lucrative agriculture, rural depopulation has been a problem for many years. Afforestation prevents migration from the countryside by providing a source of income and employment there.

Fourth, in forestry there is a long time interval between the establishment cost and the revenue which arises from the sale of timber. Due to the unusually long gestation periods which persist in forestry, some potential investors, especially the small ones, tend to stay away from it. Planting grants which are given in instalments break, to some extent, the cycle of no income as well as recovering a good part of the initial investment cost.

3.6 A PLANTING FUNCTION FOR PRIVATE AFFORESTATION PROJECTS

What are the factors that determine the size of forestry uptake in the private sector? Surely when a person invests in forestry he must do so simply because he believes that it is a good business. In a minority of cases, however, tree planting may take place for ornamental purposes. Nearly 80% of all forestation in the UK takes place in Scotland, and is mostly composed of Sitka spruce, Scots and lodgepole pines. During the last ten years the size of private sector forestry in Scotland has increased by more than 200 000 ha. Since the introduction of the 1980 forestry policy statement and enhanced rates of grant in 1981, new planting in the private sector has risen sharply.

The response of private foresters to an increase in grant levels must surely be positive otherwise there would be no point in increasing grants, or indeed in introducing them in the first place. Other variables which may have a similar impact on the uptake are likely to be tax incentives and the price of timber, which is an indication of profitability. All these variables should normally be positively correlated with decisions to plant trees.

On the other hand, there must be other variables which could discourage potential investors in forestry. The real rate of interest must be one of these factors. If this rate is increasing the negative effect will be twofold. First, the high interest rate will make the establishment cost dearer (interest rate is the cost of borrowing), thus discouraging potential investors. Second, with high interest rates, investments in time deposits will become rather more attractive, making forestry less appealing. Another negatively correlated variable is likely to be the price of land, the reason being that afforestation reduces the marketability of land. For example, in a period of rising land prices landowners may choose to sell or lease their land. This may, however, prove difficult if they are tied up in afforestation. Prices of alternative agricultural commodities, which could have been produced had the land not been put under trees, are likely to be another negatively correlated variable. A high price for these commodities would make agriculture more lucrative in comparison with forestry.

3.6.1 An economic modelling of planting function

In this section a planting function for coniferous species will be constructed in Northern Ireland, which stems from work by Kula and McKillop (1988) and McKillop and Kula (1987).

The planting function in its most general form consists of the following variables:

$$\Delta Q_t = f(PL_t, PS_t, I_t, G_t, R_t, T_t),$$

where

ΔQ_t is the change in the size of private softwood afforestation in year t;
PL_t is the price of land in year t;
PS_t is the price of imported sawn softwood in year t which dominates the price in the British Isles;
I_t is a weighted price index of selected agricultural commodities in year t;

G_t is the present value of government grants available for softwood planting in year t;

R_t is the market rate of interest, in real terms, in year t; and

T_t is the level of tax incentive.

The relevance of all these independent variables must be obvious. Land price, PL_t is included on the basis that afforestation reduces the marketability of land as has already been explained. The correlation between the dependent variable, ΔQ_t and PL_t should be negative. That is, an increase in land prices is bound to make afforestation projects less attractive for the reasons given above.

The price of imported softwood, PS_t, should be positively correlated with the dependent variable ΔQ_t as it is an indication of profitability. If the timber prices are rising, this should encourage investors to take up afforestation projects on the grounds that profits in this line of business are increasing.

The independent variable I_t, price index of select agricultural commodities, is relevant because it is a measure of the profitability in the competing agricultural sector. The correlation between this variable and ΔQ_t should be negative because an improvement in agricultural profits would, *ceteris paribus,* reduce the attractiveness of afforestation projects.

The present value of government grants for forestry, G_t, is a crucial parameter in planting functions, and one should expect it to be positively correlated with the dependent variable, ΔQ_t. This variable must be expressed in terms of net present value because of the fact that grants are given in instalments.

Finally, the market rate of interest, R_t, and an index based upon tax levels, T_t, are also included. However, neither of these proved to be significant for the reasons which will be explained below.

The planting function is specified in log-linear form, which tends to give the best results. It is also very convenient for elasticity calculations. The specification by ignoring the time lags is as follows:

Log-linear model without lags

$$\ln \Delta Q_t = \ln \beta_0 + \beta_1 \ln PL_t + \beta_2 \ln PS_t + \beta_3 \ln I_t + \beta_4 \ln G_t + \beta_5 \ln R_t + \beta_6 \ln T_t.$$

However, this model is incomplete owing to the problems of delayed response. For example, the current value of the dependent variable, ΔQ_t, is a function not only of the current grant, but also of the past, or lagged, values of it. Because of this problem a distributed lag model which was first developed by Almon (1965) and refined by Cooper (1972) is used. In this model coefficients are constrained to lie on a polynomial-shaped lag. The lag structure specified in the model was that of a polynomial degree of three. This contains the lag structure, to have at most two turning points. However, as detailed in the results, the lag structure for the majority of variables was quadratic in form.

Log-linear model with distributed lag

$$\ln \Delta Q_t = \ln \beta_0 + \sum_{i=0}^{n-1} \beta 1_i \ln PL_{t-i} + \sum_{i=0}^{n-1} \beta 2_i \ln PS_{t-i}$$
$$+ \sum_{i=0}^{n-1} \beta 3_i \ln I_{t-i} + \sum_{i=0}^{n-1} \beta 4_i \ln G_{t-i} + \sum_{i=0}^{n-1} \beta 5_i \ln R_{t-i}$$
$$+ \sum_{i=0}^{n-1} \beta 6_i \ln T_{t-i}.$$

Figure 3.4 shows the behaviour of the lag structure with respect to government grant, G_t, over a five-year period.

With regard to the relevant lag lengths, i.e. the number of years which the change in independent variables will last, experiments with lag lengths of $n = 4$ to $n = 12$ were conducted. The criteria for deciding on the optimal lag length are R^2 (the goodness of fit) and a requirement of close similarity of the weights between the optimal lag and lags of slightly longer length. On the basis of these criteria, lag lengths of the order $n = 6$ were arrived at.

The reader should note that, if the length of the lag is incorrectly specified, then the estimates will be biased.

The data sample covers the years 1948/49 to 1983/84, a total of 36 observations with explanatory variables, where appropriate, expressed in 1966/67 prices. Observations for the dependent variable, ΔQ_t, annual increase in softwood plantation, are shown in the second column of Table 3.7. Land prices in Northern Ireland, the first independent variable, are deflated by an index of agricultural product prices in order to capture the change in real terms. The price of imported sawn softwood is deflated by the wholesale price index. A composite

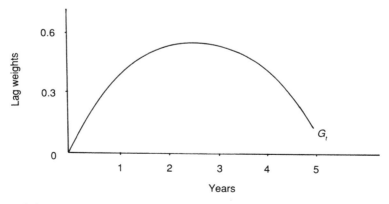

Figure 3.4   Behaviour of the government grant, G_t, with a distributed lag in log-linear planting function.

Table 3.7 Variables for forestry planting functions

Year	Increased hectares under softwood afforestation (new planting and restocking) (ΔQ_t)	Price of land per hectare at constant prices (PL_t)	Price of imported sawn softwood at constant prices (PS_t)	Index of agricultural commodities at constant prices (I_t)	Present value of grant at constant prices (G_t)
1948/49	81	44.6	79.9	100.5	74.2
1949/50	81	42.8	74.8	100.2	83.7
1950/51	68	47.9	87.4	94.1	88.9
1951/52	62	45.2	101.4	98.4	88.9
1952/53	149	33.9	93.1	105.2	84.1
1953/54	150	30.3	86.3	107.5	85.5
1954/55	162	36.1	87.0	99.4	100.0
1955/56	214	26.0	90.2	94.6	107.9
1956/57	263	19.7	91.2	95.7	102.9
1957/58	197	26.0	84.6	101.3	107.3
1958/59	173	35.7	77.7	99.4	112.7
1959/60	200	42.2	81.9	102.0	110.7
1960/61	220	50.2	88.7	105.3	108.0
1961/62	201	47.1	86.2	107.2	107.2
1962/63	215	64.4	93.7	101.6	106.7
1963/64	214	89.2	99.8	105.8	108.7
1964/65	227	100.4	104.9	107.7	103.5
1965/66	194	98.1	105.7	97.1	102.6
1966/67	187	100.0	100.0	100.0	100.0
1967/68	205	100.8	103.4	106.8	120.2
1968/69	223	100.1	110.0	112.0	137.3
1969/70	283	123.1	113.8	112.8	130.2
1970/71	236	134.3	114.6	118.6	120.3
1971/72	223	147.5	108.6	114.7	109.0
1972/73	157	256.3	115.0	116.5	93.0
1973/74	243	316.9	150.4	129.8	78.0
1974/75	136	153.3	152.1	104.3	63.9
1975/76	73	178.2	129.0	105.9	53.4
1976/77	34	211.9	136.7	109.6	47.4
1977/78	62	234.2	133.1	142.2	79.5
1978/79	24	333.5	123.5	167.3	100.5
1979/80	37	380.8	128.7	150.4	83.0
1980/81	125	248.1	121.5	144.3	71.9
1981/82	181	259.1	105.1	131.1	273.0
1982/83	34	209.6	105.3	114.4	452.0
1983/84	66	236.3	119.7	112.5	421.0

Source: Kula and McKillop (1988).

index of agricultural commodities, I_t, is deflated by the agricultural price index. This composite index apportioned a 30% weight to cattle and 70% weight to sheep prices on the grounds that the bulk of afforestation takes place on hill and upland areas suitable in the main for sheep farming.

Over the years, the grant schemes have changed many times. For example, the first grants administered by the Forest Service related to ground preparation as well as planting. Furthermore, larger grants were available for corporate bodies than for private individuals. From 1947 to 1981 the main focus for grant aid was the Dedication Scheme, offering a planting grant with an annual management grant attached. Introduced in 1981, the Forestry Grant Scheme consisted of a planting grant only and no subsequent management grant; 80% was paid on planting and the remainder after five years, subject to satisfactory establishment. To accommodate the time value of money concept it was decided to express the total grant available to an applicant in present value form with future pay-offs being discounted at the average bank base rate for the year under construction.

As regards the average annual interest rate, R_t, on an *a priori* basis one would normally expect an inverse relationship between interest rates and forestry investment. High interest rates raise both borrowing costs and the opportunity cost of locking finance into forestry. However, this variable did not prove to be significant. Its insignificance may be explained partly by the fact that initial investment costs are grant aided.

Another variable which did not prove to be significant was that of the tax incentive, T_t. Various attempts to incorporate this variable into regression equations have been unsuccessful. A possible explanation of this is the fact that financial institutions, likely to be more responsive to tax incentives than small farmers, have so far been unwilling to invest in afforestation projects in Northern Ireland. This reluctance to invest is explained when one considers that, in order to achieve economies of scale, financial institutions are interested in large-scale afforestation (100 ha and over). Such tracts of land are not available in Northern Ireland as land is fragmented into small ownerships. The author believes that a similar analysis conducted for Great Britain would result in a highly significant role for tax incentives. The observations for statistically significant variables are shown in Table 3.7.

Results

The regression results with the distributed lag are as follows:

$$\ln \Delta Q_t = 6.3 + \sum_{i=0}^{5} \beta 1_i \ln PL_{t-i} + \sum_{i=0}^{5} \beta 2_i \ln PS_{t-i}$$
$$+ \sum_{i=0}^{5} \beta 3_i \ln I_{t-i} + \sum_{i=0}^{5} \beta 4_i \ln G_{t-i}$$
$$R^2 = 0.9; \; DW = 2.2; \; F(12, 23) = 17.6,$$

where R^2 refers to the goodness of fit; DW is the Durbin–Watson statistic, which measures the level of autocorrelation; and the F statistic measures the significance of the regression as a whole.

The polynomial lag coefficients, i.e. details of $\beta 1_i$, $\beta 2_i$, $\beta 3_i$ and $\beta 4_i$ are:

$\beta 1_0 = 0.5\ (2.6)$	$\beta 2_0 = 1.2\ (2.5)$	$\beta 3_0 = -1.6\ (2.6)$	$\beta 4_0 = -0.1\ (0.3)$
$\beta 1_1 = 0.0\ (0.1)$	$\beta 2_1 = 1.1\ (4.3)$	$\beta 3_1 = -1.5\ (4.1)$	$\beta 4_1 = 0.4\ (3.1)$
$\beta 1_2 = -0.2\ (1.7)$	$\beta 2_2 = 0.9\ (3.6)$	$\beta 3_2 = -1.2\ (3.4)$	$\beta 4_2 = 0.5\ (3.6)$
$\beta 1_3 = -0.3\ (3.1)$	$\beta 2_3 = 0.8\ (3.0)$	$\beta 3_3 = -0.8\ (3.3)$	$\beta 4_3 = 0.4\ (3.3)$
$\beta 1_4 = -0.3\ (2.0)$	$\beta 2_4 = 0.5\ (1.8)$	$\beta 3_4 = -0.5\ (1.5)$	$\beta 4_4 = 0.2\ (1.2)$
$\beta 1_5 = -0.1\ (0.9)$	$\beta 2_5 = 0.2\ (1.0)$	$\beta 3_5 = -0.2\ (0.3)$	$\beta 4_5 = 0.0\ (0.2)$

The sums of the lag coefficients are:

$$\sum_{i=0}^{5}\beta 1_i = -0.4\ (2.2); \quad \sum_{i=0}^{5}\beta 2_i = 4.7\ (3.7);$$

$$\sum_{i=0}^{5}\beta 3_i = -5.8\ (5.1); \quad \sum_{i=0}^{5}\beta 4_i = 1.4\ (3.7)$$

where the numbers in parentheses refer to the t statistic, which is a measure of statistical significance.

The meaning of these results is quite straightforward. Take the grant variable, for instance. The elasticity of planting function with respect to this variable is 1.4. This means that a 1% increase in the government grant, in real terms, would generate about a 1.4% increase in private forestry uptake over a five-year period. The elasticity of the function with respect to variable PS, the price of sawn softwood, is 4.7, meaning that a 1% increase in timber prices would generate an approximate 4.7% increase in plantation. Likewise, if there were a 1% increase in I_t, the price index for selected commodities, this would generate a 5.8% decrease in afforestation.

In terms of 'goodness of fit', the model is a good representation of the underlying data. All coefficients have theoretically expected signs, that is, annual plantation is positively correlated with both timber prices and grant levels and is negatively related to the index of agricultural commodities and land prices. With regard to land prices, as detailed earlier, a negative correlation is to be expected for Northern Ireland. In other regions of the UK where afforestation is expanding, this negative relationship may not, however, pertain. As explained above, this is primarily because financial institutions, such as pension funds and investment banks which are attracted by grants, tax exemptions and a stable political climate have placed upward pressure on land prices.

The main variable of interest in this study is that of grant aid which can be employed as a policy variable to influence forestry uptake. As can be seen in Figure 3.4 the weighted distribution takes the form of an inverted U distribution: weights increase up to year 3, and then decline on moving through the remainder of the time period. That is to say the maximum impact of an increase in grant will be realized around year 3.

It should also be noted that the estimated coefficients for the prices of timber and the index encompassing sheep and cattle indicate that these variables are highly elastic over the period.

This empirical research on Northern Ireland has shown that, according to the forestry planting function described above, only four variables are of relevance: land prices, softwood prices, a weighted price index of agricultural commodities and grant-in-aid. Furthermore, the models employed emphasize the importance of the lag structure in private afforestation projects.

From a policy perspective it is evident that the level of grant can be employed to influence forestry uptake. It must be said, however, that, as uptake is marginally elastic with respect to grant, if the government wishes to encourage substantially the size of private sector forestry, a quite fundamental restructuring of the grant system and/or a revision of the subsidies payable on substitute agricultural commodities must be undertaken.

The response of the private sector to afforestation projects in Northern Ireland has, however, so far been disappointing. It is the author's view that any further restructuring of forestry and agricultural subsidy schemes to reallocate land in the province should take two added factors into account. First, the absence of a steady income from forestry discourages many potential foresters. It takes an average of about 20 years for the first sizeable benefits to arise from thinnings. Second, some Ulster farmers are conservative and may not readily accept the idea of abandoning traditional farming in favour of forestry, irrespective of the level and structure of grants available.

3.7   THE OPTIMUM ROTATION PROBLEM

One of the most interesting problems in forestry economics is the selection of the optimal time to cut a tree or an even-aged forest. An even-aged forest refers to timber stands composed of trees of only one age group. This problem has been known since the work of Von Thunen and (1826) and Faustmann (1849) in the nineteenth century.

In practice, the harvesting period depends on many factors such as the final use of the tree, e.g. whether it is for a bean stick, Christmas decoration, pulp, palletwood or sawwood. Other important factors are the cost of harvesting and the price of wood at the time when cutting is contemplated. For example, the demand for wood in the area may be very depressed at a given time. It is only common sense to postpone the felling until the price is favourable. Identifying the optimum rotation requires clear assumptions without which there can be no solution to the problem. In establishing these assumptions an attempt will be made to simplify the problem without losing the essence of the argument. The assumptions are as follows.

Assumption 1.   Future timber yields from an even-aged stand are known with certainty, or, to put it another way, the productivity of land on which trees are planted is known. When the output details are worked out, no substantial change is expected in the fertility of the land or in the management technology over time.

Assumption 2.   The trees are widely spaced and will not undergo thinning. That is, a no-thinning regime is assumed.

Assumption 3.   The trees will be sold standing, i.e. harvesting and haulage costs will not be incurred by the owner.

Assumption 4.   All future timber prices are constant and known. The price is assumed not to be sensitive to the age or volume of wood. That is, a cubic metre of wood has the same value, independent of the size and the age of the trees cut down. This is somewhat unrealistic, but at this stage divergence from this assumption will complicate the problem.

Assumption 5.   The interest rate is known with certainty and it is constant over all future periods.

Assumption 6.   Planting and replanting costs are also known and they are constant for all future periods. When replanting is involved it is done immediately after harvesting the stand. The intervals between planting and replanting are constant.

Assumption 7.   Risks from forest fires, windblow and disease are minimal and are thus ignored.

Assumption 8.   Forest land can be bought, sold and rented. This price could be very low or very high, depending on the demand and supply conditions for the land.

Assumption 9.   There is no taxation of any kind. It is well known to many forestry economists that taxation makes quite a difference to the optimum rotation decision; this will be dealt with later on.

3.7.1 Single-rotation solution

In the single-rotation system the owner tries to identify a time period to maximize his profit over a single cycle. Although the single-rotation model has some serious limitations which will be explained later it is useful to work it out at this stage as it gives an intuitive understanding of the problem.

Bearing the above assumptions in mind, let us argue that the owner has a very simple revenue function:

$$R_t = A \left(1 + t\right)^{\frac{1}{2}},$$

$$(3.1)$$

where t is time expressed in terms of years, 0, 1, 2 ...; R_t is revenue at time t resulting from the sale of timber; A is a constant.

Let the initial capital cost be C, and the interest rate or the opportunity cost of capital be r. The net present value (NPV) of this investment is given by

$$NPV = -C + \frac{1}{(1 + r)^t} R_t,$$

which can also be written as:

$$NPV = R_t e^{-rt} - C. \tag{3.2}$$

What is the felling time which maximizes the net present value of this project? To find the answer, first we differentiate equation (3.2) with respect to time t. That is:

$$\frac{dNPV}{dt} = -rR_t e^{-rt} + \frac{dR_t}{dt} e^{-rt} \tag{3.3}$$

Then equation (3.3) is set equal to zero

$$-rR_t e^{-rt} + \frac{dR_t}{dt} e^{-rt} = 0$$

or

$$rR_t e^{-rt} = \frac{dR_t}{dt} e^{-rt}.$$

To solve for r both sides are divided by $R_t e^{-rt}$, which yields:

$$r = \frac{dR_t / dt}{R_t}. \tag{3.4}$$

Equation (3.4) tells us that the marginal rate of return on capital must be equal to the interest rate. That is, the net present value of this investment is maximized when the marginal revenue from the selling of timber is equal to the opportunity cost of the capital, i.e. the market rate of interest. In other words, the trees should be felled when the incremental revenue equals the market rate of interest. This should be intuitively obvious. It is not profitable to maintain the stand when the marginal revenue becomes lower than the market rate of interest. It would then be profitable to cut the forest and deposit the timber proceeds in a bank account. In Figure 3.5 the optimum rotation is found geometrically. There are two curves in the figure: the revenue curve which results from the sale of timber, and the opportunity cost of capital (the interest

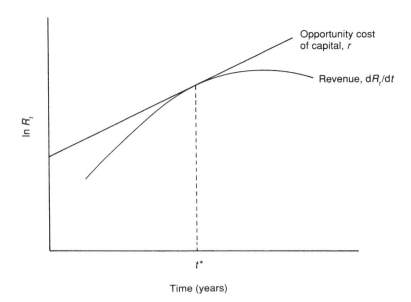

Figure 3.5   Geometric solution to the optimum rotation problem.

rate curve). Note that the vertical axis is a logarithmic scale so that continuously compounded interest appears as a straight line. Also the revenue curve becomes more compressed. The optimum solution is where the opportunity cost curve, which corresponds to the left-hand side of equation (3.4), becomes tangent to the revenue curve, or it could cut it.

Example 3.1
The revenue function $R_t = 4000(1+t)^{\frac{1}{2}}$, $r = 0.02$ or 2% and $C = 6000$. Find the optimum rotation.
Differentiating the revenue functions in terms of t, gives:

$$\frac{dR_t}{dt} = \frac{1}{2}\,4000\left(1+t\right)^{-\frac{1}{2}}$$

$$= 2000\left(1+t\right)^{-\frac{1}{2}}$$

$$= \frac{2000}{\left(1+t\right)^{\frac{1}{2}}}.$$

By using equation (3.4)

$$0.02 = \frac{2000/(1 + t)^{\frac{1}{2}}}{4000(1 + t)^{\frac{1}{2}}}$$

$$0.02 = \frac{1/(1 + t)^{\frac{1}{2}}}{2(1 + t)^{\frac{1}{2}}}$$

$$0.02 = \frac{1}{2(1 + t)}$$

$$1 = (0.02)\big[2(1 + t)\big]$$

$$1 = 0.04 + 0.04t$$

$$0.04t = 0.96$$

$$t = 24 \text{ years.}$$

3.7.2 Problems with the single-rotation model

The single-rotation period is limited to a case in which the land is available in unlimited supply and thus the land rent is zero. In other words, one planting cycle only involves cash receipts other than land rent. If the land rent is not zero, then the single-rotation model, at least in the form described above, will be a misleading one. Suppose that the land is not in unlimited supply and the forest owner has the option of cutting the stand and letting out the land. Therefore, in this more general situation, the extra cost from lengthening the rotation by one period is equal to the rental value of the land which would have been paid during this period. Equation (3.4) does not capture this argument. Many well-known economists as well as foresters (Von Thunen, 1826; Hotelling, 1925; Fisher, 1930) erred in their analyses when they generalized the single-rotation solution, which is indeed a very limited case. For an excellent review see Samuelson (1976).

Now, let us approach the problem from a quite different angle. It has been argued by some economists, such as Boulding (1935) and Goundrey (1960), that the internal rate of return method, which yields zero net present value gives the best solution to the rotation problem in forestry economics. Instead of finding the time period that maximizes the net present value, let us find a time period that maximizes the internal rate of return. By rewriting equation (3.2) we get:

$$\text{NPV} = 0 = R_t e^{-xt} - C. \tag{3.5}$$

Note that x is the internal rate of return, which is unknown. Our problem now is to find the time period in equation (3.5) that maximizes the internal rate of return, x. To do this, we first rearrange equation (3.5):

$$C = R_t e^{-xt}$$

then take the natural logarithm of both sides:

$$\ln C = \ln R_t e^{-xt}$$

and solve it for x:

$$xt = \ln R_t - \ln C$$

$$x = \frac{1}{t} \ln \left(\frac{R_t}{C} \right). \tag{3.6}$$

By using the revenue function in Example 3.1, which is $R_t = 4000(1+t)^{\frac{1}{2}}$, and the cost function, which is $C = 6000$, let us find which time period yields a maximum x in equation (3.6). For this let us try various values for t, starting from 10.

For $t = 10$:

$$x = \frac{1}{10} \ln \left(\frac{4000(1 + 10)^{\frac{1}{2}}}{6000} \right)$$

$$= \frac{1}{10} \ln (2.21)$$

$$= 0.079 \text{ or } 7.9\%.$$

For $t = 5$:

$$x = \frac{1}{5} \ln \left(\frac{4000(1 + 5)^{\frac{1}{2}}}{6000} \right)$$

$$= \frac{1}{5} \ln (1.63)$$

$$= 0.098 \text{ or } 9.8\%.$$

For $t = 4$:

$$x = \frac{1}{4} \ln \left(\frac{4000(1 + 4)^{\frac{1}{2}}}{6000} \right)$$

$$= \frac{1}{4} \ln (1.49)$$

$$= 0.10 \text{ or } 10\%.$$

For $t = 3$:

$$x = \frac{1}{3} \ln \left(\frac{4000(1 + 3)^{\frac{1}{2}}}{6000} \right)$$

$$= \frac{1}{3} \ln (1.33)$$

$$= 0.095 \text{ or } 9.5\%.$$

The optimum period that maximizes the internal rate of return in this example is four years. The maximized internal rate of return, which is 10%, is well above the market rate of interest, which is 2% in Example 3.1. However, this method of finding the optimum solution is based on an assumption that is unwarranted. The internal rate of return criterion implicitly argues that funds provided in the project can continuously be reinvested in projects with proportional expansion of scale. To make it more clear, in our example we start out with an investment of £4000. After four years, the prescribed solution to the rotation problem, the investor would reinvest:

$$£4000e^{(0.1)(4)} = £5969$$

and so on. As long as the internal rate is greater than the opportunity cost of capital (2% in Example 3.1) the present value of an infinite replication of a proportionately growing investment is infinite. Obviously, something is not quite right.

3.7.3 Multiple rotation with constant scale replication

The correct solution to the optimal rotation problem is to assume that the project can be replicated indefinitely at a constant scale. In this, once the trees are harvested, the same acreage is replanted at a constant scale over and over again. In other words, the same acreage is devoted to the afforestation project from now to eternity. This will yield a net present value formula (NPV) which is:

$$NPV = -C + (R_t - C)e^{-rt} + (R_t - C)e^{-2rt} + (R_t - C)e^{-3rt} + \dots . \quad (3.7)$$

The first term on the right-hand side of the equation is the initial planting cost which takes place at present, year zero, for which the discount factor is 1. The second term represents the discounted value of the first harvest and replanting, which take place at a future date. The third term is the second harvest and replanting, which take place at a time which is twice as long as the first one and thus it is discounted at e^{-2rt}, and so on.

Multiply both sides of equation (3.7) by e^{-rt}. This gives:

$$NPVe^{-rt} = -Ce^{-rt} + (R_t - C)e^{-2rt} + \dots + (R_t - C)e^{-(n+1)rt}. \quad (3.8)$$

Subtract equation (3.8) from equation (3.7):

$$NPV - NPVe^{-rt} = -C + Ce^{-rt} + (R_t - C)e^{-rt} - (R_t - C)e^{-2rt} + \dots - (R_t - C)e^{-(n+1)rt}.$$

All the terms on the right-hand side cancel out except $-C$; Ce^{-rt}; $(R_t - C)e^{-rt}$ and $(R_t - C)e^{-(n+1)rt}$. Since n is a very large number, then $(R_t - C)e^{-(n+1)rt}$ becomes almost zero. Then we have

$$NPV(1 - e^{-rt}) = -C(1 - e^{-rt}) + (R_t - C)e^{-rt}$$

then

$$NPV = \frac{-C\left(1 - e^{-rt}\right)}{1 - e^{-rt}} + \frac{(R_t - C)e^{-rt}}{1 - e^{-rt}}$$

which is

$$NPV = -C + \frac{(R_t - C)e^{-rt}}{1 - e^{-rt}}$$

or

$$NPV = \frac{R_t - Ce^{-rt}}{e^{rt} - 1}$$

or

$$NPV = -C + \frac{R_t - C}{e^{rt} - 1}$$

or

$$NPV = -C + (R_t - C)\left(e^{rt} - 1\right)^{-1}. \qquad (3.9)$$

To find the optimum rotation we have to differentiate equation (3.9) with respect to time and set it equal to zero (the first order maximization condition), that is

$$\frac{dNPV}{dt} = \frac{d}{dt}\left[-C + (R_t - C)\left(e^{rt} - 1\right)^{-1}\right] = 0$$

$$\frac{dR_t}{dt}\left(e^{rt} - 1\right)^{-1} + re^{rt}(R_t - C)(-1)\left(e^{rt} - 1\right)^{-2} = 0$$

multiply both sides by $(e^{rt} - 1)^2$:

$$\frac{dR_t}{dt}\left(e^{rt} - 1\right) = re^{rt}(R_t - C)$$

then :

$$\frac{dR_t}{dt} = \frac{re^{rt}(R_t - C)}{e^{rt} - 1}$$

or :

$$\frac{dR_t}{dt} = \frac{r(R_t - C)}{1 - e^{-rt}}. \qquad (3.10)$$

Example 3.2
The revenue and cost functions are $R_t = 4000(1+t)^{\frac{1}{2}}$ and $C = 6000$, and the market rate of interest r is 2%. Find the optimum rotation with constant scale replication.

Substituting this in equation (3.10) gives:

$$\frac{2000}{(1 + t)^{\frac{1}{2}}} = \frac{0.02\left[(4000)(1 + t)^{\frac{1}{2}} - 6000\right]}{1 - e^{(-0.02)(t)}}.$$

By trying various values of t we get the following results:

Left-hand side of equation (3.10)	*Right-hand side of equation (3.10)*	*Difference*

For $t = 10$:

$\dfrac{2000}{(11)^{\frac{1}{2}}}$	$\dfrac{0.02\left[(4000)(11)^{\frac{1}{2}} - 6000\right]}{1 - e^{-0.2}}$	
602.41	$\dfrac{145.6}{1 - 2.72^{-0.2}}$	
602.41	$\dfrac{145.6}{0.18}$	
602.41	808.89	+206.48

For $t = 7$:

$\dfrac{2000}{(8)^{\frac{1}{2}}}$	$\dfrac{0.02\left[(4000)(8)^{\frac{1}{2}} - 6000\right]}{1 - e^{-0.14}}$	
706.71	$\dfrac{106.4}{0.13}$	
706.71	818.46	+111.75

For $t = 5$:

$\dfrac{2000}{(6)^{\frac{1}{2}}}$	$\dfrac{0.02\left[(4000)(6)^{\frac{1}{2}} - 6000\right]}{1 - e^{-0.1}}$	
816.33	$\dfrac{76}{0.095}$	
816.33	800	−16.33

For $t = 4$:

$\dfrac{2000}{(5)^{\frac{1}{2}}}$	$\dfrac{0.02\left[(4000)(5)^{\frac{1}{2}} - 6000\right]}{1 - e^{-0.08}}$	
892.85	$\dfrac{59.2}{0.08}$	
892.85	740	−152.85

So the optimum rotation must be just over five years because from $t = 5$ to $t = 4$ the difference grows wider.

3.7.4 Factors affecting the optimum rotation

Taxes can extend or shorten the optimal rotation period or they can have no effect on it. Of course, when the optimal rotation is altered, this in turn will alter the volume as well as the grade of timber coming on to the market. There are a number of taxes that are relevant.

A sales tax on timber

A tax levied on the revenue resulting from the sale of timber means a decrease in the net price the forest owner receives. This leads to a longer interval between planting and harvesting.

Example 3.3
Let us take the above example of multiple rotation. With the imposition of sales tax the revenue function would be reduced but there are no changes in the interest rate or the cost function. That is:

$$R_t = 3500(1 + t)^{\frac{1}{2}}; \; r = 0.02; \; C = 6000.$$

Let us work out the optimal rotation by using equation (3.10):

$$\frac{dR_t}{dt} = \frac{1750}{(1 + t)^{\frac{1}{2}}} = \frac{0.02\left[(3500)(1 + t)^{\frac{1}{2}} - 6000\right]}{1 - e^{-0.02t}}.$$

By means of the iterative method in which we try various values for t to find out which one satisfies this equation, we get $t = 7$ (approximately).

Longer rotation periods mean that the trees felled will be slightly older and more voluminous.

A negative planting tax (grant aid)

Planting grants, which are subsidies or negative taxes, will reduce the initial planting costs and this in turn, all things being equal, will shorten the rotation period. To ascertain this, let us again modify Example 3.1 by reducing the planting and replanting costs by 1000.

Example 3.4
The revenue and cost functions are

$$R_t = 4000 (1 + t)^{\frac{1}{2}}, \; r = 0.02 \text{ and } C = 5000.$$

By using equation (3.10)

$$\frac{dR_t}{dt} = \frac{2000}{(1 + t)^{\frac{1}{2}}} = \frac{0.02\left[(4000)(1 + t)^{\frac{1}{2}} - 5000\right]}{1 - e^{-0.02t}}$$

The figure that satisfies this equation is $t = 3$ (approximately).

Afforestation projects in the British Isles are heavily subsidized by governments to encourage private entrepreneurs to take up such projects. Note that the improved grant would affect optimum rotation on the new as well as the existing plantations. The subsidy would affect the existing plantations via the reduction in the cost of replanting.

A site utilization tax

The government may impose a tax per hectare each time land is brought into forestry use. This would be equivalent to an increase in the establishment cost, which would increase the profit-maximizing rotation period. As a result the trees cut down would be older and more voluminous.

A licence fee for afforestation

A licence fee implemented on a per hectare basis would have a similar effect to a site utilization tax as it would, effectively, increase set-up costs. For example, the government may enforce a licensing requirement, at a cost, on foresters. This would be on the basis of per hectare or per year. If the licence fee were levied per year rather than per hectare, then the optimal rotation interval would be unaffected.

A profit tax

A fixed percentage tax on profits resulting from the sale of timber would not change the optimal rotation interval because it would have no bearing on the parameters that determine the optimum interval.

Changes in interest rates

Since the interest rate is a dominant parameter in equation (3.10), any change in this will alter the final result. It must be intuitively obvious that a high interest rate will shorten the optimum rotation point. Going back to the argument regarding a single rotation, in Figure 3.5 a high interest rate would mean a steeper line for the opportunity cost of capital and hence the tangency point will be nearer to the origin. Of course, a high interest rate will also make investors reluctant to tie up their money for a longer period of time in forestry projects.

Example 3.5
The revenue and cost functions are $R_t = 4000(1 + t)^{\frac{1}{2}}$, $C = 6000$ and $r = 5\%$. Find the optimum rotation.

$$\frac{dR_I}{dt} = \frac{2000}{(1 + t)^{\frac{1}{2}}}.$$

By using equation (3.10) we get:

$$\frac{2000}{(1 + t)^{\frac{1}{2}}} = \frac{0.05\left[(4000)(1 + t)^{\frac{1}{2}} - 6000\right]}{1 - e^{-0.05t}}.$$

Again, by using the iterative method in which we try various values for t to find out which one yields the desired solution, we get $t = 4$ (years) as opposed to $t = 5$ (years) in Example 3.2 above.

Therefore, other things being equal, the higher the interest rate the shorter the rotation.

3.8 FORESTRY POLICY IN THE EUROPEAN COMMUNITY

In many European Community countries forests meet an important need for industrial materials, providing economic activity and employment for a large number of people, and supporting activity in wood-processing industries. For example, in the German state of Baden-Württemberg it is estimated that forests provide employment for about 250 000 people. Forests also play a crucial role in maintaining the ecological balance and contributing to environmental quality by preventing soil erosion and desertification. Furthermore, forests provide a base for recreational and leisure activities for urban as well as rural dwellers.

The needs of the 12 member states for wood greatly exceed the currently available output from the Community forests and as a result the European Community is the world's largest importer of forest products. Net imports into the Community currently exceeds 20 000 million ECU (European Currency Units) and the situation will not improve much in the near future. This huge deficit is largely due to the insufficiency of the wooded area. Only about 20% of the total land surface in the European Community is under trees. Another factor which contributes to the problem is the underutilization of existing forests, some of which are totally unproductive. Despite its overall deficit the European Community exports about 2 million tonnes of paper and board per year and is a net exporter of furniture. The Community believes that an increase in the supply of wood is likely to sustain more activity in the wood-processing industries within the European Community.

Whereas it has a Common Agricultural Policy, the European Community does not currently have a formal forestry policy. However, the European Community is in the process of preparing a Community action programme. There are four main reasons that have compelled the initiation of such a programme. First the persistence of the agricultural surpluses can only be a temporary phenomenon. Second, the inevitable reduction in agricultural output will lead to a search for alternative crops, including forestry products. The Community's huge trade deficit in wood gives an added incentive to increase timber output. Third, urgent action is needed to stop the destruction of European forests by atmospheric pollution and fire. The former is particularly acute in the north and the latter in the south. Fourth, there is a need to maintain and expand economic activities and employment in the rural areas.

In the past the European Community has made a number of proposals for the development of forestry. In 1979 the Commission proposed a resolution on the Community's forestry policy and the setting up of a Standing Forestry Committee, and in 1983 it proposed objectives and lines of action for Community policy regarding forestry and wood-based industries. So far no decision has been taken by the Council on any of these proposals.

Some areas of marginal agricultural land in Europe provide ideal growing conditions for trees. For example, in the west of Ireland large areas are not suitable for lucrative agriculture but are good for forestry. Ironically, most of the rural population in those areas are engaged in traditional types of farming such as sheep and cattle, and the income generated by these means is not enough to sustain families. In parts of counties Galway, Mayo and Sligo the most rural communities rely heavily on social security payments. Because of the poor farming conditions rural depopulation has always been a major problem in that part of Ireland. Since 1982, with the help of the European Community, the Irish government has been actively promoting afforestation projects in the west to prevent further depopulation. There is considerable scope in these areas for short-term cropping with a view to press-board production as well as for traditional forms of timber production.

It is a well-publicized problem that, in the European Community, agricultural surpluses have been growing steadily since the introduction of the Common Agricultural Policy (CAP). Currently, the Community is considering possible measures for the development of forestry as an alternative to agriculture. In this respect, the cost of supporting forestry production has already been seen in relation to the cost of agricultural support and that of other measures for taking land out of agricultural production.

Despite the absence of a clear forestry policy, in the past the European Community invested heavily in forestry in the context of its other policies. Between 1980 and 1984 about 470 million ECU were committed to forestry projects in Europe and in developing countries. For example, the Irish Western Package is one of many such investments. The European Community has also been promoting forestry in Africa, the Caribbean and the Pacific region, countries which supply the Community with tropical wood. The Community has co-operation agreements through the Lomé Convention with some of the exporting countries. Overexploitation of tropical forests has grave consequences for climate, agricultural production and supply of wood. The European Community has a vested interest in all these matters.

The extension of the forest area in the Community in an environmentally acceptable manner appears to be a priority issue. This is particularly important at a time when the Community has huge agricultural surpluses and a large deficit in timber. The European Community believes that an incentive package including planting grants as well as tax reliefs may be a good way to start the ball rolling. It has been observed that similar packages that already exist in some member states give satisfactory results: Scotland is a good example. However, this can only be a short- and medium-term strategy. In the long run, the objective of agro-forestry action must be to develop an activity which is self-sustaining and does not require substantial subsidies. Another problem that must be solved is the absence of a steady income in forestry. Indeed, there is a considerable time gap between planting trees and receiving income from thinnings and felling. This is quite a deterrent factor, especially for low-income farmers. In order to lure them into forestry they must be provided with a regular income similar to that from farming.

The improvement of productivity in existing forests is another key area. A considerable proportion of the Community's forests are unproductive. The European Community believes that a large increase in output can be achieved with relatively small additional investment. About 60% of the Community's forests held in the private sector are characterized by problems of small size: the average holding is about 8 ha. Forests under public sector management consist of larger plots, and the infrastructure facilities are better than those of the private forests. Extraction of timber from the remote forests is costly, and private owners are unwilling to harvest timber in such places. Another factor is the lack of demand for wood in remote areas, so extracted material requires costly transportation to places where demand exists. This situation could be improved by the creation of forestry associations for the owners. The European Community is ready to help with the infrastructure problems and the creation of woodland associations where they are needed. In aiming to improve productivity it is also necessary to support the expansion of wood-processing industries. Measures to improve standardization of forest products would be a useful step to encourage the development of the timber trade.

The European Community believes that expansion of the Community's forests and the improvement of productivity in the existing ones must be carried out in an ecologically acceptable manner. There is some evidence of strong feeling in some regions at the spread of conifers at the expense of broadleaved forests. If a grant scheme is to be implemented to encourage afforestation, it should make provision for greater planting of deciduous trees. In any forestry policy the production of timber should not be the only objective. It must be accompanied by the safeguarding of the long-term fertility of the soil, the regulation of the water cycle, the conservation of the fauna and flora, the preservation of the landscape and the provision of recreational facilities for the general public. Unrestricted access to forests may help to bring about greater public understanding of woodlands and nature in general. However, it also increases the risk of damage. The Community believes that the aim of a forestry action programme must be to promote the opening up of forests to the public with good management to reduce the risk of damage to trees.

Another area that requires urgent action is the damage caused to woodlands by atmospheric pollution. The destruction has reached considerable proportions, especially in Germany and France. About 50% of the German forests and 40% of the woodlands in eastern France are damaged. The future economic development of these regions is dependent on forestry and tourism. Although the causes of the damage and the mechanics involved are not completely understood, it is generally agreed that atmospheric pollution is a major factor. The pollution is caused by emissions from car exhausts and coal-fired power stations. The European Community has already made proposals to reduce pollutant emission from cars and it envisages making further proposals on speed limits and on emissions from diesel-powered cars and from trucks. It is believed that implementation of these proposals would considerably reduce atmospheric pollution and the damage to forests. A Community proposal to establish a regular inventory of forest damage and a network of observation posts for measurement is making only slow progress in the Council.

The Community has also proposed measures aimed at reducing other risks such as fire and insect damage to the Community forests. In relation to the former, the European Community is willing to improve the fire-fighting capacity of member states. However, it is doubtful that this issue will become a priority case in the Council. Biotic damage cause by insects or disease poses another challenge to forests, and the Standing Committee is consulted whenever a potential danger is perceived. One important measure would be the adoption of a code of conduct on genetic impoverishment, a factor which is partly to blame for the weakening resistance of European forests.

Development of a Community action programme for forestry will require additional efforts in forestry research, statistics and information. A considerable research effort has already been made in the Wood as a Renewable Raw Material Programme, which has been extended for another five years. Other programmes exist in the areas of agriculture, energy and environment which include forestry. However, the European Community believes that the existing research effort is inadequate in view of the ground that needs to be made up and the scale of the requirements. Likewise, statistics and other information on forests are also inadequate for the needs of a Community action programme. All in all, efforts are needed to provide the Community with research findings, data and information in all relevant areas.

3.9   FORESTRY IN THE UNITED STATES

Before colonial times forests covered the entire eastern half of the United States in an almost unbroken tract from the Atlantic right up to the central plains. When the explorers landed, they saw America as a sea of trees. The view from the mountaintops was overwhelming; trees stretched in every direction as far as the eye could see. The early Spanish explorers coming to the northwest from Mexico were also astonished by the vastness of the American forests. These early settlers cleared some land of trees in order to make room for crops, but such clearances were mere nibbles at the edges of the great forests. Trees provided ready building material and fuel, which were as important as crops to the early pioneers. As the nation expanded westwards lumber from the forests provided a major part of the building materials. West of the Great Plains, forests occurred in more isolated groups, chiefly in the mountainous areas where rainfall is heavier. Towards the middle of the nineteenth century, the area of forest cut annually was about 800 000 ha. This figure went up to about 4 million ha at the turn of the century. A considerable part of the cleared land eventually reverted to trees but there was a substantial time lag. The forest area of the United States has been stabilized since the First World War. Today, about 270 million ha of forests cover the continental land mass, amounting to 34% of the land surface. It was estimated by Clawson (1979) that around the mid-nineteenth century this figure was 50%. In Alaska, about 60 million ha of the land is forested.

   The present-day forests contain about 800 native species. This variety is due to the extremely varied conditions of climate, soil type and altitude. About 60% of the nation's forests are in the eastern United States. In the northeastern corner, spruce, fir and pine predominate. The area from New Jersey to Texas, the southern pine region, is dominated by pines. This area supplies nearly 60% of the pulp and more than one-third of the timber cut in the United States. The region between New England and the Mississippi Valley abounds with various species of oak and hickory. Maple, birch and beech are widespread in the lake states and parts of Pennsylvania, New York and New England. Parts of the lake states are dominated by aspen, which is usually a short-lived species and may ultimately be displaced by maple and fir. In contrast, in the bottom lands of the south there are dense hardwood forests of oak, sycamore and maple and a variety of other trees and shrubs. In the west, high up in the Rocky Mountains up to 3000 m, and in limited areas of Oregon and Washington, spruce forests are common. Commercial uses for these western forests have so far been limited, though they are of great importance for watershed protection and for recreation. Among the more famous forests are the redwoods of northern California, which can be found in a narrow belt about 160 km long and 30 km wide. Some stands contain more than 5000 m³/ha – the heaviest stands of timber in the world.

   Wood was the basic fuel throughout the US until well into the late nineteenth century, and it remained so on most farms right up to the beginning of the Second World War. One reason for its prolonged use on farms was the fact that most of them lacked electricity until the end of the 1930s. Lumber, another large consumer of wood, was dominant in the late nineteenth and early twentieth centuries. The major activity which it supported was construction, which became very depressed in the 1930s. Since the end of the Second World War, lumber output has increased again and has remained fairly high. Pulp, plywood and veneer have become major wood-using sectors in the last few decades. When all forms of manufactured wood in the US are considered, an irregular but significant upward trend over the last 200 years is evident. Of course, different uses of wood over time require different kinds of wood. For example, lumber and plywood are made today from species whose log qualities and sizes were considered quite unsuitable only a couple of decades ago. Much of the wood used as fuel in earlier times would be considered suitable for manufacture today. Currently in the US the annual lumber output is about 11 million m³. Wood for plywood, veneer and paper constitutes another 140 000 m³, most of which is used domestically. Under 56 000 m³ of wood is consumed for fuel and other purposes. These figures make the US the largest wood-consuming nation in the world.

As for the ownership of forests, the situation has changed over the years. Until the turn of the last century, ownership was mainly public, federal, state or other, and the land policy was to convert the lands in the public domain to private ownership. Around 1900, concern for future timber and water supplies led to lands being reserved for national forests and to basic changes in other land policies. In the early years of this century, many federal governments purchased lands back from the private sector, which increased the size of the public forest areas. In the 1930s the federal and state governments repossessed millions of acres of private forests, mainly for non-payment of taxes. The public/private division of forest ownership has stabilized during the past few decades. Today more than 50% of all US forests are privately owned. Some of the private forests are owned by the forest industry firms which have increased their holdings steadily since the turn of the century. During the last few decades there has also been a decrease in farm forest acreage.

In the United States the national forests are managed by the Forest Service under a decentralized system of operation calling for 11 regional foresters, 150 forest supervisors, and 750 district rangers. Each district ranger administers timber, water, wildlife and recreational resources in an area that frequently exceeds 100 000 ha. The objectives of the Forest Service are to achieve the maximum potential for the national forests, to provide wood, to protect wildlife habitats, to provide and protect a water supply, and to provide facilities for outdoor recreation. The Forest Service is not anticipating any major shift in land use, at least up to the early part of the twenty-first century. The withdrawal of land from forests is also likely to be very small and mainly for transportation and urban redevelopment. On the other hand, abandonment of farmland is also likely to be very small, and some land may be shifted back into forestry. The Forest Service expects that the volume of standing timber will increase modestly over the next few decades. This trend will include some liquidation of old-growth timber, but this will be more than offset by increasing inventory from younger stands which will meet the increasing demand in sectors such as pulp and plywood without any contraction in timber stocks. The total consumption of lumber may have more or less stabilized but there may be some increase in firewood consumption.

In addition to wood supplies the contribution of forests to wildlife, recreation and watershed values is substantial. These aspects of forests are widely appreciated today, particularly as Americans have become more affluent and more urban. In the mid-1920s total recreational visits to the national forests were about 6 million; today the figure is over 200 million. For many years recreational use of forests increased at a rate close to 10% per annum. From the early 1930s to the late 1970s numbers of big game killed in the national forests increased by about 2.5 times. Over the same period the amount of forage consumed by wild animals increased by about six times. The amount of water flowing off the national forests has probably remained steady, but the water stored in dams or used for irrigation has increased substantially.

As was mentioned earlier, there are two distinctly different kinds of privately owned forests in the United States: those owned by wood industry firms and those owned by others. The former own about 14% of the entire commercial forests in the country. Most of the firms are vertically integrated and their operation is extensive in all major forested regions; some even have subsidiaries in other countries. The other owners are mainly smallholders who have no manufacturing facilities to process their wood. The forest industry firms have achieved a much higher growth rate per hectare than other owners, mainly because of better management and the advantage they have in large-scale operation.

One problem which the private owners are facing is that of externalities. Both industrial and non-industrial owners are unable to capture the full benefits of their forests, especially with regard to water supply, protection of wildlife, recreation and aesthetic enjoyment. Most forest owners suffer from trespass for these uses. Trespass for the purpose of cutting trees for fuel is on the increase. It is possible to impose an entry fee, but no charge can be made for hunting because the wild animals in forests and elsewhere belong to the state. Likewise a forest owner is unable to capture the value of the water flowing out of the forests. Policing the estate to prevent unauthorized use is a very costly matter in most cases and this makes many forest owners helpless. To make matters worse, an attitude has grown up in many localities that access to private forests should be free to the public. In the past some forest industry firms closed their property to public access; the public reaction was so hostile that today almost all firms make their lands freely available to hunters and other leisure seekers. Without a monetary reward for providing non-wood outputs, owners have no incentive to invest in the production of such outputs, nor do they have any inclination to preserve the existing facilities.

3.10   FOREST DESTRUCTION IN THE TROPICS

Tropical forests run like a girdle around the equator and cover about 1050 million ha, 9% of the earth's land area. The bulk of the tropical forests are in Latin America, with the rest shared about equally between South-East Asia and

Africa. The Latin American forests are largely in Brazilian Amazonia but they also stretch into Peru, Bolivia, Venezuela, Colombia, Guyana, Surinam and Ecuador. Most tropical forests grow on very poor soil, primarily through the evolution of symbiotic associations between trees and microflora which enables a remarkably efficient recycling of minerals.

There is a growing awareness that tropical forests are essential to the ecological well-being of the earth because they are a major source of the world's biological diversity, perform important – but not fully understood – climatic functions, and act as a carbon bank which helps to counter the 'greenhouse effect'. For all these reasons most environmental scientists call for complete preservation of these forests. However, governments that have jurisdiction over the area see the forests and the land they cover as resources to be developed. Such development projects include felling for tropical wood, agricultural expansion, mining, hydroelectric dams and basic road infrastructure to open up thus far untouched areas. They yield marketable benefits such as timber, crops and meat, minerals, hydroelectricity and tourism. But perhaps the largest benefits of tropical forests are in non-marketed products such as climatic functions, biodiversity and carbon storage. Unsustained development projects are at the expense of non-marketed global benefits.

Forest destruction in the tropics has been so extensive in recent decades that it has become a major concern throughout the world. Annual deforestation now amounts to over 8 million ha per year and the tropical forests in some countries such as Nigeria and the Ivory Coast are now virtually gone, cut down in the last 50 years. Over one-third of this annual destruction is taking place in Brazil (Table 3.8).

At the heart of these 'development' projects lie poverty, overpopulation, ignorance and greed. In Brazil over the last few decades millions of landless peasants have moved into Amazonian forests. Landlessness has been a major problem in Latin America for many years; 93% of arable land is held by 7% of the population. Every main road brings in more settlers; every dam encourages the formation of townships on the shores of its reservoirs; every mining project attracts scores of migrant workers. In Indonesia the situation is just as bad. More than 4 million landless peasants have been settled into the densely forested outer island as part of the country's transmigration programme and it has been estimated that about 1 million ha of forest land has been converted to agricultural land as a result of this transmigration policy (Barbier, 1991).

Stripped of its forest cover by slash and burn agriculture, the land rapidly degenerates, leading to erosion, the silting up of rivers and loss of fertility. The farmers then abandon their plots in search of more productive areas. On many occasions this process has actually been encouraged by government policies. For example, in Brazil state subsidies and other policy distortions are estimated to have accounted for at least 35% of forest conversion (Barbier, 1991). Export subsidies, rural credit for cattle ranching and mechanized agriculture, small farmer settlement allowance and capital investment incentives in livestock and

Table 3.8 Deforestation in tropical countries

(Regions) countries	Total forest area (mn ha)	Annual deforestation ('000 ha)
(Amazonia)	(613.5)	(4186)
Brazil	347.0	3200
Peru	73.0	300
Bolivia	55.5	60
Venezuela	42.0	150
Colombia	41.1	350
Guyanas	27.2	63
Surinam	15.2	3
Ecuador	12.5	60
(Central Africa)	(167.1)	(325)
Zaire	103.8	200
Congo	21.1	22
Gabon	20.3	15
Cameroon	17.1	80
Central African Republic	3.6	5
Equatorial Guinea	1.2	3
(South-East Asia)	(167.3)	(1707)
Indonesia	108.6	1315
Papua New Guinea	33.5	22
Malaysia	18.4	255
Philippines	6.5	110
Brunei	0.3	5
Other areas	97.7	2300
Total	1045.6	8518

Source: Food and Agricultural Organization (1990).

wood production are all taking their toll on the country's forests. In Indonesia and Malaysia, government policies to switch from the export of logs to processed timber products has led to the establishment of inefficient wood-processing industries. The result is overexploitation by individuals after quick profit.

Pearce (1991) contends that slash and burn agriculture has no financial rationale; its existence depends on substantial subsidies which should be abolished. However, Winpenny (1991) concludes that tropical agriculture, especially cattle ranching, has gathered a momentum which will not be halted by removal of financial incentives to large companies. 'There are powerful political vested interests in the current trend, and the forces fuelling land speculation originate deep in Brazil's present economic and social situation. Nevertheless, policy reforms would take some of the force out of the motives to deforest' (Winpenny, 1991). Land title should not be issued for areas with poor soil and, when property rights are given, those proposing sustainable agriculture and Indian tribes should be given priority.

At a global level, policy options are currently being discussed. Some environmental pressure groups suggest that the world's tropical forests should be exchanged for the Third World debt and held in trust in perpetuity by an international body, possibly the United Nations. In addition, it is proposed that the World Bank should have a global environmental fund specifically to aid sustainable projects in the tropics. The proposed establishment of a biodiversity fund turned out to be the major issue at the 1992 Rio Conference. Finally, it is clear that the International Tropical Timber Organization should become much more effective in discouraging the sale of timber cut down on an unsustained basis.

4
Agriculture and the environment

In the case of cultivated land, a man is not entitled to think that all this is given to him to use and abuse, and deal with as if it concerned nobody but himself.

J.S. Mill

In most countries the landscape we see today – the shape of the land and its vegetation – has been moulded more by human activity than by the forces of nature. Take the British Isles for instance. As explained in Chapter 3, up until the Middle Ages most of Britain was clothed with trees; from then on a relentless process of forest clearance took place for the purpose of agricultural expansion and extraction of timber and firewood. As a consequence, by the turn of this century the once forest-rich lands had become one of the baldest spots in Europe.

As the forest clearance was intensifying in the nineteenth century, Britain introduced a succession of Corn Laws designed to ensure plenty of home-grown food at steady prices. These became increasingly protective, restricting imports and causing prices to rise, until 1846 when they were abolished and British agriculture was exposed to world conditions (Court, 1967). Around that time the enclosure movement was almost complete, forming many of the farm boundaries in existence today. The establishment of private property rights on land resulted in a rapid increase in agricultural production as farmers were able to take risks and try out new farming methods. Between the end of the Corn Laws and the First World War agricultural productivity in England grew almost fivefold.

During the First World War Britain started to pay the price of deforestation as home-grown timber, which was a strategic commodity at the time, became scarce and importation proved to be difficult due to conditions imposed by the German Navy. The war also created widespread food shortages. After the war there was a half-hearted attempt to afforest the countryside (Kula, 1988b), but then came the Second World War and timber and food shortages reappeared with a vengeance. From then on, the British government decided to support both agriculture and forestry on a sustained basis.

The 1947 Agriculture Act established guaranteed prices for many farm products and deficiency payments made up the difference between the market price and the guaranteed price. This was designed to provide food security and the guaranteed prices were set annually by negotiations between the Ministry of Agriculture, Fisheries and Food and the National Farmers' Union. During the 1950s and 1960s a technological revolution took place on farms throughout the UK and the Western world. Widespread use of farm machinery, chemical fertilizers and pesticides, improved plant varieties and animal husbandry and increases in specialization encouraged by guaranteed prices resulted in a substantial increase in farm output. However, Hanley (1991) and Hodge (1991) contend that these changes in the organization and technology of farming were beginning to undermine the complementary relationship between agriculture and the countryside, leading to a decline in the quality of the landscape in many regions.

In 1973 the UK joined the European Community, alongside Denmark and Ireland, and thus British farming became a part of the Common Agricultural Policy (CAP). The objectives of the CAP are defined in Article 39 of the Treaty of Rome. They are:

● to increase agricultural productivity by promoting technical progress and by ensuring the rational development of agricultural production and optimum utilization of all factors of production, in particular labour;
● to ensure thereby a fair standard of living for the agricultural community, in particular by increasing the individual earnings of persons engaged in agriculture;
● to stabilize markets;
● to provide certainty of supplies; and
● to ensure supplies to consumers at reasonable prices.

Some additions were made to these objectives in 1958 at the Stresa Conference.

● to increase farm incomes not only by a system of transfers from the non-farm population through a price support policy, but also by the encouragement of rural industrialization to give alternative opportunities to farm labour;
● to contribute to overall economic growth by allowing specialization within the Community and eliminating artificial market distortions; and
● to preserve the family farm and to ensure that structural and price policies go hand in hand.

Two years after entry into the EC the British government published a White Paper entitled *Food from Our Own Resources* (1975) in which it was argued that continuing the increase in agricultural output was in the national interests. In 1979 another White Paper, *Farming and the Nation,* revised this view by arguing that continued expansion over the medium term was in the national interests.

It is not only the European Community that has a protective agricultural policy; similar policies are pursued in most industrialized countries, even in those like the United States and Canada who appear to have strong comparative advantage in agriculture. Protective policies can take various forms, such as the deficiency payments that were popular in the UK before she joined the European Community, import controls, export subsidy for domestic producers in order to reduce the quantity of home-grown produce reaching the domestic market, income support for marginal farmers so that they can stay in business and contribute to the domestic production, and non-price/non-income measures such as health standards on imported produce, which is a popular measure in Japan. About 90% of Japanese agricultural imports face health controls as opposed to 50% and 60% in Switzerland and Australia, respectively. Table 4.1 shows the nominal protection coefficients for various agricultural products in a number of countries and regions. The figures show the percentage by which the internal prices received by domestic growers exceed world prices. Winters (1989) points out that since 1980–82 Japan has liberalized somewhat, but Europe and the US have become more protectionist.

Supported by protectionist policies, continuous expansion and intensification of domestic agriculture have created a number of environmental problems in many countries. Pollution by nitrogenous fertilizers and pesticides, the disposal of livestock wastes, removal of landscape features, loss of wildlife and soil erosion have become major environmental issues.

There must, at least at a theoretical level, be an optimum level of intensity in agriculture allowing for factors such as environmental pollution, wildlife, amenity, rural employment, etc. (Traill, 1988). To explore the trade-offs between intensity of agriculture and various factors of value to society, Figure 4.1 plots the intensity of production, per hectare, along the horizontal axis and agricultural output, pollution levels, values of landscape, wildlife and rural employment along the vertical axis, again on a per hectare basis. Theoretically,

Table 4.1 Farmer protection coefficients (percentage excess of producer price over world price), 1980–82

Countries	Wheat	Sugar	Rice	Dairy products
EC-10	25	50	40	75
EFTA-5	70	80	0	140
Australia	4	0	15	30
Canada	15	30	0	95
US	15	40	30	100
Japan	280	160	230	190

Source: Anderson and Tyers (1986).

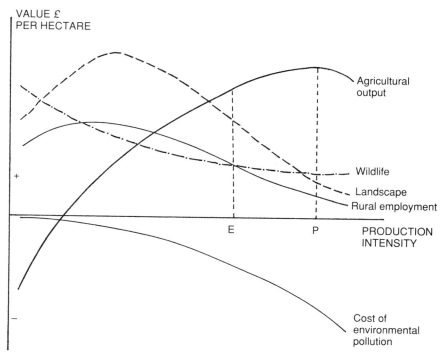

Figure 4.1   Value of land using activity against production intensity.

it is not difficult to envisage the nature of the relationships between agricultural intensity and these factors, in order to find a socially optimal level of production. At P the value of agricultural production is at a maximum but poor performance in the other factors makes it socially suboptimal. At E the combined values of all aspects of land utilization are maximized.

A number of factors could change these curves. A shift towards more environmentally friendly methods of production would reduce the downward slope of the environmental variables. Such a shift would also contract the agricultural output schedule. Farming techniques are always changing, an improvement in farming technology, *ceteris paribus*, shifts the product curve upwards. Present farming support schemes have already shifted point P to the right along the horizontal axis. Of course, different regions have different curves reflecting their circumstances and conditions. Harvey (1991) contends that current agricultural support policies in the European Community and elsewhere have encouraged an excessive intensification and capitalization of agriculture which has in itself created problems for the environment. Artificially high prices distort farmers' decisions in favour of production of supported farm products and away from the supply of environmental goods. Furthermore, the lack of a functioning market for environmental goods means that they are frequently treated as free inputs by farmers and others.

4.1   INORGANIC NITRATE POLLUTION

At the turn of this century the chemical industry managed to convert atmospheric nitrogen to ammonia. This heralded the arrival of the 'age of nitrogenous fertilizers'. Nitrogen, which is necessary to plant growth, is not abundant in most natural soils and this tends to limit the growth of plants and microorganisms. Nitrogen enters the soil in a variety of ways such as fixation by some plants, through plant and animal residues, in rain, and by direct application of nitrogenous fertilizers. After the Second World War plant breeders developed new, durable, and more productive species of grain which require high levels of nitrogenous fertilizers to yield effectively. A substantial capital investment took place in the chemical industry to meet the growing demand for inorganic fertilizers. For example, in 1938–39, 50 000 tonnes of nitrogenous fertilizer were used in England and Wales; this had increased to 1.2 million tonnes in 1980–81, and has now stabilized at around 1.6 million tonnes per annum. A similar situation has occurred in other parts of the world, making the manufacture of nitrogenous fertilizer a multi-billion pound industry.

Nitrogen in the environment is cycled between soil, water and air. Some of the available nitrogen in the soil is taken up by plants as they grow, some remains in the soil, some is lost to the atmosphere by volatilization and denitrification, and finally a good deal is leached into soil water which drains into streams, rivers and eventually the sea. The amount of nitrate lost through leaching depends on the vegetation and the climate. Leaching is lowest in forested areas and is slightly higher in permanent grasslands, although as stocking rates go up, leaching increases (Ryden et al., 1984; Croll and Hayes, 1988; Conrad, 1988). Leaching is highest during times when rainfall is high, and evaporation and crop demands are low.

The release of nitrogen from soils is a slow process and where the huge quantities of nitrogenous fertilizers used in intensive farming have been applied, there will be an unnaturally high levels of nitrogen on the land for many decades to come. As it leaches out slowly into streams, rivers, lakes and seas, these waters will also contain unnaturally high levels of dissolved nitrogen. Potentially this is harmful to health and serious drinking water problems have already emerged in many regions. The presence of excess nitrates in drinking water has been linked to cancer and other diseases in humans, although the evidence is not conclusive. In the 1980s it was discovered that a number of German water companies were supplying drinking water with nitrate concentration exceeding European standards.

Some ingredients of nitrogenous fertilizers lead to excessive growth of algae if allowed to accumulate in surface waters. These algal blooms smother and kill fish directly – some are even toxic – but also form food for microorganisms which deplete the oxygen levels in the water. Many species have already disappeared in parts of the Great Lakes in North America and in many rivers like the Thames and Tyne in England. Burnett (1990) argues that even if the use of nitrogenous fertilizers was stopped now, high concentrations would still persist for many decades. Hanley (1990) reports on concentration of nitrate in water courses and groundwater in southern and eastern England where arable agriculture is dominant and rainfall is relatively low. In these areas the water contains nitrate levels in excess of European Community and World Health Organization upper limits. Some of these problems may have been created by ploughing-up of pastures during the Second World War to increase cereal production.

What are the policy options to minimize the damage caused by nitrate pollution to human health and the environment? Some of the options available are:

1. Regulation of nitrogenous fertilizer use by various means such as direct control and permit systems.
2. Taxation to reduce the use of inorganic fertilizers by making them more expensive. The effectiveness of a nitrogen tax would largely depend upon the price elasticity of demand for nitrogenous fertilizers. Some studies reveal that the price elasticity of inorganic fertilizer is low, in the range of -0.08 to -1.20, with an average of -0.62 (Burrell, 1989). Therefore, in order to be effective such taxes would need to be very high. For example, it was estimated by Dubgaard (1989) that to achieve a 30% reduction in nitrogen use, a tax rate equal to 150% of the current price would need to be imposed in Denmark. Also, the highly locale-specific nature of the problem makes a nitrogen tax a very blunt instrument.
3. Reduction in agricultural subsidies, especially for grain, which will curb output and hence reduce demand for fertilizers. Some studies argue that taxation of nitrogenous fertilizers would be a more effective policy than reduction in farm subsidies or a cut in cereal prices (England, 1986). For example, for winter wheat, an 8.6% cut in nitrate use can be achieved by a 100% levy on nitrogenous fertilizers, whereas a 30% cut in wheat prices yields about a 2% fall in nitrate applications.
4. Change in farming patterns, in particular a move to winter-sown cereals where the ground will not remain empty over the winter. Avoidance of ploughing-up of pastures would also help.
5. Change in land use patterns in favour of afforestation, a shift away from high nitrogen demanding crops and the introduction of set-aside schemes. The last item has already been in operation in North America and some countries of Europe and is due to be expanded in the coming years. For example, in 1993 in Britain some 1.5 million acres (600 000 ha) were taken out of crop production voluntarily and in return farmers received £84 on every productive arable acre they farmed. The effect of a much expanded set-aside scheme on the British and northern European countryside will be profound. The current 'manicured' look will gradually give way to ragwort, thistles and,

eventually, birch, ash and sycamore. Some argue that if just abandoned and left to its own devices, this land will become derelict. Others believe that natural succession will create havens for wildlife.

6. Treatment of polluted water. A wide range of methods have been suggested for water treatment, including a process known as blending, in which high-nitrate with low-nitrate water are mixed. This method is restricted by the availability of low-nitrate water in a blending area. Other methods include biological denitrification, electrodialysis, ion-exchange and reverse osmosis (Croll and Hayes, 1988). Various steps could be taken to prevent nitrate reaching water courses, such as reservoir storage, relocation and lining of boreholes. In England, Anglian Water, who operate in a high nitrate region, use blending, closing boreholes and treatment. In France, ionizing appears to be a preferred method.

4.1.1 The UK policy on nitrate pollution

As a member of the European Community the UK must comply with the minimum environmental standards set by the Commission. The European Commission Drinking Water Directive of 1980 limits nitrate in drinking water to 50 mg per litre, a limit which has not been achieved in all cases. The Commission's Draft Directive 88 requires a number of further management actions to reduce nitrate levels in surface and groundwaters by way of stocking limits, set-aside, afforestation and cutting down on application to land. Hanley (1991) contends that depending on the interpretation of Directive 88, as much as 4 million ha of arable land could be designated as vulnerable zones with respect to rivers and streams, with about 30 000 ha designated with respect to groundwaters.

In response to Draft Directive 88 the British government introduced a nitrate sensitive pilot project in ten areas of England, amounting to 16 400 ha (Figure 4.2). In addition, nine areas have been designated as advisory regions. Participation in these schemes is voluntary, but those who take part in the nitrate sensitive areas receive a payment of £55–95 basic rate per hectare for their co-operation. Farmers participating in this pilot project agree to use no nitrate in the autumn, maintain a winter crop, maintain hedgerows, ponds and woods, and apply no more than 120 kg of nitrate per hectare at any one time. In order to qualify for payment farmers must enter into an agreement for at least five years.

There is also a premium rate scheme available to farmers who register on the basic rate scheme. In order to benefit from higher rates of pay farmers must switch land into one of the following options:

- transfer of arable land to grassland with woodland;
- transfer of arable land to unfertilized and ungrazed grassland;
- transfer of arable land to unfertilized land with grazing;

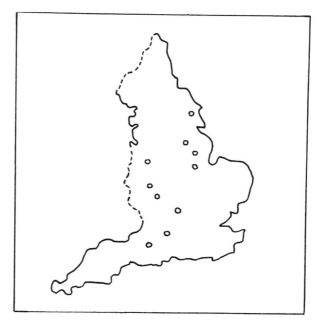

Figure 4.2   Areas designated for participation in the UK nitrate sensitive pilot project.

- transfer of arable land to grassland, where limited fertilizer is used and grazing allowed.

The rates vary across the nitrate sensitive areas.

4.2   PESTICIDE POLLUTION

The application of pesticides to crops has been remarkably successful in increasing agricultural yields. Powerful pesticides such as DDT were first developed in the late 1930s and used on non-edible crops such as cotton and tobacco. After the Second World War the desire to increase domestic food supplies in Western countries initiated research to develop a range of effective pest control agrochemicals. This resulted in a variety of insecticides, herbicides and fungicides which are widely used throughout the world.

   However, this extensive application of pesticides in modern agriculture created a number of serious side-effects, such as damage to flora, fauna, contamination of groundwater, surface water and seas. Some pesticides are highly persistent and their effect extends to species other than the target population and in this way they could affect human health. A quantity gets washed off by rain, which eventually ends up in water supplies and finally seas (Kahn, 1991). Perhaps the most worrying aspect of pesticides is the way in which they accumulate and concentrate in the food chain. When farm animals eat contaminated plants, the chemicals pass into the gut bacteria and the body fat. When humans eat the farm animals, the chemicals tend to accumulate in them, especially in children.

The use of DDT on non-food crops in the United States had already contaminated much of the country and the surrounding seas by the early 1950s. Burnett (1990) points out that from this contamination a diversity of animals as far afield as penguins in Antarctica, were affected. Many species of animals, large and small, birds and fish, fall victim to pesticide pollution. For example, it takes only seven mice fed on treated grain to kill a kestrel. A further concern is that pesticides affect the microscopic life of the soil on which much of its fertility depends. In this way sustained use of pesticides can become counterproductive, leading to a reduction in crop yields. The frequency and intensity of application as well as the ingredients of pesticides aggravate the situation.

The most effective policy to remedy the situation is a complete ban on the most harmful substances. In effect, this has been done in many countries. However, even after their abolition the effects of some chemicals continue for a considerable period of time. Quantity regulation by authorities in the form of a quota on active ingredients as well as application is also a policy option. However, variations in pesticide and methods of application make the administration of the quota system complicated. Variations in regions as well as climate are some other factors which make the operation of the quota system difficult.

As an economic instrument, taxation could be an effective way of reducing the application of pesticides in agriculture. This is a flexible method of control which could be changed from year to year in response to pest-infestation levels, climate variations, agricultural patterns and other variables. However, it has been argued by Dubgaard (1991) that it is virtually impossible to devise a socially optimal taxation level on pesticides because their harmful effects cannot easily be quantified. Furthermore, the existing variety in types, active ingredients and prices add to the difficulty.

If taxation is substantial it can create an incentive to improve pest management systems, including the composition of chemicals and the efficiency of spraying equipment. The most desirable effect of taxation, or indeed a quota system, could be the discovery and adoption of safe, natural biological controls. For example, in some crops such as tomatoes, white fly have been controlled by breeding their known predators. Winter moth in Canada is controlled by the introduction of its natural predators. Apart from administrative problems, one further adverse effect of pesticide tax is the reduction of farm incomes. Dubgaard (1987) calculates that a 120% increase in pesticide prices would reduce land rent by around 10% on good soil and more on marginal areas in Denmark. The most affected crops would be winter wheat, winter barley, potatoes, sugar beet and, to a lesser extent, spring barley. However, this problem could be overcome by reimbursement of the tax revenue in the form of lump-sum payments based upon a flat rate per hectare.

In 1972 the United States government introduced the Federal Insecticide, Fungicide and Rodenticide Act for regulation of pesticides and herbicides used on crops, on application as well as on production. According to this Act no new pesticide or herbicide can be sold in the US without the approval of the Environmental Protection Agency (EPA). Furthermore, permission must be sought for any new use of an already approved agrochemical. For example, let us say that pesticide X was initially approved for use on barley, widening its application to corn would require EPA approval. All new products must pass a number of tests designed to identify hazards to human health and the environment. The tests are normally carried out by the manufacturer. Regulations cover application dose, frequency, ingredients and labelling. A product can be removed from the market if a harmful effect becomes apparent, even long after its introduction. In such cases, however, the burden of proof falls on the EPA.

In the European Community it has been recognized that there is a need to cut down on pesticides used on Community farms. The Community is constantly reviewing measures to make agriculture more ecologically agreeable, which includes the prohibition of dangerous pesticides. On the protection of flora and fauna from pesticide and other damage the European Community is working to establish, by at least the year 2000, a comprehensive network of protected areas in all regions of the Community. One of the priorities in the 1990s is to control pesticides that could harm human health and the environment. The Community plans uniform rules for the approval of pesticides with a view to minimizing the use of dangerous substances (European Community, 1989).

In some countries such as Denmark there are proposals to reduce the use of pesticides drastically. The 1986 Danish Action Plan to Reduce Pesticide Application calls for a 50% reduction in pesticide use by the mid-1990s compared to pesticide use in the period 1981–85 (Dubgaard, 1991).

4.3 ANIMAL WASTE AND POLLUTION

The intensification of the meat, poultry and dairy industries, especially in the smaller countries of Europe, has led to an increase of polluting emissions into their small but densely populated territories. Furthermore, the emission of ammonia from livestock contributes to acid rain, and the emission of methane from livestock contributes to the greenhouse effect, both of which are global problems. The intensification process gathered pace during the 1960s and 1970s and was most marked in Holland, which became the most densely farmed region of Europe. Consequently, Holland has the most acute environmental problems stemming from animal waste. Table 4.2 shows the number of pigs, poultry and cattle per hectare in a number of regions in Europe.

When the waste is put onto the ground some of it gets washed away into the surface water, while the remainder penetrates the soil. Some of the minerals are taken up by the plants. Nitrates quickly leach into the groundwater. In saturated soils other minerals such as phosphate and potassium eventually end up in the

Table 4.2   Number of animals per hectare

Country/region	Poultry	Pigs	Cattle
Holland	45.1	6.1	2.6
Brittany, France	45.0	2.8	1.6
Lombardia, Italy	40.2	2.8	1.6
Belgium	15.4	3.9	2.2
Yorkshire and Humberside, UK	9.4	1.5	2.0
Denmark	5.4	3.2	0.9

Source: Stolwijk (1989).

groundwater. The full effects of animal waste pollution are not completely understood as a good deal of research is still in progress. However, it is well-known that a build up of minerals, especially nitrate and phosphate, causes a reduction in the number of species. Furthermore, high concentrations of these minerals in drinking water adversely affect human health (Tamminga and Wijnands, 1991).

In Holland, most farmers are smallholders who were under pressure to diversify in the 1960s and 1970s. Intensification of livestock farming was the only means of staying in business as cheap feed imports were available despite the European Community's farm policy. Intensification of livestock created large quantities of animal waste which was then put on the land. In this way farmers hoped to both dispose of the waste and increase the fertility of their land at the same time. However, as the years passed it became clear that the high concentrations of manure were creating serious pollution problems. The social cost of nitrate and phosphate pollution is estimated to be 200–760 million Dutch guilders per year (Baan and Hopstaken, 1989; Tamminga and Wijnands, 1991). Even if animal waste related pollution stops now, the cost will continue due to persistence. For the year 2010 this cost is expected to be 150–580 million Dutch guilders.

Annually, cattle produce about 55 million tonnes; pigs, 20 million tonnes; and poultry, 5 million tonnes of manure in Holland. About 80% of this is disposed of in farms of origin, the rest is transported to other farms. The poultry waste contains the highest concentration of minerals and it is transported over long distances for disposal. The Dutch government estimates that the average cost per farm is in the region of 4000 florins. The concentration of poultry farms is highest in the southeastern part of the country.

Holland is a low-lying country with sandy soils. This exacerbates the environmental problems created by animal waste as minerals that exist in the waste leach more easily into groundwater on sandy soils. Furthermore, Holland has plenty of surface water into which minerals get washed. The European Commission requires that nitrate should not exceed 50 mg per litre in drinking water. In the sandy regions of eastern Holland this standard is exceeded by 70% in the upper section of the groundwater.

Animal waste became an issue of debate in Dutch politics in the early 1980s. In 1984 parliament decided that stringent measures which could hurt the economic position of farmers should be avoided and passed an interim law restricting poultry and pig farming. As a result the setting up of new farms in sensitive regions was prevented and the growth of stock on existing farms was restricted. Three years later the Manure Law was passed, the basic features of which are: prohibition of expansion and starting of new farms; restrictions on the relocation of manure-creating activities in sensitive regions; levy on manure surpluses and creation of a national manure bank. In addition, the Dutch government passed a law on soil protection, the main characteristics of which are: restriction on the application of manure to the soil; regulation of the spreading of manure on fields; and regulation of methods of putting manure on the ground.

4.4   SOIL EROSION

Soil erosion is the oldest and probably the most persistent environmental problem associated with agriculture. History provides us with many examples of soil erosion. North Africa, once the fertile and highly productive breadbasket of the Roman Empire, has lost much of its natural fertility and some countries there nowadays depend upon imports to feed their population. Even today, literally millions of acres of cropland in Africa, the Middle East, Asia and the Americas are being abandoned each year because of severe soil erosion.

In the United States of America the droughts of 1890 and 1910 brought seasons of poor crops and accelerated erosion, especially on the Great Plains. During the Great Depression, a particularly severe drought occurred in Kansas, Oklahoma and Texas and destroyed much of the native vegetation. The dry ploughed lands were swept by strong winds which carried off the topsoil. Some of the blown soil piled up high as dust dunes, some was blown from hundreds and even thousands of kilometres away. This was the dust bowl, one of the turmoils of the Great Depression, which led to the migration of farmers as described in John Steinbeck's *The Grapes of Wrath*. In dry climates the role of plant life in preventing the soil from being blown away is crucial. Before the arrival of agriculture, in vulnerable regions of the United States the delicate soil structure was topped by stabilizing grass and roots. When farmers began ploughing, this disrupted the structure of the soil and reduced its capacity to hold water. After the above-mentioned droughts a good part of the affected areas have remained unused. Soil scientists argue that some of those areas should never have been ploughed and planted in the first place.

In the Middle East, once rich soils have been rendered useless because of overuse. Due to relentless grain farming, topsoil in a good part of Iraq, Syria and the Lebanon was washed away centuries ago. In India between 1950 and 1990 a 175% increase in agriculture was achieved as a result of the 'green revolution'

(Douthwaite, 1992). During this period many farmers were encouraged to replace their traditional methods of farming with modern techniques using deep ploughing, irrigation and the use of inorganic fertilizers. With traditional methods, light ploughing and harrowing was used on a soil made friable by the pre-monsoon heat. Rain then washed the surface soil into the deep cracks, exposing the new layer as a seedbed. The application of a mixture of wet straw, leaves and mulch to the soil then provided nutrients to increase the soil bacteria and retained moisture in the soil.

The use of deep ploughing and chemical fertilizers instead of mulch increased the rate of soil erosion rapidly. It has been estimated that soil loss on fields without mulch is about 233 tonnes/ha per annum, while on mulched fields it is only 8 tonnes. Forty-two per cent of the rain runs off unmulched land but only 2% if mulch is applied. In this way less rainwater penetrates the soil to replenish the water table.

Overgrazing is another major cause of soil erosion. Heavy grazing by cattle and sheep pulls out plant roots causing exposure of the soil which is then vulnerable to rain and wind. A good part of South Africa and Australia has been overgrazed and transformed from grasslands into bushy, half-bare plains. An increase in population in developing countries is often accompanied by an increase in livestock numbers. As sheep, cattle and goat populations increase they denude the countryside of its natural grass cover and prepare conditions for soil erosion. This process is most reckless in parts of Africa. In densely populated regions such as Pakistan, India, China and Central America severe water and wind erosion of the soil caused by overgrazing has become a major environmental concern.

Deforestation is another precursor to soil erosion. When networks of live tree roots no longer exist, the run-off of rainwater is greatly increased which washes away the soil. Especially after heavy rain, fast flowing water races downhill to erode gullies in meadows or farmlands located at lower levels. After the destruction of trees there is usually a great increase in the amount of sediment in streams. Silt, once part of wooded soil, clogs irrigation networks and interferes with the agriculture. Where areas of tropical forests have been destroyed for the twin purposes of timber and planting of cash crops, the soil is first disturbed by deforestation and then by ploughing the soil. Within a few years, the soil is washed away in tropical rainstorms and agriculture becomes impossible.

In Turkey, a country with harsh winters, demand for fuel and construction material led to the decimation of the original forests. Once cut over, large areas became grazing grounds for sheep and barren stretches expanded steadily as topsoil was washed and blown away. In the Sahel region of Africa the desert has been expanding steadily as the land has been cleared of trees and overgrazed by animals. Since fossil fuels are not available to these communities, wood and animal wastes are the only fuels available.

In the United States soil erosion was recognized as one of the earliest environmental problems (Eliot, 1760). In 1930 the Congress recognized the seriousness of the problem and commissioned a study on the subject. In 1935 the Soil Conservation Service was established under the auspices of the Department of Agriculture. The Service carries out comprehensive and periodic research to estimate soil erosion rates in the United States (Table 4.3). The United States conservation policy has been influenced by historic events such as the dust bowl of the 1930s (mentioned above), the aftermath of the Second World War and expansion of agriculture in the 1970s. There are a number of policy options, some of which are as follows:

Cost-sharing practices. This is one of the oldest conservation programmes in which the federal government shares farmers' costs related to soil conservation.
Technical assistance. In this the Soil Conservation Service provides technical assistance to farmers through about 3000 conservation districts.
A combined cost-sharing and technical assistance. The 1956 amendment to the Soil Conservation Act established practices which allowed Great Plains states to combat wind erosion. In this the authorities and farmers agree on a long-term technical assistance and cost-sharing on soil conservation.
Set-aside scheme. This is a voluntary programme of reduction in crop acreage to conserve soil. In return farmers receive payments from the Department of Agriculture.

For further details of these and some other policies see Heimlich (1991).

Table 4.3   Gross erosion rates on cropland in the United States

Inventory	Area ('000 acres)	Erosion rate (tons per acre, per annum)	
		Wind	Sheet and rill
1977			
Cultivated	367 825	N/A	5.1
Not cultivated	45 452	N/A	1.3
1982			
Cultivated	376 516	3.4	4.7
Not cultivated	44 866	0.4	0.8
1987			
Cultivated	376 916	3.6	4.1
Not cultivated	45 925	0.7	0.9

Source: Heimlich (1991).

4.5   SALINATION

A traditional method of increasing agricultural productivity, especially in dry regions, is irrigation. However, even this tried and tested farming aid is not without its problems. Mesopotamia, once the hub of civilization, suffered severely from salination of its soils brought about by irrigation and it is thought that this was a major contributory factor in its decline and fall. Today, salination, coupled with the contamination of irrigation water by chemical fertilizers, threatens the success of many modern agricultural projects.

The problem is that the irrigation of arid lands needs enormous quantities of water due to the high evaporation rates. In many areas it is necessary to build large, capital-intensive dams and reservoirs to supply this water and many of these are filling with silt while communities downstream suffer from reduced water supplies. In addition to this, because the soil in dry areas tends to be low in humus and nutrients, large amounts of inorganic fertilizers are needed to grow crops successfully, especially in new agricultural territories.

When the land is irrigated with these large quantities of water, the naturally occurring salts in the water, such as sodium chloride and calcium carbonate, tend to concentrate in the surface layer of the soil. In humid conditions these salts are leached away into surface and groundwaters, but in dry conditions the leaching process is less efficient and the high evaporation rate means that the salts concentrate to harmful levels in the topsoil. In Imperial Valley, California, intensive irrigation has produced widespread damaging soil salination, and in India 25 million ha of irrigated land have been affected by growing salinity.

The successful growth of crops on irrigated lands depends upon maintaining the fertility in the soil. Encouraged by agricultural subsidies, irrigation puts water supplies under pressure, undermining the long-term viability of irrigation projects and thus many currently irrigated regions face serious problems in the future. Water management techniques adjusted to local climate and hydrological conditions are necessary alongside prevention of excessive salination.

4.6   MULTIPURPOSE LAND USE

The countryside is, essentially, a public asset. When a farmer owns a plot of land where he is engaged in a commercial farming activity, say sheep farming, his land is still a part of the national heritage. As I mentioned earlier, the countryside in many parts of the world has been shaped by the farmers, and this is most conspicuous in Europe. The land in almost all European countries is predominantly occupied by agricultural use. For example, in the UK, agriculture claims about 65%, forestry 10%, urban uses 10% and the rest is in a natural or semi-natural state. Whitby (1991) contends that it is difficult to be precise about these figures, but the dominance of agriculture in land use is obvious. The main reason for the dominance of agriculture is the extensive and generous agricultural support policies, coupled with numerous research establishments which have been in existence for a long time. However, the pattern of land use is changing. Edwards (1986) estimates that between 1985 and 2000 up to 6 million ha of agricultural land will be surplus to requirement. For more moderate projections see North (1990). According to some land economists, farming contributes to visual amenity as it has produced a distinctive and much appreciated countryside in many parts of the UK (Hodge, 1991).

Apart from agriculture, there are many aspects of rural land use such as wildlife conservation, outdoor recreation, forestry, education, etc., which are of value to individuals. Many of these aspects exhibit characteristics of public goods which do not usually bring revenue to owners and thus there is little incentive for private landholders to supply them. Since there is a wide range of interest in the countryside then the government ought to take steps to enhance its capacity to meet the demand. Colman (1991) argues that one way of conserving vulnerable artifical and natural features of the landscape is public purchase. An account of the history of public ownership of land in the UK is given by D.I. Bateman (1989). Even prime agricultural land is a finite depletable national asset and thus government must take an interest in it with a view to conserving it (Buddy and Bergstrom, 1991).

The provision of countryside benefits presents many difficulties. Conservationists do not seem to agree on what they want; some advocate spatial separation of farms from wildlife areas whereas others prefer some form of integrated approach. Another problem is that countryside values are site-specific, depending, amongst other things, on the size, taste and income levels of the community. For example, a woodland may not have enormous value in a remote countryside location, but it would be highly valued in a green belt. Some countryside benefits can be achieved at little cost, whereas for others large amounts of money are required. In some Western countries there is a clear objective to commit lands to specific uses by taking into account the cost of service and demand. In Holland, for example, land reclaimed from the sea at a considerable expense is carefully planned to cope with competing uses. In countries with abundant space such as Canada, the United States and Australia, land use history has been dominated by a shifting frontier as new land is brought into use. Whitby (1991) notes that in the United States about 40% of the land surface is in public ownership compared with 5% in Holland, which gives the former considerable convenience and flexibility to plan.

Hodge (1991) contends that conditions relating to the supply of and demand for countryside benefits clearly imply a role for the government. The minimum would be to provide information and education to land managers about the supply of countryside services. Various financial incentive/disincentive packages can accompany the information and education process. For example, in England the Peak District Rural Development Experiment allows payments to farmers on the basis of the number of wildflower species found in their fields. Control of rural activities by way of planning permission and public ownership of key sites can also be policy options.

4.7   FUTURE PROSPECTS

The recent GATT negotiations and the resulting reduction in farm support policies of 1992 will not lead to great improvements to the environment of the European Community. Environmental improvement may come about as a result of isolating greater areas of land from intensive agriculture and by encouraging farmers to adopt less profitable but environmentally friendly practices. This may require a package other than the currently implemented set-aside schemes. Rapid technological changes and new policies are likely to improve agricultural productivity for strong farmers, whereas weaker farmers are bound to get pushed out of business in the face of declining farm support policies.

Harvey (1991) contends that society as a whole would be better off without the farm support policies, provided that some form of Kaldor–Hicks compensation scheme is implemented to aid losers. However, complete elimination of support packages may turn out to be politically difficult; powerful interest groups are likely to prevent such a move. Therefore, it is probable that some form of an agricultural support policy will be in operation in the future. What shape or form could this policy take? Harvey proposes a mixture of quota and subsidy package called the Producers' Entitlement Guarantee (PEG). Under this scheme eligible farmers get a price support only for a fixed quantity of production. Output above the supported quantity would be sold on the free market. The PEG would allow a base level of support on the basis of eligible farmers' past production. According to Harvey, PEG could assist in promoting the development of an environmentally friendly agriculture by combining specific environmental policies with farm support programmes. Furthermore, as long as PEG is independent of farm assets, land prices are bound to fall, making alternative land use activities easier and cheaper. Present support policies encourage an increase in the size of fields, the removal of hedgerows, the use of capital-intensive farming techniques and the use of minimal labour. The PEG can be implemented by deliberately targeting small farms or it can even be made conditional upon particular farming practices deemed to be environmentally friendly.

Developments in genetic engineering relating to agriculture could have environmentally beneficial effects. The production of pest-resistant species, for instance, would reduce the need for pesticides. Developments in this area depend upon the identification of the gene components in cultivated or wild plants which contain genetic advantages and devising methods for transferring the genetic code into host plants. This makes it even more urgent to prevent the extinction of plant and animal species and thus preserve as large a gene pool as possible. Today it is possible, for example, to engineer tobacco and tomato plants with resistance to some viruses. Tomato plants have also been engineered to yield slowly ripening fruit which are easier to market and distribute. Similar developments in genetic engineering are taking place which could have highly beneficial results from both production and environmental viewpoints.

5

Economics and policies in mining, petroleum and natural gas

> A farm, however far pushed, will, under proper cultivation, continue to yield forever, a constant crop. But in mining there is no reproduction and the produce once pushed to the utmost will soon begin to fail and sink to zero.
>
> W.S. Jevons

The economic analysis for mining, petroleum and natural gas is fundamentally different from the analysis of agriculture, manufacturing and service sectors. The reason for this is that mine, petroleum and gas deposits are destructable resources. Fossil fuels, e.g. oil, coal and natural gas, are non-renewable whereas metal deposits are, normally, recyclable. Agricultural output, fabricated goods and services can be supplied over and over again. In other words, the production of these resources, under proper management, can be a continuous process. In the mining, petroleum and natural gas industries, however, a given stock of resource can be supplied only once, i.e. extraction of a unit is a once and for all affair. If we make the assumption that the resource owner, just like any other entrepreneur, is a rational profit maximizer, then he/she must consider some additional factors which are unique to mining, petroleum and natural gas.

In other sectors, profits will be maximized by operating at a supply level where the marginal cost equals the marginal revenue. Consider a case in which a firm is facing a horizontal demand curve which implies that demand for its product is very large at a given market price. This demand curve is also the marginal revenue for the firm. By superimposing the marginal and average cost schedules we get the profit maximizing output level OQ_1 (Figure 5.1) where the long-term average cost is at its lowest point. Wunderlich (1967) argues that this policy has been the basis of the 'maximum efficiency recovery' programmes widely used in the regulation of petroleum extraction in the United States of America, where production is recommended at the minimum long-term average cost. However, as will become obvious below, this policy would be appropriate only in a special case where the interest rate is zero.

In deciding whether to extract now, the resource owner must be satisfied not only that current profits are increased, but that this increase outweighs the reduc-

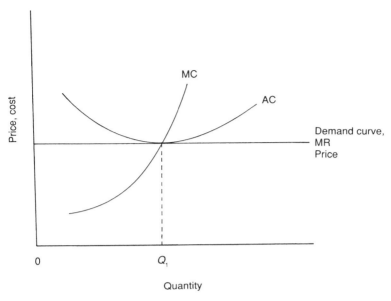

Figure 5.1 Profit maximizing output for the non-extractive sector under competitive market conditions. MC, marginal cost; AC, average cost; MR, marginal reserve.

tion in future profits, had the extraction been postponed. Here there are two cost components: the marginal extraction cost, arising from the current extraction, and the marginal user cost, arising from forgone future profits. The former relates to the extraction activity which uses up scarce resources such as labour and capital. The latter relates to future profits which could have been earned by a decision to postpone the operation – the time element which also has an opportunity cost. This decision to wait is not always well understood by everybody. For example on 3 February 1977, a couple of years before the second energy crisis of the 1970s, a United States journal ran a front-page article entitled 'The waiting game: sizeable gas reserves untapped as producers await profitable prices' (*Wall Street Journal*, 1977). A few weeks later the United States Secretary of the Interior warned companies holding federal oil and gas leases that they must show reasons why they were not producing or face the termination of their leasing arrangements.

The economics of extraction dictates that the revenue which would arise from current extraction should be high enough to cover the marginal extraction cost as well as the marginal user cost. This leads us to a rule which is known as the 'fundamental principle' in the mining, petroleum and natural gas industries (Gray, 1916; Hotelling, 1931). That is, for the extraction to be justified the net market price of the resource (net of extraction cost) must rise in line with the market rate of interest. This principle can easily be grasped by imagining two situations:

1. Assume that the market price of the resource minus extraction cost is rising at a rate which is less than the market rate of interest. What is the appropriate course of action for the resource owner? If he/she is a rational profit maximizer, *ceteris paribus*, he/she should extract and sell his/her stock as quickly as possible and invest the proceeds elsewhere, say, in a time deposit.
2. Market price minus extraction cost is rising at a rate higher than the market rate of interest. What should be the best decision? The owner should leave the resource in the ground as it would represent a superior investment to alternatives.

Therefore, a net price rising at the same pace as the market rate of interest is the equilibrium condition.

The yield on a non-renewable resource deposit is sometimes called the resource rent. If this rent does not increase over time at the same pace as the discount rate people would be unwilling to purchase the deposit as the rate of return on alternative assets would be more valuable. Furthermore, the owner of an existing deposit would speed up the extraction so that he/she can invest the proceeds elsewhere. Let us assume a coalmine owner who is under a contractual obligation to extract 100 000 tonnes of coal per year. It is technically feasible for him/her to increase the annual extraction beyond this level. Also assume that the market rate of interest is 10% and the current market price of coal is such that each additional tonne, on top of 100 000 tonnes, gives the owner £10 profit. If he/she increased current output by 1 tonne, he/she can put the resulting profit, £10, in the bank which will become £11 next year. Suppose that he/she is anticipating more favourable prices for the next year as he/she believes that his/her net earning on each additional tonne extracted will go up to at least £12. Obviously, he/she will make more money by deferring extraction of the extra tonne until the next year. Indeed, during 1979–81 when the price of crude oil was rocketing, many Arab producers stated quite plainly that in a period like this it would be better to 'leave the stuff in the ground'.

The fundamental principle can be obtained formally by making a number of simplifying assumptions:

1. The resource owner has a fixed non-renewable stock, say, oil and he/she wants to deplete this at a rate which will give him/her the maximum profit, i.e. he/she wants to maximize the present value of his/her revenue resulting from extraction over time.
2. The quality of oil is uniform at all points of extraction, i.e. there is no difference between the first and the last barrel of oil extracted.
3. The cost of extraction is constant.

The profit function which is to be maximized is:

$$\pi = P(0)Q(0) - C + [P(1)Q(1) - C] (1 + r)^{-1} +$$
$$[P(2)Q(2) - C] (1 + r)^{-2} \dots + [P(T)(Q(T) - C] (1 + r)^{-T} \tag{5.1}$$

subject to stock constraint

$$Q(0) + Q(1) + \dots Q(T) = \overline{Q} \tag{5.2}$$

or in short

$$\text{Maximize} \sum_{t=0}^{T} \pi(1 + r)^{-t} = \sum_{t=0}^{T} \underbrace{[P(t)Q(t) - CQ(t)]}_{\text{profit}} \underbrace{(1 + r)^{-t}}_{\text{discount factor}} \tag{5.3}$$

Subject to

$$\sum_{t=0}^{T} Q(t) = \overline{Q}, \tag{5.4}$$

where
π is the discounted profit function;
$P(t)$ is the price of oil at time t;
C is the cost of extraction, constant;
$Q(t)$ is the quantity extracted at time t;
t is time in terms of years;
T is the number of years that the deposit will be worked out, i.e. the planning horizon;
r is the interest rate; and
$\overline{Q}$ is the total stock.

By using the Lagrangian multiplier method the augmented function becomes

$$L = \sum_{t=0}^{T} [P(t)Q(t) - CQ(t)](1 + r)^{-t} + \lambda \left[\overline{Q} - \sum_{t=0}^{T} Q(t) \right]. \tag{5.5}$$

By differentiating with respect to $Q(t)$ and setting it equal to zero we get

$$\frac{\partial L}{\partial Q(t)} = [P(t) - C] (1 + r)^{-t} - \lambda = 0$$

which is

$$[P(t) - C] (1 + r)^{-t} = \lambda$$

or

$$P(t) - C = \lambda(1 + r)^{t}. \tag{5.6}$$

The left-hand side is the net price of the deposit (net of extraction cost) and the right-hand side is the resource rent. Equation (5.6) shows that price minus extraction cost must increase in line with market rate of interest.

5.1   DETERMINING EXTRACTION LEVEL AND RESULTING PRICE PATH OVER TIME

One objective in Hotelling's work was to examine the optimal extraction ratio for non-renewable resources from the viewpoint of the government who wanted to maximize the social welfare from exploitation of these resources. Then he showed that a competitive industry, which consists of many firms, facing the same extraction costs and demand conditions as the government, and having information about resource prices, will arrive at exactly the same extraction path. That is, the optimal extraction rate determined unilaterally by each single firm in a competitive world will yield the socially optimal result.

Let us assume that the entire extraction industry is facing a downward sloping linear demand curve as shown in Figure 5.2. Obviously, the greater the industry's output, the lower the price for output at any given point in time. There are a number of issues which must be emphasized at this point. First, the industry must reduce the quantity extracted in each period of time in order to secure higher prices as implied by equation (5.6), i.e. net price must go up over time in line with interest rate. This will happen when the level of extraction contracts. Therefore, extraction next year must be less than this year's output to ensure that price goes up just enough to satisfy equation (5.6). Second, with this type of curve there is a price level at which no one is willing to buy the output. In Figure 5.2 this level is $\bar{P}$ where the demand curve cuts the vertical axis. This is also called the choke-off price, meaning that demand for the commodity becomes zero at that price level. Third, if there is going to be any stock left in the ground

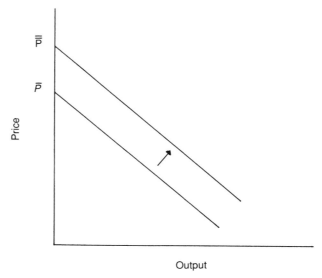

Figure 5.2   Demand for the extractive industry's output and the choke-off price.

at the time when the price hits the critical level, from the industry's viewpoint, there will be waste because the left over can never be sold at that price level or beyond. It can, of course, be disposed of at a discount price – in which case the industry will suffer a loss.

In view of all these the owners will want to make sure that their resource will deplete completely over time before the choke-off price is reached. They will work out a plan to decrease the level of output at each point in time, which in turn will determine the price path. In this, the planning horizon and the output levels are to be worked out simultaneously. Figure 5.3 shows the situation. The choke-off price on the left becomes the ceiling for the diagram on the right, i.e. the price level which will reduce the demand to zero. The deposit must therefore be exhausted just before this point is reached. The justification for this idea is that when the price becomes prohibitively high customers will choose a substitute commodity. The cost of extraction, which is constant, is illustrated by the horizontal line c; ab is the price path which is achieved by reducing the level of output at each period of time. The difference between price line (ab) and cost (c) is the resource rent which increases at a rate equal to the market rate of interest in the perfectly competitive economy. If the demand curve on the left shifts to the right then the ceiling on the right will rise, allowing the owners to acquire much higher rent over time.

It is possible to argue that given a certain value of ca, which is the initial rent, a path similar to ab can be established by way of a trial and error process. If the path reaches the choke-off price earlier than the stock is exhausted there is clearly scope for a higher price at each point in time and owners will see the possibility of receiving higher royalties. If, on the other hand, owners feel that

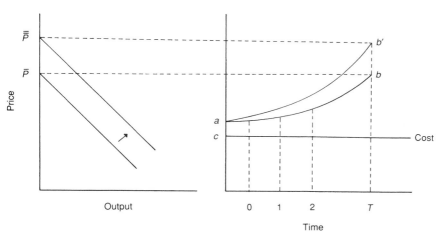

Figure 5.3 Output–price path for the extractive industry.

some stock will be left in the ground at the choke-off price then they will be forced to reduce their royalties. In this way a path more or less conforming to *ab* will be discovered by a process of iteration.

At this stage it is worth mentioning that for any study of fossil fuel for energy, or for any other energy source for that matter, it is essential to be aware of two laws of thermodynamics. The first states that the sum of energy in all its various forms is a constant. Physical processes change only the distribution of energy, never the sum. In other words, energy can neither be created nor destroyed. This first law tells us only that energy is transferred and does not explain the direction in which the process operates.

The second law specifies the direction taken by physical processes; that is, heat transfer takes place from a hot body to a cooler body. Therefore, it is impossible to construct a system that will operate in a cycle, extracting energy from a source and doing an equivalent amount of work in the surroundings.

The second law explains why 100% conversion efficiency cannot be achieved in any process. Part of the energy will inevitably be lost as unavailable heat. For example, the chemical energy which is locked up in a barrel of oil may be burnt to create heat to transform water into stream which may then be used to turn a turbine and hence to generate electricity. During each of these transformations there is always a loss of energy as heat. Likewise, there will always be fossil fuel left in the ground as waste which is not worth recovering; and also during the extraction there will be some seepage.

5.2   FACTORS AFFECTING DEPLETION LEVELS

A number of factors will affect the price–output path worked out by the competitive extractive industry, namely: fluctuations in interest rate; fluctuations in extraction cost; and the introduction of taxes by the government. Some of these, such as taxation and interest levels, can be employed as policy variables by the government to influence the level of extraction in mining, petroleum and natural gas industries.

5.2.1 Change in interest rate

Fluctuation in the level of interest will have a powerful effect on the price–output path of the extraction industry. To begin with let us assume that the market rate of interest rises. This would mean that rate of return on alternative investment projects, say, time deposits, increased. If the owners do not make any alteration on the previously worked out plan the stocks will be earning a suboptimal rate of return over time. The way to avoid this loss is to shift production to the present. That is, the owners will extract and sell more now which will drive down the current market price. Thereafter less will be extracted so that the net price on the remaining deposits can rise at the higher rate. This means that the deposits would be exhausted in a shorter space of time than the one worked out before the rise in interest.

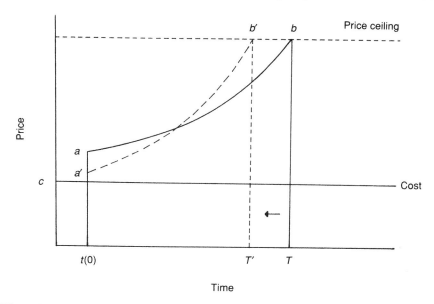

Figure 5.4   Effect of a rise in interest rate on the price–output path and depletion time.

Figure 5.4 illustrated the situation. The curve *ab* is the output–price path before the rise in interest rate. Immediately after the rise owners should make an adjustment by increasing the level of output and then the starting price will fall to $t(0)a'$. For the rest of the time less will be extracted so that the rent on the remaining deposit is growing at a higher rate. This will shorten the depletion time from T to T' The new price–output line, $a'b'$, will be steeper than the previous one, *ab*.

If the interest rate comes down exactly the opposite will happen. The starting price will go up as owners shift output towards the future by reducing current output. This is because falling interest rates make stocks more attractive assets to alternatives so the present output will contract. This should also be evident by the fact that a lower interest rate would indicate a slower growth path than before. This means that the time to depletion must rise as shown in Figure 5.5.

5.2.2 Fluctuations in extraction cost

Let us first assume that the extraction cost has gone up. This can happen for a variety of reasons, such as a shortage of skilled manpower, increasing the wage cost in the industry, or a declining resource base as the owners begin to extract from less easily accessible deposits.

A rise in cost will reduce the level of current extraction and hence increase the starting price but lower later prices. This, in turn, will reduce the quantity

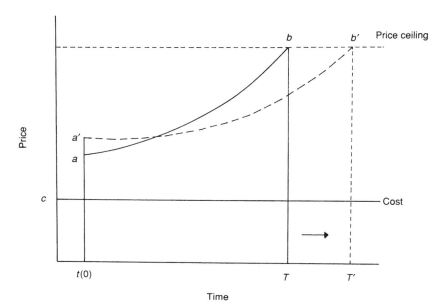

Figure 5.5   Effect of a fall in interest rate on the price–output path and depletion time.

demanded in the earlier periods and increase it later. The net effect will be an increase in depletion time. This situation is depicted in Figure 5.6. As the cost of extraction goes up, the rent will be squeezed. In response owners will reduce current output and this will increase the initial price from $t(0)a$ to $t(0)a'$ and the new price–output path will be $a'b'$.

A fall in the extraction cost will have the opposite effect by increasing the value of the initial rent. If no adjustment is made, this would create a situation in which the choke-off price would be touched earlier than the desired time, leaving owners with unsold stocks. In order to avoid this, the owners must reduce the starting price. The outcome will be that when the extraction cost comes down the immediate output level will go up which in turn will reduce the starting price and depletion time (Figure 5.7).

5.2.3 Taxation

Taxation can have a powerful effect on the utilization policies of extractive industries. There can be a number of tax cases some of which are as follows:

Excise tax

A levy on the output of extractive industries will increase the cost, which will have an effect similar to that described in Figure 5.6. For the mine owner a tax on output is a cost which will reduce the level of current extraction and increase the depletion time. Furthermore, this type of tax will make companies postpone

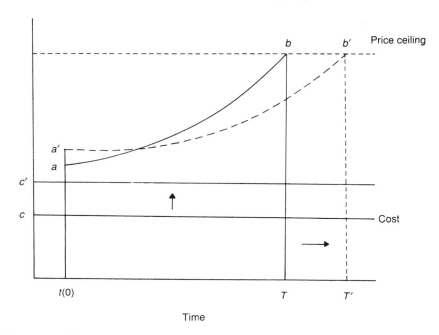

Figure 5.6   Effect of a rise in extraction costs on the price–output path and depletion time.

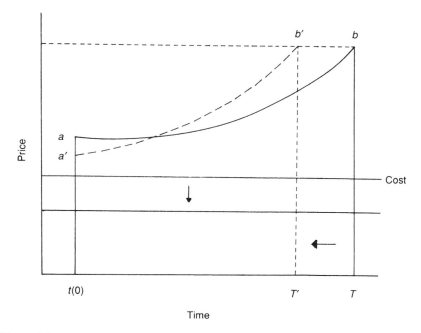

Figure 5.7   Effect of a fall in extraction costs on the price–output path and depletion time.

extraction so that they can defer their tax liabilities. They prefer to keep deposits in the ground – in which case there is no tax payment.

Ad valorem *tax*

This is a tax levied on the price of each unit of output, usually a given percentage of the value of extraction. The effect of this tax will be a reduction in the rate of depletion and an increase in the time to depletion. There is a difference between the affects of *ad valorem* and excise taxes; that is, the strength of reduction in depletion rates will be less strong in the former. Let us say that owners are thinking about delaying their tax liabilities by lowering extraction rates. They will see when *ad valorem* tax is in force that as future sales will be at higher prices tax payable on their sales will therefore be higher. Therefore, in the case of *ad valorem* tax, lowering the rate of depletion is less likely to be such an attractive proposition compared with excise tax.

The difference in strength between specific and *ad valorem* taxes can have an important bearing on policy-making. If the government feels that the nation's natural resource deposits are depleting too fast then a strong measure in the form of an excise tax may look quite appealing to moderate the depletion rates. A somewhat less-effective form of action will be to go for *ad valorem* tax as opposed to excise duty. This is one reason why hard-line conservationists tend to argue for an excise duty on extraction.

Property tax

This will shorten the time to depletion. It is inherent in equation (5.6) that the value of stocks in the asset market is the present value of the future net profits from extracting and selling them. This value in equilibrium will increase over time at the market rate of interest, thus providing owners with the incentive to hold them. *Ceteris paribus*, an annual tax on resource value will greatly reduce this incentive since the longer a deposit is held the greater will be the tax paid on it. One way to avoid paying this tax in all future periods would be to extract as fast as possible and invest the proceeds in areas where there is no similar tax.

5.3 FURTHER POINTS

The operation of the fundamental principle will be restricted by a number of real world constraints. For example, consider fluctuations in the market rate of interest: if it goes up *ceteris paribus* the pace of extraction should increase; conversely, if the rate comes down there will be a slow down in extraction. It is well known that the market rate of interest can go up and down quite briskly in a short space of time. Should an automatic adjustment in output levels be expected every time the interest rate moves? Table 5.1 shows the fluctuation in the base rate of interest in the UK since 1968; at no time did the rate stay the same two years running.

Table 5.1   Real rate of interest (after inflation) in the UK, 1968–88

Year	Real rate of interest (%)
1968	4.6
1969	4.3
1970	–0.5
1971	–5.3
1972	–2.8
1973	2.0
1974	–4.9
1975	–15.0
1976	–1.9
1977	–2.0
1978	–1.5
1979	3.6
1980	–4.0
1981	2.6
1982	1.9
1983	4.4
1984	4.8
1985	5.4
1986	7.6
1987	4.8
1988	6.2

Source: Bank of England Quarterly Bulletin and Economic Trends.

It is quite unrealistic to expect an automatic response by resource owners to changes in interest rates. Let us say that the market rate of interest has risen quite considerably, compelling owners to increase the pace of extraction so that proceeds can be invested in high return back deposits. Normally, an increase in output levels in the mining, petroleum and natural gas industries requires an expansion of capacity which takes time. Furthermore, the high interest rate period may not last long which will make owners think twice before engaging in a costly exercise of capacity expansion.

Similar problems can arise with regard to taxation. National tax policies in many countries change with the government in power. Therefore, resource owners cannot be certain about the longevity of a particular tax policy. If capacity and output levels in extraction industries are based strictly upon current tax legislation, when this is suddenly changed owners may be left with excess or inadequate structures to operate optimally.

Another important factor is the change in technology with regard to exploitation of natural resources. A technological breakthrough may reduce the dependence and hence the demand for a particular deposit. For example, compare solar energy with the use of fossil fuel. A rapid technological development in tapping solar power may reduce, substantially, the demand for fossil fuel. This type of uncertainty is always in the minds of resource owners when they work out a depletion plan for their deposits. Needless to say, in addition to the fundamental principle owners are also guided by a rule which dictates: 'Sell your stocks at a time when there is demand for them'. When the depletion time (T in the present analysis) is long then this problem becomes serious as the likelihood of a technological breakthrough increases.

Last, but not least, impatience can have a powerful effect on the depletion of natural resource deposits. For various reasons a resource owner may be desperate for cash which can be acquired either by selling the property rights or by hastening the extraction regardless of what the fundamental principle says. When deposits are owned publicly, selling the property rights may not always be politically feasible, and the government is left with one option: deplete resources fast for early cash. In effect this happens quite often in the real world. Take, for example, two major oil producers, Iran and Iraq, who were engaged in a long and costly war for most of the 1980s. During the hostilities both nations were in need of hard currency to maintain their war efforts and this was achieved, mostly, by selling their oil.

Mexico provides another example. Having borrowed heavily from the international financial community in the 1970s, the debt servicing became an enormous problem requiring large sums of money. For many years to come the nation is likely to rely on her oil wealth for hard currency to meet her financial commitments to her creditors. When resource-owning nations are in urgent need of cash the wisdom of the fundamental principle is unlikely to be a powerful guiding force when they work out their extraction plans.

5.4   A TEST OF FUNDAMENTAL PRINCIPLE

In Chapter 1 it was pointed out that the prices of many natural resource-based commodities have been declining over a long period of time (Potter and Christy, 1962; Barnett and Morse, 1963; Barnett, 1979; Smith, 1979). There were some exceptions, such as timber which showed a definite upward trend and the price of oil which went up quite rapidly between 1973 and 1982, although between 1982 and 1988 it fell (Figure 5.8). On the other hand, the fundamental principle makes it very clear that *ceteris paribus* prices of mine and fossil fuel deposits should increase in line with the market rate of interest. Then a question must arise, has there been a contradiction between the economic theory of natural resources and the situation observed in the real world? It is worth pointing out once again that in Hotelling style models we consider the rise in the net price of mine and fossil fuel deposits over time, that is, market price minus extraction costs, all in real terms. Still, we know that except for brief periods the real rate of interest has been positive in many countries throughout history. So has there been a sustained reduction in extraction costs which may explain the actual price trends in the face of the Hotelling rule? This has been tested by Slade (1982) who attempts to reconcile the theoretical

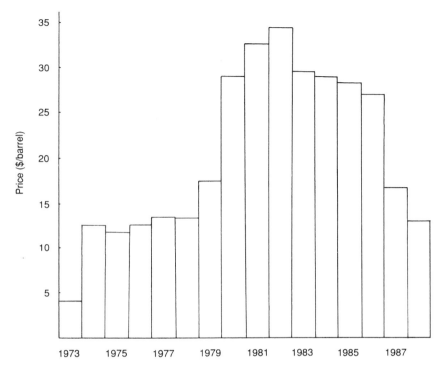

Figure 5.8   Crude oil prices (Arabian light), yearly average. (Source: OPEC, various years.)

prediction of increasing real prices for natural-resource commodities with the above-mentioned empirical findings of falling prices.

The Slade model assumes exogenous technical advance and indigenous change in the grade of deposits and these are used on price trends for all the major metals and fuels in the United States. If equation (5.6) is modified slightly by allowing the cost of extraction to change over time, then

$$P(t) = C(t) + \lambda(1 + r)^t.$$

Slade permits a fall in price by arguing that although $\lambda(1 + r)^t$, which is rent, is normally rising, if the technological advance is substantial, then $C(t)$, i.e. the cost of extraction, may fall substantially and could yield a declining price trend. At early stages of operation, a decline in cost may outweigh the increase in rent, but later on its power may diminish yielding a U-shaped price trend. For example, between 1900 and 1940 the real price of copper in the United States fell because of the advent of large earth-moving equipment, which made possible the strip mining of extremely low-grade ore bodies, and the discovery of froth flotation, which made concentration of low-grade sulphide ores very economical and thus overall cost fell. By 1940 the switch to new technology had reached its natural limits and since then the decline in unit cost has become moderate, giving rise to an increasing price trend (Figure 5.9).

Slade's data consist of annual time series for the period 1870–1978 for all the major metals and fuels with the exception of gold. Prices were deflated by the US wholesale price index in which 1967 was taken to be the base year. Adjusted for inflation, the prices of these resources show some very interesting trends. Two price paths were estimated for each resource – one where the price is a simple linear function of time, i.e. a straight line, the other where price is a U-shaped function of time. Twelve commodities yielded statistically significant results. Figures 5.10–5.12 show the price trends for coal, petroleum and natural gas, respectively.

As well as being intuitively more appealing, the U-shaped curves tend to fit well to the data. At the early stages of extraction the fall in prices can be understood by assuming that the rate of technological change offsets ore grade decline and costs fall. From then on prices may stabilize because the two factors, i.e. falling ore grades and advancing technology, go at the same pace. At the rising cost stage the rate of technological progress is no longer high enough to offset the decline in ore grade and consequently costs and prices increase.

Slade notes that her model is simple and naïve. It neglects many important aspects of the extraction industry such as environmental regulations, tax policy, price controls and market structure.

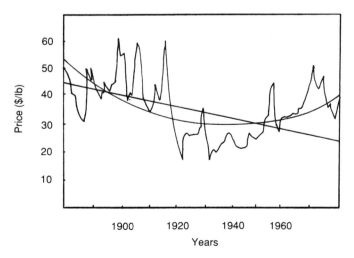

Figure 5.9   History of deflated prices and fitted linear and U-shaped trends for copper. (Source: Slade, 1982.)

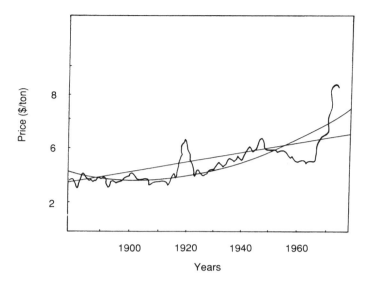

Figure 5.10   History of deflated prices and fitted linear and U-shaped trends for coal. (Source: Slade, 1982.)

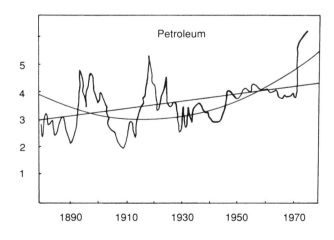

Figure 5.11   History of deflated prices and fitted linear and U-shaped trends for petroleum. (Source: Slade, 1982.)

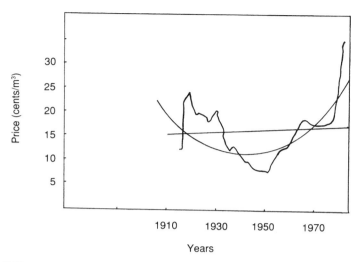

Figure 5.12   History of deflated prices and fitted linear and U-shaped trends for natural gas. (Source: Slade, 1982.)

5.5   MARKET STRUCTURE AND RESOURCE USE

It has been argued that market imperfections, especially the polar case of monopoly, are the conservationists' best friend (Hotelling, 1931). It should be pointed out that monopoly can exist in extraction as well as in manufacturing, effecting the depletion ratios and hence costs/prices of metal and fossil fuel deposits.

5.5.1 Monopoly in manufacturing

Economic theory normally assumes that the objective of any firm, e.g. perfectly competitive, monopolistic, oligopolistic or monopolistically competitive, is to maximize profits by equating the marginal revenue (MR) with marginal cost (MC). Although this objective may be common for all types of firms the market conditions are different in each case. For example, a perfectly competitive firm is faced with a horizontal (or near horizontal) demand curve which implies that the demand for its product is infinitely large at a given price. The monopolist, at the opposite end of the spectrum, faces the market demand curve.

Now let us assume two situations; in one case all branches of the manufacturing sector are dominated by monopolists, in the other all markets are perfectly competitive. We must also bear in mind that the industry is the major user of metal and fossil fuel deposits for raw material and source of energy. Figure 5.13(a) shows the level of output OQ_m in one branch of manufacturing when the market is dominated by a monopolist. In Figure 5.13(b) the market is perfectly

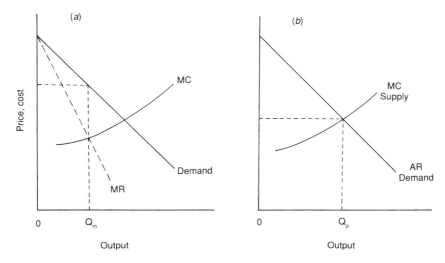

Figure 5.13   Price and output levels in manufacturing industry with (a) monopoly and (b) perfect competition.

competitive and contains a large number of firms where the output level will be OQ_p; the entire industry's marginal cost, which is also the supply schedule, intersects the demand curve. In this, each firm takes the market price as given and the price equals marginal costs.

A comparison of Figures 5.13(a) and (b) shows that the monopoly output, OQ_m, is smaller than the competitive level, OQ_p. This means that a perfectly competitive market structure in manufacturing will call for more raw material and energy resources from the extractive sector compared with a situation in which the market is dominated by a monopolist. Consequently, metal and fossil fuel deposits will last longer in a monopoly situation.

5.5.2 Monopoly in extraction

In what ways would a monopolist behave differently from a perfectly competitive firm in the extractive sector? Once again the objective for all firms will be the same; that is, to extract in a manner which will maximize the present value of profits over time. When profits are maximized over a period of time the market rate of interest will be one of the determining factors for a monopolist as well as for a perfectly competitive firm. This is inherent in the fundamental principle. Let us assume, to begin with, that a monopolist is the sole owner of a fixed stock of deposit which he/she can extract at a zero cost. The market demand curve, which the monopolist is facing, remains stationary over time. His/her problem is to find an extraction path which will yield maximum discounted profits over a period of time until the deposit is exhausted. The increase in his/her marginal revenue between two consecutive time points must be:

$$\frac{MR(t + 1) - MR(t)}{MR(t)} = r, \qquad (5.7)$$

which says that the percentage change in marginal revenue over time equals the rate of interest, r, or

$$MR(t)\,(1 + r) = MR(t + 1), \qquad (5.8)$$

that is, at each time period the monopolist's marginal revenue rises at the market interest rate.

Figure 5.14 shows that output at any period is chosen so that equation (5.8) holds. In Figure 5.14(a) an output level for time $t(Q_t)$ is determined which corresponds to price level P_t. The next period, $t + 1$, the price and hence the marginal revenue must go up in line with the market rate of interest which can only be achieved by cutting back output, that is,

$$Q_{t+1} < Q_t.$$

When discounted at the market rate of interest the marginal revenue is the same between time periods. The last unit extracted will yield the highest undiscounted marginal revenue, which corresponds to the ceiling price p^*.

As the monopolist is moving up its marginal revenue curve the competitive extraction industry, which contains a large number of firms, will be moving up the demand curve in each period of time. In both market structures the fundamental principle must be satisfied given the same interest rate. Since the marginal revenue curve is steeper then the demand curve, i.e.

Slope MR > Slope D

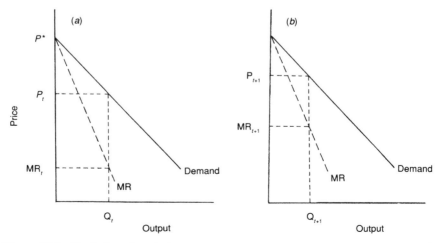

Figure 5.14   Depletion of stocks in monopoly with (a) time = t and (b) time = $t + 1$.

the monopolist will have to decrease output less than a competitive industry in each period of time. Thus the monopolist will extract its deposits more slowly than the competitive extraction industry.

As for the price path under two different market conditions, since the monopolist's output is initially lower than that of the competitive industry, the starting monopoly price must be higher. As the marginal revenue is less than the price (Figure 5.13(a)) then the price path which belongs to the monopolist will be flatter than that of the competitive industry, i.e. the price is rising more slowly. This is shown in Figure 5.15. In a monopoly, initially prices are higher but the rate of increase is slower. Assuming that there is no change in the demand then with monopoly the stocks will last longer compared with a situation in which the extractive industry is competitive.

It has to be pointed out that when the depletion time (T) is long there can be a lot of change with regard to the technology of extraction and resource use which is likely to affect the demand curve. It could well be argued that since in perfect competition the price is rising fast, this is likely to encourage users to seek alternatives. In a monopoly, however, demand for natural resource-based commodities may not shirk as users get accustomed to slowly rising prices and hence maintain a stable demand. With stable demand the ultimate depletion will take place whereas in perfect competition, if users gradually turn away from the commodity in question, a part of the stock may be left in the ground.

It is also inherent in the above argument that the monopoly rent, which includes both resource rent and excess profits, exceeds the resource rent accruing to the competitive industry. Therefore, it is quite understandable that

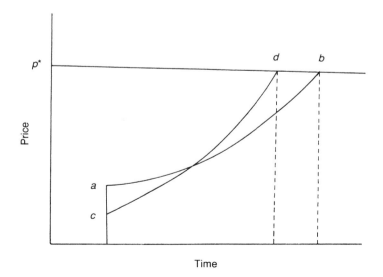

Figure 5.15   Price path in monopoly (*ab*) and perfect competition (*cd*).

independent resource owners would wish to form a cartel in which they can behave as a collective monopoly. Indeed, a cartel is a group of independent owners attempting via collusive agreement to behave as one firm. In a cartel situation each owner agrees to produce less than it would have produced under competitive conditions. The anticipated effect of a cartel is to drive the market price up so that producers can make excess profits.

Figure 5.16 shows output levels in both competitive and cartel conditions. For the sake of simplicity we assume a zero interest rate in this analysis. As regards to the former, Figure 5.16(a) shows the competitive price, OP_p, and output OQ_p, established with the intersection of market demand and supply schedules. A competitive firm takes the market price as given and produces at a level where its marginal cost equals price which then becomes the marginal revenue and demand schedule for it. Its market share is only a small part of the total sales.

Now let us assume that in order to achieve excess profits all these competitive firms get together to form a cartel. They cut back output to, say, OQ, so that the market price can go up to OP_c. Note that the ability to supply is not diminished as output levels are reduced artificially. Each firm in the cartel is given a quota so that the reduced market output level can be maintained. The individual firm in Figure 5.16(b) is told to lower its output to, say, OQ_c.

In the new situation there is strong temptation to cheat. The individual firm, by reducing its price by a small amount (just below the cartel price) can sell up to OQ_d at the expense of the other members. At OQ_d the firm's marginal cost equals the new marginal revenue, MR', and the shaded area shows the super

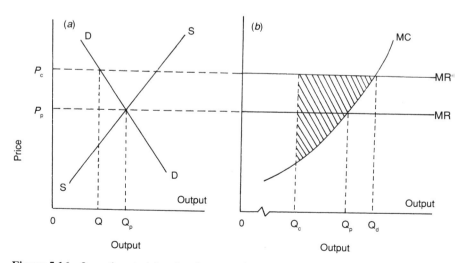

Figure 5.16   Incentives to 'cheat' under a cartel.

excess profits which can be acquired by the cheating firm. But the fact of the matter is that all firms are tempted in the same way. If they all cheat, then the market output will be much greater than the pre-cartel level and consequently the price will be much lower. In other words a cartel can also bring disaster to its members.

The conclusion in this analysis is that cartels can only raise the price by cutting back output. However, at higher prices members who make up the cartel are tempted to produce even more than the competitive equilibrium. The more successful the cartel the greater the incentive to cheat. In order to succeed a cartel requires some form of policing to ensure that each member is sticking to its quota. In terms of resource depletion, in a cartel situation, output extracted each period of time will be much less than the one which is dictated by competitive conditions and consequently stocks will stay longer in the ground.

5.5.3 Oil cartel

The Organization of Petroleum Exporting Countries (OPEC) formed in 1960 is the best-known cartel. Before that time the world oil market was dominated by a few large oil companies such as Royal Dutch Shell, Mobil Oil, British Petroleum, Standard Oil of New Jersey, etc. At the time these companies were accused of acting collectively in order to fix the price of oil – an accusation that was largely unsubstantiated. In 1960 the price of crude oil began to go down which led the oil exporting countries to form OPEC. Ironically, the United States government gave a lot of encouragement to these countries, hoping that the cartel would be helpful to them in their development struggle.

OPEC comprises 13 countries, seven Arab and six non-Arab; Algeria, Ecuador, Gabon, Indonesia, Islamic Republic of Iran, Iraq, Kuwait, Libya, Nigeria, Qatar, Saudi Arabia, United Arab Emirates and Venezuela. In the 1970s and early 1980s OPEC became a powerful economic force and gained a tight control over the price of crude oil and as a result some member states accumulated vast sums of money.

During the 1973 Arab–Israeli war, OPEC countries acting in solidarity with the Arab world decided to use oil as a political weapon against the West hoping that Western countries, especially the United States, would put pressure on Israel. The price of Saudi crude was $2.12 per barrel on 1 January 1973, and rose to $7.01 one year later. On 1 January 1975 the price went up to $10.12. OPEC attempted to convince the world that the rapid price increase was largely due to the aggressive oil companies. The fact of the matter was that before the first sharp price increase the oil companies' share from 1 barrel of oil was about $0.60 by which they covered their operating costs and profits. This figure did not change much after the rise.

From 1975 onwards the price rose more slowly, reaching $13.34 in January 1979, an increase of about 30% which was below the rate of inflation experienced in many parts of the world. For example, between 1975 and 1979 the purchasing power of the US dollar fell by 38% which meant that there had been a decline in the real price of oil. In 1979 events in Iran, a major producer, caused a substantial contraction in the output level and as a result the price began to rise again. In 1981 the price of Saudi crude exceeded $35 per barrel. As an excuse for such a high price some OPEC members mentioned rising world inflation and also argued that the cartel was looking after the interests of future generations.

From 1981 onwards the OPEC oil revenue began to decline steadily (Figure 5.17). Following the first oil shock of 1973 industrialized countries gradually increased their efforts to reduce the consumption of oil by way of conservation and the use of substitutes such as coal and gas. Furthermore, the world recession of 1982–84 reduced the demand for oil. Increased production by non-OPEC countries such as the Centrally Planned Nations, Britain, Norway and Mexico contracted further the share of OPEC exports in the world market. Figure 5.18 shows the steady erosion of OPEC's share.

The discovery of substantial oil deposits in the North Sea in the 1960s and their successful exploitation has made the UK a major oil producer. For example, in 1984 the UK became the fourth largest oil producer in the non-communist world after the United States, Saudi Arabia and Mexico. There were

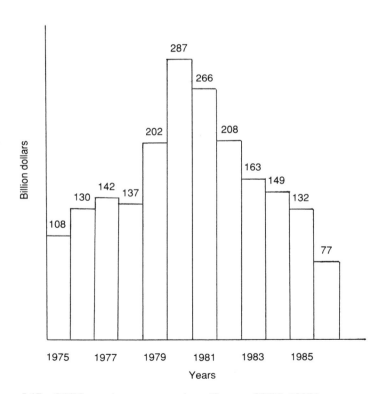

Figure 5.17   OPEC petroleum export values. (Source: OPEC, 1987.)

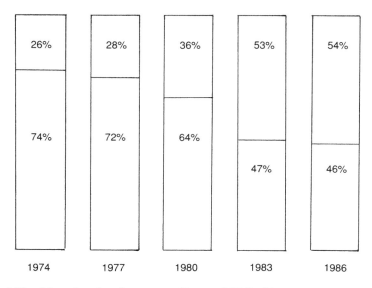

Figure 5.18   Oil market share by exports. (Source: OPEC, 1987.)

some suggestions that the country should join OPEC but, of course, she declined to do so. There were a number of reasons for non-membership. First, as explained on p.164–5, in a cartel situation members must restrict their output so that high prices can be maintained. Those who stay outside the cartel will have a very substantial degree of flexibility to determine their own level of output. Once a price level is agreed by the cartel, non-members can, if they wish, undercut.

Second, cartels are illegal in the UK. In 1948 the creation of the Monopolies Commission gave this body warrant to examine trade practices which are not in the public interest. The 1956 Restrictive Trade Practices Act gave the Monopolies Commission authority to investigate mergers and cartel practices. For example, in 1959 the Restrictive Trade Practices Court dismantled the cotton yarn spinners' agreement on the grounds that the proposed price fixing was against the public interest.

In the European Community cartels are also illegal. Article 85 of the Treaty of Rome relates to uncompetitive behaviour by two or more undertakings and prohibits agreements or other practices which distort competition and adversely affect fair trade between member states. Price fixing, agreements on market shares and production quotas are illegal. Article 85 also spells out the consequences of uncompetitive behaviour. That is, such behaviour is null and void from the outset. The Community has the power to order the parties to terminate such illegal behaviour otherwise they will be liable to a fine of up to 1 million ECU (£1 = 0.75 ECU at the time of writing).

OPEC, realizing the unfeasibility of Britain's membership, requested a number of times that she should co-operate in oil production agreements. The cartel knows that Britain now plays an important part in the world oil market and in certain instances might be able to destabilize the world oil price. In March 1983 OPEC sounded out Britain on this matter but the official British response was negative. In 1986 when oil prices were sliding downwards Saudi Arabia's oil minister, Sheik Yamani, announced that there would be no limit to the downward price spiral. He warned that such a drop would entail adverse and dangerous consequences for the whole world economy. By contrast, the prospect of cheap oil brought unsuppressed glee to importing nations such as Japan, West Germany, France, Italy, Spain and Brazil.

However, would a further fall in oil prices be good for a consuming and producing country like the UK? Beenstock (1983) argues that despite the nation's oil-producer status, lower oil prices would be beneficial on the whole. His argument is that a fall in oil price will stimulate the gross domestic product on a sustained basis on two considerations. In the short run, the fall in energy prices reduces inflation, which raises aggregate spending in the economy. Second there are beneficial implications for profitability of lower energy costs. Increased profitability raises entrepreneurial incentives and output is permanently stimulated. As for the offshore economy, it is likely to suffer from lower oil prices. The pound weakens in real terms because the contribution of oil production to the balance of payments is now lower. However, the offshore economy is small in relation to the onshore economy, so the benefits to the latter dominate the losses to the former and as a result the British economy gains as a whole.

Although in recent years the effectiveness of OPEC in manipulating the price of oil has diminished substantially, it may be misleading to conclude that its future prospects are grim. Figure 5.19 shows how the world crude oil reserves were distributed in 1987 and Figure 5.20 gives the estimated stocks for the 13 member countries. In the face of these facts it will be hard to dismiss OPEC as irrelevant.

5.5.4 Uncertainty

Uncertainty about market conditions can have a powerful effect on extraction policies. Unlike academics who devise their depletion models in the security of their tenure, when extraction companies work out their plans they consider all type of genuine hazards which exist in the real world. In addition to technological uncertainties there is the ownership uncertainty when companies operate in a politically unstable region where nationalization may be on the agenda. This is likely to make them very nervous indeed and can even happen in well-established democracies especially during the reign of socialist governments. For example, in Britain the coal and gas industries were nationalized in the late 1940s.

The fear of expropriation by the state is bound to influence private owners' extraction policies. Extraction companies who operate in politically volatile countries will almost certainly want to exhaust stocks as quickly as possible and get out.

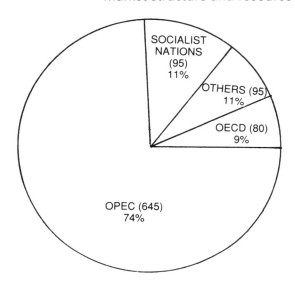

Figure 5.19   World crude oil reserves, 1987, in billion barrels. (Source: OPEC, 1987.)

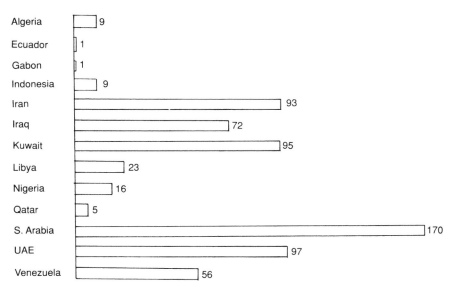

Figure 5.20   OPEC countries' crude oil reserves, 1987, in billion barrels. (Source: OPEC, 1988.)

Another type of uncertainty arises from lack of co-ordination between various sectors of the economy. When a businessman decides to undertake an investment project he estimates future demand and prices. In many cases there is a lack, or even complete absence, of co-ordination between sectors of the economy which makes these estimates difficult. For example, investment in iron ore extraction may not match investment in the steel industry. At some point in time the demand for iron ore may fail to match the supply at anticipated prices. If future demand and prices had been correctly anticipated, this would have created quite a different level of investment and output decisions in ore extraction than the one in which decisions are made in isolation. Therefore, due to lack of this kind of co-ordination, extraction decisions in the mining and fossil fuel industries may be suboptimal.

Problems of multiple ownership in general were discussed in Chapter 2. When an extractive sector is shared by a number of different companies, the resulting investment and output levels will be quite different from the one when the sector is dominated by one firm. Suppose that the oil industry in a country is under a single ownership in which a number of wells are drilled and a certain level of output is realized. The same level of investment and extraction will never be achieved when the oil industry is split into a large number of companies each bearing its own risk. That is to say that under a single ownership the entrepreneur will be able to make a greater number of drilling decisions than is likely under a fragmented ownership. This is because under a single ownership the risk of failure to find oil will be spread between a large number of projects.

5.6   SOME TRENDS IN FOSSIL FUEL USE

Figure 5.21 shows world consumption levels for oil, natural gas and coal. These are substitute energy resources; when there is an increase in the price of oil, there is a tendency to increase the consumption of natural gas and coal as users switch whenever it is technically possible. Furthermore, an increase in the price of oil leads to an increase in efforts to find new oil deposits. After the first price shock in 1973 a substantial increase in exploration for new oil deposits occurred all over the world. Exploratory drilling activities increased not only in regions where conventional oil deposits had been found in the past but in new locations such as the North Sea and the Arctic. Figure 5.22 shows the new discoveries during the last three decades.

Increases in oil supply are largely determined by fresh efforts in the fields of discovery and technology and the price of oil. There may be considerable deposits beneath the ocean floor and in some hostile regions of the world such as the Arctic. Extraction from these areas is a costly operation. Improvement in extraction technology and an increase in the price of oil may bring these stocks to the margin of profitability in the future. Furthermore, there is the possibility of exploiting oil shales and tar sands. The former are sedimentary rocks containing matter than can produce oil. Deposits, some of which are very large, exist in

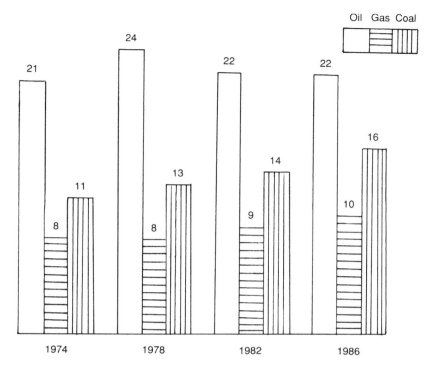

Figure 5.21   Primary energy consumption by type, % in total, for selected years. (Source: OPEC, 1987.)

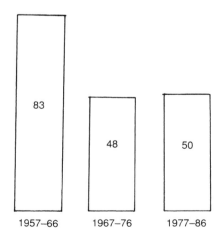

Figure 5.22   New oil discoveries, world total, in billion barrels. (Source: OPEC, 1987.)

the United States, Canada, Brazil, Zaire, China and the Soviet Union. Unfortunately there are some problems in the exploitation of oil shales. First, the cost of producing oil from shales is quite high. The estimates for the cost per barrel output range from $12 to $23 (1978 dollars) (Griffin and Steele, 1980). Second, the extraction and processing of oil shale involves some serious environmental problems. It must be strip mined and requires large quantities of water during processing, which leads to air and water pollution.

The tar sands, which are in thick liquid form, are found in large quantities in North America and Venezuela. At present some processing plants are in operation in Canada. Problems that are prevalent in shale oil also exist in tar sands. With the current technology production is very costly. Strip mining and sulphur emissions from processing plants create serious environmental problems.

Coal is the most abundant fossil fuel energy resource on earth. Huge reserves are available in some parts of the world and it provides about a third of the world's energy. Although, in some regions, its overall contribution as a source of energy has been declining in recent years with the increasing use of oil, gas and hydroelectricity, in some oil-deficit countries its share has gone up, particularly since the two oil shocks of the 1970s. In China, for example, which has vast coal reserves, the contribution of coal to its energy needs increased from 53% in 1978 to 80% in 1988. Turkey, the second most coal-intensive country, burnt more than four times as much coal in 1988 as in 1978. Electricity generation accounts for most of this increase. In contrast, consumption of coal fell in the Soviet Union, Britain and France during the 1980s. Norway is the least coal-intensive country; it has abundant oil and water for hydroelectricity, and coal accounts for only 1% of its energy consumption (Figure 5.23).

In the past coal was the main source of energy which powered the industrial revolution in Britain. Politically coal has always been a sensitive commodity in this country. Policies regarding the depletion of coal deposits have always been strongly influenced by political considerations. In the 1950s there were doubts about the coal industry's ability to meet the rapidly growing demand for energy. Then the government encouraged the use of oil in power stations which proved to be a very effective measure in moderating the demand for coal. However, a substantial reduction in coal supply was politically impossible due to the anticipated militancy of coal miners and the government ordered power stations to use coal again. Also, imports from the United States were discouraged. During that time the government felt that demand forecast for coal, oil and natural gas had become necessary and a set of five-yearly forecasts were published in the mid-1960s (UK Government, 1965a). It was later proved that the overall forecast for energy was remarkably accurate but the demand for coal was over-estimated.

In the 1960s when Britain was a major oil importing nation there were doubts about the reliability of supplies from the Middle East and oil imports were becoming a burden on the balance of payments account. Since coal was an indigenous fuel it made sense to boost coal output to minimize the reliance on

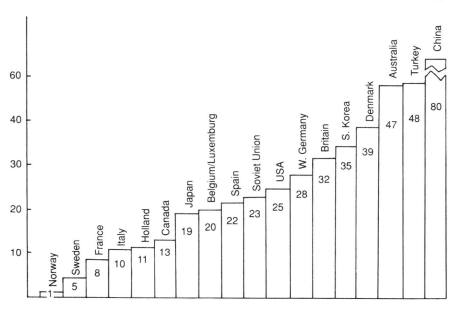

Figure 5.23   Coal as a percentage of total energy consumption, 1988. (Source: *The Economist*, 1988.)

imports. One problem was that coal was becoming very expensive in relation to oil and something had to be done to reverse the trend. In 1965 the government announced measures to assist the coal industry (UK Government, 1965b). These included: closure of uneconomic pits for which funds were allocated to the industry for redundancy and redeployment of labour; writing off about £400 million owed by the Coal Board to the Treasury; and the further use of coal in power stations and in public buildings for heating.

Later in 1965 substantial deposits of oil and natural gas were discovered in the North Sea. Many also believed that widespread commercial use of nuclear energy was imminent which changed the energy picture almost completely. The government realized that a decline in the demand for coal was inevitable and that this could create unpredictable political problems. They thus decided to support the industry. In the 1970s the situation was different again. On the one hand oil was becoming a very expensive commodity and on the other hand the UK became a major oil producer. High oil prices favoured the use of coal as a cheap source of energy.

In the late 1970s the government finally realized that there should not be a reliance on a single fuel for the nation's energy requirements. Instead, there should be a flexible strategy which would be altered in the face of changing events. Furthermore, there was a genuine need for a detailed forecast for various fuels up to the year 2000 and beyond (Energy Consultative Document, 1978).

The broad objective of this document was the provision of adequate and secure supplies of energy at minimum social cost.

In the 1980s the fortune of the coal industry in Britain was changing again. Environmental pressure was building up due to the emission of carbon dioxide and sulphur dioxide from the nation's coal-fired power stations. The increased use of natural gas in power stations and the development of international trade in coal began to threaten this commodity in its principal market. Following the famous 'Plan for Coal' of 1974, which was intended to revive the industry, there were moves to open up new pits such as Selby, the Vale of Belvoir and Margam. The central objective of the plan was to change from a high-cost to a low-cost industry by closing down uneconomic pits, opening up new fields and increasing the productivity of miners. However, these policies were not supported by miners. By 1984, ten years after the Plan for Coal, the workforce in the industry had decreased by 72 000 and 70 pits were closed. The morale in the industry was low and the historic hostility between management and workers' unions was at boiling point.

In October 1983 the National Union of Mineworkers (NUM) imposed an overtime ban and in March 1984 the year-long miner's strike began. The strike ended with the defeat of the NUM which lost considerable sums of money in fighting the strike; its membership shrank when the Nottinghamshire miners formed a breakaway union, and many miners accepted redundancy and left the industry. The Coal Board lost 70 million UK tons of deep-mined output, 73 coal faces were gone and many others damaged. The result was a great political victory for the ruling Conservative government.

In 1988 the Coal Board reported positive changes in the industry since the 1984 strike, i.e. the workforce had been reduced by about 100 000 without any compulsory redundancies, 79 pits had been closed or merged, productivity had more than doubled and the 1987–88 accounts showed an operating profit of £190 million before interest payment, for the first half of the year, giving the best financial performance for 20 years.

Since the 1947 Nationalisation Act coal has been an extremely turbulent and overprotected industry in Britain. There are small areas which belong to the private sector. Some privately owned deep mines produce about 800 000 tons a year and employ some 200 workers; private open-cast sites turn out about 1.5 million tons a year and 15 million tons of open-cast output are produced by contractors working for the Coal Board. In the United States, Australia, Germany, and many other countries coal production is in the hands of the private sector. Following the privatization of the UK electricity industry in 1990/91 the Conservative government argued that privatization of the coal industry would be the ultimate privatization. Bailey (1989) points out that despite recent improvements the coal industry in Britain is still unattractive to the private sector as it is still incurring net losses: £17 million in 1986/87.

At an international level rich coal deposits exist over a wide geographical area. Figure 5.24 shows proven recoverable coal deposits in 1986. Currently Australia,

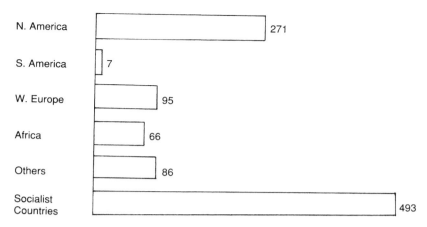

Figure 5.24   Proven recoverable coal reserves including bituminous and sub-bituminous coal, anthracite and lignite, billion tons, 1986. (Source: OPEC, 1987.)

Poland, South Africa, the United States and Colombia are offering cheap coal to many consumers throughout the world. Bailey (1989) argues that these exports are not sustainable as few producers in these countries are making any profit. It is quite difficult to argue that coal will ever regain its dominance in energy markets due to its bulk and pollution problems. However, there are techniques that can turn coal into oil and gas. Widespread commercial exploitation of these techniques depends on further innovations to reduce the production costs.

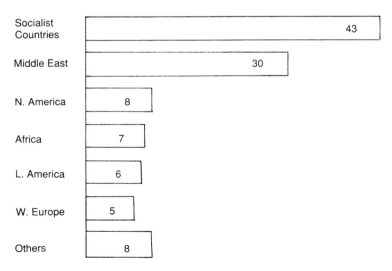

Figure 5.25   Proven global gas reserves, thousand billion standard m^3, 1986. (Source: OPEC, 1986.)

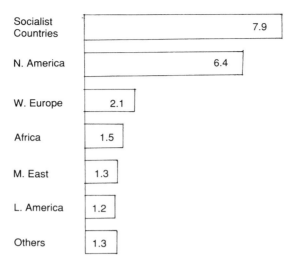

Figure 5.26   Natural gas production, hundred billion standard m^3. (Source: OPEC, 1986.)

As for natural gas, its exploration increased dramatically since the mid-1970s, as did new discoveries and additions to reserves of existing deposits. Figures 5.25 and 5.26 show the proven reserves and production in 1986, respectively. One major constraint in tapping the natural gas is the location of reserves. If the oil price were to go up again in the future, which is more than likely, then much of the gas will be brought to the market.

6

Economics of environmental degradation and policies

> To defend and improve the environment for present and future generations has become an imperative goal for mankind.
>
> The Declaration of the UN Conference on Human Environment, Stockholm, 1972

The issue of environmental degradation is similar to that of the problems of fisheries (Chapter 2). As in fisheries, environmental problems came about largely as a result of the market failure to define and enforce property rights. However, environmental problems as a whole are much more complex and consequently harder to deal with than the problems of fisheries.

As indicated in Chapter 1, environmental problems became prominent in economic literature only in the second half of this century. Long before that Marshall (1890) made the first approach to economic analysis of environmental degradation by introducing the concept of external economies. Although Marshall had in mind only the benefits that accrue to the economic identities through general industrial development, the concept of externalities contains the key to the economic analysis of environmental problems. The advantages referred to by Marshall are enjoyed by businesspersons without payment and outside the market.

Later, Pigou (1920) realized that the concept of externalities is a double-edged sword, containing not only benefits but costs as well. As an example of negative externalities, Pigou uses the case of woodlands damaged by sparks from railway engines. He makes it clear that not only can the production conditions of third parties be influenced outside the market but also the welfare of private persons can be seriously affected both in cost and benefit terms.

The first substantial treatment of externalities was by Kapp (1950) who anticipated the far-reaching adverse consequences of economic growth on the environment. The social cost, which is defined as all direct and indirect burdens imposed on third parties or the general public by the participants in economic activities, is the central point in Kapp's analysis. He explicitly mentions all costs emanating from productive processes that are passed on to outsiders by way of air and water pollution, which harms health, reduces agricultural yield, accelerates corrosion of materials, endangers aquatic life, flora and fauna and creates problems in the preparation of drinking water.

6.1 EXTERNALITIES

Externalities are also called side-effects, spillover effects, secondary effects and external economies/diseconomies. They come about when activities of economic units (firms and consumers) affect the production or consumption of other units and where the benefits or costs which accrue to these units do not normally enter into the gain and loss calculations. In other words, these effects, although noticed, are left unpriced and hence the bearers are, normally, uncompensated in the private market environment. If externalities are priced and the bearers compensated then they are said to be internalized.

Some economists (e.g. Bator, 1958) emphasize that externalities are merely a result of market failure. Baumol and Oates (1975) point out that market failure is a very broad issue which occurs in many areas of economics. They favour the approach taken by Buchanan and Stubblebine (1962) in which externalities are defined not in terms of what they are but what they do. That is, they violate the conditions for optimum allocation of resources in the economy.

There are a number of cases of externalities involving environmental degradation: congestion, which could be urban, industrial or recreational; noise; land, water and air pollution. In this section environmental degradation will be discussed mostly in the context of air pollution because it is the most topical case. However, what will be said could be applicable to other forms of environmental issues although their analyses may require some slight modification.

6.1.1 Technological and pecuniary externalities

Technological externalities occur when the production or the consumption function of the affected is altered. In the former, less output is obtained for a given level of input because of the external diseconomies. In the latter, one less utility is acquired from a given level of income because of the externalities. Numerous examples can be given for technological externalities. An enlarged volume of smoke in a locality will no doubt increase the cost of laundry service as more powder, manpower and machine time will be required for washing compared with the previous situation when the smoke was less intense. The smoke will also affect, adversely, the enjoyment capacity of leisure seekers in the area. Less time will be spent sitting in the garden and furthermore the reduced leisure activity will be less enjoyable due to increased disturbance.

Pecuniary externalities result from a change in the prices of some inputs or outputs in the economy. That is to say that one individual's activity level affects the financial circumstances of another. For example, an increase in the number of handbags sold raises the price of leather and hence affects the welfare of the buyers of shoes. Pecuniary externalities do not normally affect the technological possibilities of production and should not create misallocation of resources in a competitive economy.

The distinction between technological and pecuniary externalities is important in the sense that the former reflect real gains or losses to the parties involved. The latter, however, reflect only transfers of money from one section of the community to another, via changes in relative prices. Pecuniary externalities would be highly relevant if the economic analysis aims to capture the distributional issues involved in a particular case. They do not constitute any real change in the efficiency of the productive process viewed as a means to transform inputs into outputs.

6.1.2 Private versus public externalities

The distinction between private and public externalities is drawn by Hartwick and Olewiler (1986). A private externality is typically bilateral, or involves relatively few individuals. In this case one agent's actions affect the actions of another agent, but there is no spillover on other parties. The key characteristic of a private externality is that the external effect must be fully appropriated by the agents involved. For example, if a chemical firm dumps some mild toxic material into a pond in a residential district, those who live around the pond are the only ones affected.

A public externality arises when a natural resource is used without payment and the utilization by one agent does not normally reduce the quantity available to others. However, the quality of the natural resource may be affected due to the use-as-you-please principle. Air and water pollution are examples for this kind of externality as they emerge in the form of 'public bad' – something consumed by a lot of people.

The distinction between private and public externalities is important for the internalizing process. Sometimes, due to large numbers involved in public externalities, it may not be possible to internalize them through private actions. In private externalities the role of negotiation by the parties involved or the emergence of a market can be very helpful in internalizing them.

6.2 THE OPTIMUM LEVEL OF ENVIRONMENTAL DEGRADATION

From a purely economic viewpoint, complete elimination of externalities is neither practicable nor desirable. There is an optimum level of environmental degradation and this is not at a zero level. To see this let us look at Figure 6.1, where marginal gain and loss are measured along the vertical axis and the scale of industrial activity, which is degrading the environment, along the horizontal axis. Two parties are involved: gainers and losers. The former are those who

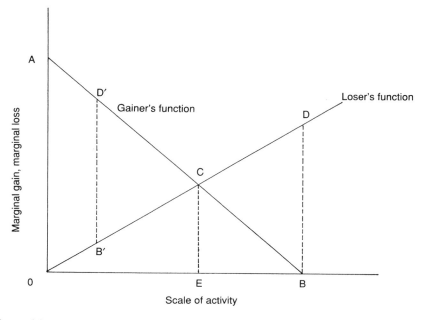

Figure 6.1   Socially optimal level of environmental degradation.

benefit from the industrial activity, e.g. wage earners and profit takers in the industry. Losers are, say, the public at large who suffer from the external effects generated by the industry.

It is clear from Figure 6.1 that the two groups have conflicting interests. From the gainers' viewpoint OB is the best situation. They will want to push the level of activity to B at which the marginal gain is zero. On the other hand, the best point for the losers is the origin, point O, where the loss is zero.

In a situation when gainers do not have to pay compensation to the losers the scale of activity is likely to be expanded to point B.

Here we would have:

Benefit (gainers)   = OAB
Cost (losers)   = ODB
Overall gain (community) = OAD'B': OAB minus ODB.
                           Since OCB is common to
                           both areas we have to
                           subtract CDB from OAC.

The society gains overall and OAD'B' represents the scale of gain with the maximum level of industrial activity.

If, however, the gainers were to reduce the scale of activity from B to E, the gain to society would increase quite considerably. At point E:

Benefit (gainers)   = OACE
Costs (losers)   = OCE
Overall gain (community) = OAC

OAC is greater than OAD'B' by B'D'C. So there is a social advantage when the scale of industrial activity, which is causing environmental deterioration, is reduced to E, which is the socially optimum level.

It is highly unlikely that the socially optimum level of externalities will be attained in a free market situation. The task for the economist is to identify what is socially desirable and then inform the decision-making authority, who may wish to take steps to move towards it.

Clearly the socially optimal scale does not require a complete elimination of external effects. These can be reduced drastically by firms in the industry adopting an externality-free technology. This may or may not be desirable depending upon the cost of 'friendly' technology.

Figure 6.2 shows a hypothetical case in which there is complete elimination of externalities by way of a costly technology. Due to the installation of expensive devices the gainers' function shrinks to A'B' and the scale of activity is reduced to OB'. The loss function disappears due to absence of externality and the net gain to society becomes OA'B'. In comparison with the net gain in the previous situation, OAD'B', where the scale of activity was OB, society is clearly worse off with a zero externality which is achieved by a costly technological shift. OAD'B' in Figure 6.1 is greater than OA'B' in Figure 6.2.

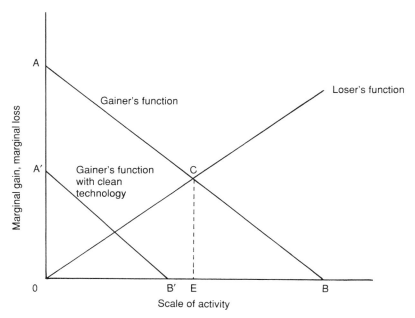

Figure 6.2   Socially optimal level of activity with a costly pollution-free technology.

6.3   METHODS OF OBTAINING OPTIMUM LEVELS OF POLLUTION

6.3.1 Bargaining solution

A typical reaction of the polluter to criticism would be to point out that his/her firm is attempting to benefit society by producing goods which it wants at the lowest possible cost. If, at some stage, an environmental problem was to be discovered and associated with his/her activities, then improvement should not be done at his/her expense. He/she should be compensated for the cost he/she going to incur in meeting the required environmental standard. Advocates of the virtues of the free market tend to argue that a polluter who is concerned to maximize his/her own profits will not voluntarily offer compensation to the victim. It is up to the victim to enter into a negotiation with the polluter. In this way a bargain can be struck over the environmental issue in the same way as it is done over the sale of commodities in the free market.

The bargaining solution is particularly favoured by the non-interventionist school who draw attention to the 'Coase theorem' (Coase, 1960). This suggests that given certain assumptions a desirable level of environmental degradation can be achieved by negotiations between polluter and polluted. If the polluter has the property rights the polluted can compensate him/her not to pollute. If the polluted has the right the polluter can compensate him/her to tolerate damage. In this there is no requirement or restriction on the nature of the deal done; it could be a bribe as well as compensation. The assignment of property rights would solve the problem (Olson and Zeckhauser, 1970; Farrell, 1987).

However, there may be a serious conceptional problem associated with the bargaining solution. Pearce (1977) points out that this solution might be extended to thieves and murderers: if we agree to pay them not to steal from or murder us, they could capitalize on a very lucrative activity in blackmail. However, one defence of the polluter in this respect could be that there can be no comparison between the thief or murderer, and the polluter who, in the first place, is engaged in a legitimate and beneficial activity of producing commodities for consumers. Therefore, it could be argued that there is nothing wrong with the idea that the polluted should initiate a debate with the polluter with a view to coming to a financial agreement which would satisfy the parties involved.

One important question in relation to the bargaining argument must be: will the bargaining process lead to the socially optimal outcome as shown in Figure 6.1? In the bargaining process, the victim will be willing to pay any money less than the suffering he/she would otherwise have to bear. The culprit, on the other hand, will accept any money higher than his/her benefit curve for a unit reduction in the level of activity. The outcome will depend very much on the relative bargaining strengths of the polluter and the polluted.

The feasibility of the bargaining solution can also be challenged by pointing out the fact that in a modern industrialized economy there are a very large number of victims and culprits involved in environmental matters. The identification problems, mentioned above, can be enormous. Even if the victim and culprits are clearly identified, due to the large numbers involved it would be extremely difficult for each group to establish a bargaining strategy. Furthermore, the transaction costs can be enormous during a bargaining process. It is inevitable that there will be different interest groups each trying to fight his/her corner in a bargaining situation. The free rider problem may also weaken a satisfactory deal between the whole group of individuals who are affected. If there is imperfect competition the bargaining becomes more complex involving a large number of consumers as well as polluter and the polluted (Buchanan, 1969).

The whole idea of bargaining is also objectionable on moral grounds. If the hardest hit are the poorest members of society, which may well be the case, it would not be morally right to expect the victim to pay the offenders, who are likely to be well-off members of society, for improvement in the environment. This issue is important in Anglo-Saxon tradition which tends to uphold the view that the weak and the poor should be protected in the community. Furthermore, in the case of intergenerational environmental problems it is not clear in Coase's theory who would be bargaining on behalf of future generations, especially the distant ones.

6.3.2 Common law solution

Since environmental problems arise as a result of failure to define property rights, some economists, at least at a theoretical level, advocate a complete definition of these rights and their enforcement in law courts. To them the root cause of almost all environmental problems is explained either as a consequence of an incomplete set of property rights or by the inability or unwillingness of the government to enforce public or private property rights.

The legal structures existing in the Anglo-Saxon world share the English common law heritage which recognizes only a limited and qualified ownership of property. That is to say, nobody should be permitted to use his property in a manner which would inflict harm on others. If harm is done then injured parties have recourse to the common law doctrines of nuisance to correct damages caused by pollution, noise, odour and congestion. Seneca and Taussig (1979) point out that such legal remedies can be traced back for centuries in England and have also been used in the United States because of an inherited English common law tradition.

In the light of this law, in most cases the courts should be able to internalize environmental externalities. The criteria for this would be: (1) to ascertain whether a substantial violation of property rights has taken place; (2) to identify the culprit; and (3) to assess the amount of liability. In this way the courts can internalize all the external costs created by the environmental culprits. That is, the polluter's private cost of, say, production becomes identical to the social costs of production after an appropriate common law solution.

However, there would be several problems associated with any systemic reliance on the common law. First, courts may be reluctant to find in favour of environmental culprits when suits have been brought against activities that have long gone without challenge. If, for example, there was a long time interval between the start of an environmental problem and the court case then a doubt may arise about the validity of the victims' claim. Why did they wait so long to bring a case against the defendant? Furthermore, some claimants may have moved into the area with the knowledge of persisting environmental problems. That is to say that the victims came willingly to the nuisance and knew full well what to expect.

Second, in Anglo-Saxon law the burden of proof is on the claimant. In controversial cases the legal costs of proof can be prohibitively high, far beyond the means of ordinary citizens. The culprit, on the other hand, would use every legal channel, such as strategic delays and right to appeal, to discourage any claim. In the end, from the claimant's viewpoint, the cost of an environmental case may turn out to be too high.

Third, in some cases it may be extremely difficult to identify a culprit in environmental grievances. In modern environmental problems such as urban air pollution the causes are many and they are scattered over a wide area. The common law remedy is most effective when the environmental problem involves only a few identities and the damage is immediately felt and clearly traceable to a single source. Furthermore, the pollution damage may be chronic in nature, e.g. cancer induced by poor air quality or illnesses associated with a damp climate, which makes the identification problem extremely difficult.

Finally, modern pollution often crosses numerous legal boundaries. Mercury poisoning, for example, may only be manifest after crossing hundreds of kilometres, across regional and political jurisdictions. In such cases determining the appropriate court that has jurisdiction for any common law suit becomes a significant problem.

6.3.3 Pollution taxes

In view of the above-mentioned problems associated with the common law and bargaining solutions, many economists advocate pollution taxes to achieve a desirable level of environmental quality; these are usually called Pigovian taxes. These taxes should be compatible with socially optimum levels of environmental degradation (p. 180) (they do not eliminate the pollution completely). Taxing the polluter is quite an appealing method of dealing with the environmental problems because it brings culprits, although indirectly, into a forced agreement with the representative of the public at large, the government.

A tax on pollution levels activates the polluter's self-interest. The normal response of self-regard will make the culprit seek ways of reducing his/her tax liability and with it the costs he/she inflicts on the rest of society. His/her optimum level of activity will contain a solution to minimize his/her outlay for a given level of production after tax payment, a case which is demonstrated in Figure 6.3. We assume that the cost of pollution is clearly identified, measured and attributed to various sources. An analysis of this type also requires substantial information on the abatement procedure and the technology involved. The marginal damage cost (MDC) is defined in broad terms to represent the burden on society at large, which is similar to the cost curve in Figure 6.1. The marginal control cost (MCC) is attributed to controlling the pollution by the polluter. This cost will be zero at point N1, where no abatement takes place. If the entire pollution is done away with the abatement cost to the industry will be OC. Needless to say, the more emission there is the smaller the abatement cost becomes. In other words, MCC is negatively correlated with the level of pollution. MDC, which is incurred by society, is a rising curve with the level of pollution.

Now let us assume that the government identified OC_e, as an appropriate unit tax levied on the pollution created by the polluter. With this the polluter will reduce their emission level from ON_1 to ON_e, where they abate N_1N_e and emit ON_e. In this way the polluter will incur the least cost given the tax level and the shape of the marginal control cost curve. Note that pollution taxes bring about a mixed system of taxes and abatement and also make the polluter pay all costs, i.e. emission cost, by way of taxes plus an abatement cost. With the tax revenue the government will be in a good position to improve the environment. Pollution taxes are based upon the polluter pays principle (PPP).

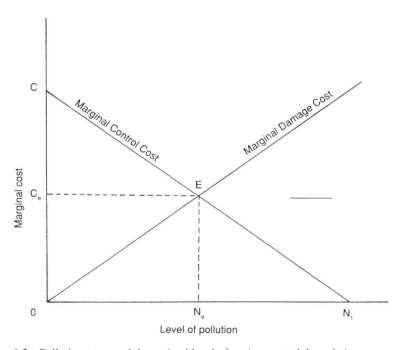

Figure 6.3 Pollution taxes and the optimal level of environmental degradation.

6.3.4 Arguments for pollution taxes

From an administrative point of view pollution taxes can be a relatively easy and cost-effective way of dealing with environmental problems. They can be handled by tax offices which already exist in all parts of the country. There may not be a need to set up independent public organizations and all the administrative and other structures that go with them for monitoring and enforcement. Some additional space and recruitment into tax offices may be sufficient to deal with pollution taxes.

From the polluter's point of view, being incentive-based schemes, taxes have cost saving properties, especially when firms have varying abatement costs. Firms should be able to make cost-minimizing decisions as to how much they should treat their emissions and thus save on pollution tax (Baumol and Oates, 1971, 1975, 1988; Hanley et al., 1990). All firms would adjust their emission levels until their marginal treatment cost was equal to tax and in this way minimize their pollution cost and thus their overall cost. Pollution taxes also encourage firms to do research in pollution control methods and install new abatement equipment as they become available.

It is also possible to manipulate the level of pollution in an area by increasing and reducing taxes. For example, let us assume that a tax which was thought to be appropriate to achieve the optimum level of pollution in a region was introduced. However, events have shown that the discharge exceeded the desired level and environmental contamination in the area is much higher than the optimal level. In the next financial year the government can increase the levy until the desired level is obtained. Conversely, if the tax was too high more abatement will take place than the intended level. Consequently the tax rate can be reduced to allow for more emission. It must be noted that this iterative method of achieving the optimum level of pollution will be satisfactory as long as sufficient precautions are taken to prevent the discharge of highly toxic waste which may create an irreversible problem.

Overall, pollution taxes encourage interested parties to use market forces. They are a flexible method of achieving the desired level of discharge and they could be relatively cost-effective for both industry and the regulator. Taxes can achieve a full cost principle in industry by bringing the operation cost of firms closer to social cost which includes environmental damage.

6.3.5 Arguments against pollution taxes

There are a number of arguments against pollution taxes. Such taxes, if implemented by one country alone, would have undesirable effects on the competitiveness of firms operating in that country. This is because they would push up costs and make goods less competitive in world markets. Firms that operate in countries where there are no pollution taxes would be at an advantage.

Figure 6.4 explains the situation by comparing two countries. In Country A a pollution tax is in force whereas in Country B there are no taxes but pollution is

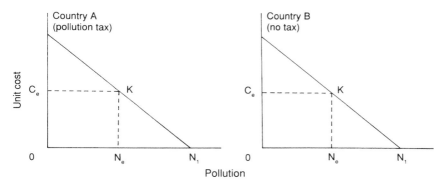

Figure 6.4   Industry's environmental cost in two different countries.

controlled directly by the government. Assume that the structure and the size of the industry in these two countries are very similar. The cost of abating pollution in the industry does not differ much between the two countries and hence the abatement cost curves are the same. Also assume that the communities have similar tastes regarding the quality of the environment. That is to say that the optimum level of environmental degradation is the same in the two countries, ON_e. Then the industries' costs will be:

Country A: Pollution Tax		Country B: Direct Control	
Abatement Cost:	N_1KN_e	Abatement Cost:	N_1KN_e
Pollution Tax:	OC_eKN_e	Pollution Tax:	0
Overall Cost	OC_eKN_1	Overall Cost	N_1KN_e

The firm in Country B is told not to emit beyond ON_e otherwise it will be liable to a hefty fine. Therefore, the cost consists of abatement expenditure only. In Country A, the firm incurs abatement cost plus tax, so the cost to industry will be OC_eKN_1, much greater than the cost in Country B. It follows that countries who implement pollution taxes in isolation will lose out. This, indeed, opens the way to controlling pollution internationally. However, in order to reduce industry's costs some countries may operate a hybrid system of taxation and subsidy. For example, a pollution tax may be in force in one country but government may provide pollution control equipment at a subsidized price, or even for free.

Pollution taxes require extensive information for both marginal abatement and marginal damage cost functions so that precise rates can be established or modified in the event of changing environmental circumstances and communal tastes. The control agency is required to know firms' marginal abatement costs to attain an aggregate abatement function; this could be very difficult to achieve. Even if this is done at one point in time, with changing production and emission treatment technologies the aggregate function becomes volatile. Imposing a single tax could be more expansive than uniform control standard (Seskin *et al.*, 1983). Marginal damage cost estimates could provide even greater problems than marginal control cost estimates. The high degree of scientific uncertainty associated with damage function makes precise valuation of cost estimates almost impossible. It is quite likely that as soon as one group of experts turns out a set of damage estimates they will be challenged by another group. Pearce and Turner (1990) contend that it is unrealistic to argue that an optimal environmental tax can be calculated precisely. What we need is to obtain some overall feel for the level of environmental damage.

There is also the fear that environmental taxes, once introduced, could go beyond the optimal level. Especially when the Treasury is in need of revenue, pollution taxes could become a target. Pezzey (1988) argues that although the industry may tolerate a pollution tax based upon a realistic polluter pays principle, it is unlikely to put up with yet another excuse to increase the tax burden. There may even be cases in which although an overall tax would be realistic for the entire industry, it may cripple some firms who are operating under unfavourable conditions such as spatial remoteness, high labour costs, etc. In the case of the tax levied directly on polluting fuel, say, carbon emission from power stations, this would treat all burners alike without differentiating between high and low polluters and would therefore penalize high efficiency plants. If a carbon tax was levied on domestic as well as industrial fuel this would be likely to be regressive as poor households spend a higher proportion of their income on energy than well-off members of the community. Furthermore, since the rich are keener on environmental quality than the average income family, across-the-board taxes would be a way of imposing the will of the rich on the rest of the community.

One further point is that although a fuel tax can go a long way towards rectifying some of the current distortions in resource pricing, it would do little to satisfy those who require specific targets on emissions. Indeed it would be very difficult to establish a set of tax rates that could guarantee a specific environmental quality.

The advocates of pollution taxes contend that despite their shortcomings pollution taxes are credible methods of regulating environmental quality. However, they note that governments have not taken much notice of their suggestions in adopting such taxes. They are still the exception rather than the rule. Why is it that the cry of taxation economists is largely falling on deaf ears? Pezzey (1988) and Hanley et al. (1990) offer a number of explanations. First, they cite ignorance. The idea is simply that policy-makers do not know about the existence of ideas regarding pollution taxes. Even if they do, it is difficult for them to grasp the intricacies of the taxation debate which is normally conducted by a select group of economists who use extensive mathematical models and impenetrable jargon as they argue the problem amongst themselves. In such debates most articles are not written to inform policy-makers. As Beckerman (1975) puts emphasis on the ignorance factor, Hanley et al. (1990) points out (with respect to British policy-makers) that this is not the case.

A second factor is that of theoretical problems, especially those relating to lack of knowledge regarding marginal damage cost and marginal emission cost functions as explained above. Without a sound knowledge of those cost curves it is impossible to identify the optimum level of taxation. A third factor is institutional or cultural considerations. Many regulatory bodies in Western countries were established more than a hundred years ago in response to legislation on public health and safety. Pollution taxes are frowned upon in regulatory circles, partly because they wish to know for sure why the existing system is inadequate, partly because they see the new system as a challenge to their position and authority. For a good review see Hanley et al. (1990).

6.3.6 The proposed European carbon tax

The idea of taxing carbon dioxide emissions to reduce the rate of increase of greenhouse gases in the atmosphere is being studied in various parts of the world. Norway, Sweden, Holland and Finland have already introduced carbon taxes and the Danish parliament has passed a law calling for one. In Britain the idea was taken up by Nicholas Ridley in 1989 when he was Secretary of State for the Environment. His successor, Chris Patten, also hinted that a carbon tax may be on the cards. At the same time the Energy Secretary, John Wakeham, declared that the price of fuels should reflect environmental costs.

The European Community has been studying the concept of a carbon tax for some time as a major step forward to moderate the emission of greenhouse gases. One objective of the European Community is to stabilize carbon dioxide emissions by the year 2000 at their 1990 levels. Originally, the carbon tax was intended to be levied in 1993 at $3 per barrel of oil then increased by $1 per year until it reached $10 per barrel in the year 2000. The tax revenue would belong to each member state to be used exclusively for environmental improvement.

The Toronto Conference of 1988 recommended that emissions be reduced by 20% from their 1988 level by 2005. The British government contended that it would stabilize its emissions at 1990 levels by 2005, provided that other countries did too. It was widely believed that a carbon tax could be employed to achieve these target figures. Although in 1992 the European Community shied away from the introduction of a community-based carbon tax, the UK and US governments and the European Commission are still studying theoretical as well as practical implications of a carbon tax with a view to introducing it at a future date.

Carbon dioxide (CO_2) is not the only greenhouse gas but it is particularly important both because it is emitted in large quantities and because it persists in the atmosphere for a long time. Figure 6.5 shows the relative influence of the different greenhouse gases. In the 1980s there was growing concern about the effect of CO_2 in aggravating global warming. A spate of publications were released showing the results of atmospheric tests. The conclusions of a typical study were that the surface air temperature of the earth has increased on average by 0.3–0.6°C during the last 100 years; global sea levels have risen by 10–20 cm; and the seven warmest years occurred since 1980. Bolin (1992) suggests that a doubling of the carbon dioxide concentration in the atmosphere

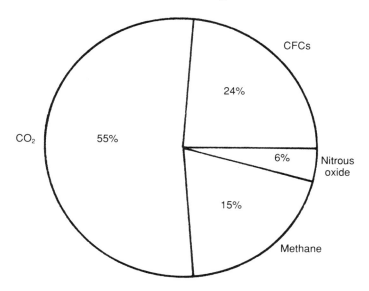

Figure 6.5   Contribution of gases to the greenhouse effect.

would increase the mean global surface temperature by 1.5–4.5°C and that this could happen by the middle of the next century. Environmental scientists are generally agreed that emissions of greenhouse gases, especially CO_2, must be stabilized or even reduced in order to prevent an environmental disaster (see also Chapter 7).

Although the greenhouse effect is a global problem there are major players. For example, the United States contributes 21% of the global emission of greenhouse gases; the figure for the European Commission is 14%. The UK's individual contribution is 3%. It is not surprising that the carbon tax became a major issue in the United States and the European Community. Furthermore, some European Community countries and parts of the United States are low-lying and thus a rise in sea levels will hit such regions particularly hard.

There are three broad objectives of a carbon tax: (1) to establish an incentive for fuel substitution away from the most carbon-intensive fossil fuel (coal) to less carbon-intensive ones (gas); (2) to reduce the overall energy demand by curtailing normal and wasteful use; (3) to raise revenue.

Figure 6.6 shows the theoretical position of the carbon tax. The horizontal axis represents percentage abatement. At 100% level the marginal cost of abatement is high, C_{100}. When the abatement rate is zero, the cost is also zero. Let us say that zero abatement is the current situation. Cost rises as abatement increases. To reduce emission by, say, 20%, we need to convert, say, all coal power stations into gas-powered units. To achieve even higher levels of abatement we

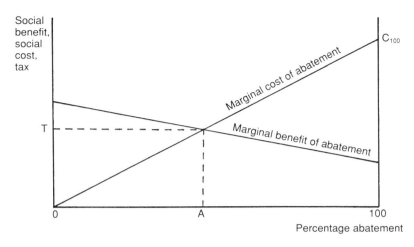

Figure 6.6 Optimal carbon tax.

need a mixture of nuclear power, hydroelectricity, gas, solar power, etc. Overall, the marginal benefit would probably fall with the level of abatement, as shown in the figure, although it is quite conceivable that the marginal benefit function could increase over some intervals, especially if CO_2 and CFC emissions were eliminated. For the sake of simplicity we assume a continuous and declining marginal benefit schedule. The optimum carbon tax corresponds to abatement level A where marginal costs and benefits are equal. If the tax is higher than T more abatement will take place and thus the marginal cost of abatement would exceed the marginal benefit.

Nordhaus (1989) computed a crude global marginal abatement cost curve for the year 2050 and a marginal benefit function for the year 2100. The difference in timing is a rough attempt to allow for the delay between emission and environmental response. According to Nordhaus the optimal tax would be about $3 per US ton of carbon with 9% CO_2 abatement. However, he warns that the error of estimation could result in a tax as high as $37 per US ton with 28% abatement.

There are a number of arguments in favour of a carbon tax. First, efficiency. It has been suggested that carbon tax is a cheap and efficient method of pollution control. It provides a direct incentive to users to burn less polluting fuels to reduce their tax liability. Second, it could stimulate research and development in environmentally friendly technologies. Third, carbon tax will make environmental culprits pay a part of the cost that they are imposing on the rest of society. Finally, it will raise revenue which can be used to improve the environment.

As for the arguments against carbon tax, a number of points could be made. First, tax revenue might be used by governments on matters other than environmental improvement. It has been estimated by Barrett (1990) that a carbon tax equivalent to $10 per barrel of oil in the European Community would raise fiscal revenue equivalent to 1–1.5% of GNP, or about 3% of the tax receipts. The UK's share would exceed £5 billion per annum. The then British government could use this money for, say, direct tax cuts. For a survey of various studies on carbon tax and their effects see Barrett (1991).

Second, carbon tax is likely to hit the poor. Since the poor spend a greater proportion of their income on heating and lighting than the rich, the increased price of energy would be regressive. However, it has been suggested that the government could overcome this problem by an increased benefit scheme to the vulnerable group. Third, a carbon tax could create adverse macroeconomic consequences, especially by reducing fossil fuel-using industries' competitiveness in the world market. Furthermore, a carbon tax could reduce output and income in some sectors in a country. Finally, a carbon tax must be levied globally, in all polluting countries. A global tax is difficult, if not impossible, to achieve. If the UK, for example, were to implement a carbon tax unilaterally, the tax would be borne by the UK but the environmental benefits would be spread globally. In other words the UK citizens would be unable to capture the benefits of the tax unless everybody does the same.

How large a tax? In order to answer this question one needs to look at the price elasticity of various fuels. The proposed European carbon tax is likely to be levied on fossil fuel and other sources of energy such as nuclear power. It could have two components: carbon content and energy content. Tax on nuclear power would relate to energy content only; tax on fossil fuel would relate to both energy and carbon content. Coal is the least environmentally friendly fuel; roughly speaking it is 25% worse than oil and oil is about 33% worse than natural gas. So we need a higher tax on coal than oil, and a higher tax on oil than gas.

Some studies suggest that the price elasticities of demand for coal, oil and gas are very low in the short run. That is, in the short run the demand function is unresponsive to a change in price. Barret (1990), using UK Department of Energy estimates, tried to ascertain the effect of a 1% increase in the price of three fuels.

Short Run	Gas	Oil	Coal
Gas	–0.25%	0.05%	0.10%
Oil	0.05%	–0.20%	0.10%
Coal	0.05%	0.10%	–0.30%

Long Run	Gas	Oil	Coal
Gas	–1.30%	0.85%	1.15%
Oil	0.05%	–2.45%	0.65%
Coal	0.85%	1.60%	–2.25%

In the short run a 1% increase in the price of gas will lead to a 0.25% decrease in its consumption but 0.05 and 0.1% increases in the consumption of oil and coal, respectively. In the long run (about ten years) the decrease in gas use would be 1.3% but the oil and coal consumption would go up by 0.85 and 1.15%, respectively.

How heavily should each fuel be taxed in order to reduce emissions by a certain percentage? Barret (1990) made some tentative recommendations.

	Reduction in emission of CO_2		
Short Run	*20%*	*35%*	*50%*
Gas	40%	70%	100%
Oil	54%	94%	134%
Coal	67%	117%	167%
Long Run	*20%*	*35%*	*50%*
Gas	14%	25%	35%
Oil	19%	33%	47%
Coal	24%	41%	59%

That is, in order to a achieve a 20% reduction in CO_2 we need to tax gas by 40%, oil by 54% and coal by 67% in the short run.

6.3.7 Direct control

Under a direct control system firms are notified by the public agency responsible for the environment not to emit beyond the socially acceptable, or optimal, level. Those who exceed the limit pay a hefty fine. In practice, standards tend to be set as political decisions in response to pressure by various lobbies. But this does not mean that they are to be ignored by economists as arbitrary measures which stem from the political process. In effect, the arguments and counter-arguments on standard setting contain considerable information about the costs and benefits which fall on the interested parties. The debate on standards is usually conducted in terms of tolerable levels of nuisance and commercially agreeable costs of emission.

There are some problems associated with this method of environmental control. First, direct control can be a costly process, consisting of standard setting, monitoring and enforcement. Second, enforcement has often been proven to be unreliable and uncertain. It can be unreliable as it depends on the vigour and vigilance of the responsible public agency which may fluctuate from time to time, especially during election times and when the government is expected to change. Enforcement can also be uncertain because it may be difficult to prove in the courts that the firm exceeded the legal limit. Third, direct control can be time consuming. When there are a number of court cases, some may take a long time to conclude.

6.3.8 Propaganda

Whereas direct control is based on punishment, a propaganda campaign aims at changing attitudes and appeals directly for solidarity with society. However, a campaign such as 'Keep Britain Clean' may be effective only in very limited cases. Sometimes there may not be enough time to enforce direct control or levy taxes, and the authorities may appeal to the public to co-operate. For example, the quality of air may deteriorate quite suddenly in a region. In order not to aggravate the situation any further the regional authority may appeal to the public to restrict the use of cars in order to reduce exhaust fumes. The public may respond very quickly to such an appeal but, unfortunately, the response tends to deteriorate as time goes by.

6.3.9 Marketable permits

A number of economists contend that by issuing marketable pollution permits the regulator can achieve a desirable level of environmental quality (Dales, 1968; Montgomery, 1972). In this the pollution control authority allows only a certain level of pollution discharge and issues permits which can be traded on the market.

In order to implement a marketable permit system the regulatory authority must first work out the geographical boundaries of the market. In most cases this is determined by political boundaries where various government departments have jurisdiction. Then the type and the level of discharge must be ascertained. In order to distinguish between various polluters the regulatory authority may split the market into a number of groups, such as industrial plants, cars, lorries and households. Some polluters may have a negligible affect on the environment and thus the regulator may decide to exclude them from its scheme. Should there be an initial price that permit holders must pay or should rights be given free of charge? Naturally, polluters would tend to favour free distribution which would give them a form of rent. In some cases the authority may decide to auction pollution rights in order to raise money to be used to improve an already deteriorated environment. And finally, the authority must devise a system to enforce its scheme and deter violations. Penalties may involve fines, removal of permits and even shutdown.

In the United States the 1979 amendments to the Clean Air Act, see below, created some possibilities for market transactions in pollution. One amendment, known as the bubble policy, allowed an increase in discharge by one firm if other sources of the same pollutant reduced their discharge by the same amount in a given geographical area, subject to the approval of the Environmental Protection Agency. The bubble policy was originally intended to transfer within a given industrial unit. Since then it has been extended to different plants and firms. One of the most significant features of the bubble policy is that it marks a shift away from the United States policy of specific standards on each discharger.

One other amendment to the Clean Air Act was the emission offset policy which allowed new firms to enter into an environmentally saturated area provided that they make existing firms reduce their emission. This confirmed a kind of property rights on existing firms for which they can extract payment.

Marketable pollution rights have some advantages in pollution control over alternatives. First, they can provide a flexible strategy for the regulator. For example, if a given number of permits is creating an excessive amount of discharge then the regulator can buy back some rights. Alternatively, if given rights are creating a situation in which the level of pollution is less than the desired level then more permits can be issued. Marketable pollution rights can be envisaged as a vertical pollution supply schedule which can be shifted by way of market operations.

Second, marketable rights can give flexibility and low-cost efficiency to polluters. When firms have different abatement costs there will be an automatic market in which low-cost polluters will sell their permits to high-cost polluters to minimize the cost of pollution control. Newcomers as well as existing firms can gain from the market transactions in permits. A new firm will buy rights if it is a high abatement cost industry, otherwise it will invest in pollution control equipment. Sometimes abatement costs are lumpy; to reduce the discharge it may become necessary to invest in a new type of abatement process which may turn out to be costly. The firm can reduce the cost by buying permits on the market.

On the negative side, if administered too rigidly, permits may give the existing firms an unfair monopolistic advantage over the newcomers. The holders can use their rights to deter new entry into the area or even into a particular industry. The barrier to entry can become quite severe in locations where a small number of firms hold all the pollution rights.

There are a number of types of marketable pollution permits. In the ambient permit system the regulator determines the areas where the pollution is received – the receptor points. Then the emission allowed in each receptor point, consistent with ambient air quality standards, is determined. Standards may change from one receptor point to another. Under this scheme rights can be obtained from different markets and probably with different prices. From the regulator's viewpoint the ambient-based system is relatively easy to operate. The regulator does not need to gather information about the abatement costs for each polluter, nor does it need to compute marginal damage and marginal benefits of pollution, whereas in a tax system it is necessary to know these. From the polluter's viewpoint the ambient-based system can be extremely cumbersome. Each polluter would have to have permits for every receptor area that its pollution is affecting. There would be as many markets as there are receptor areas and polluters might find it quite costly to conduct transactions.

An emission-based marketable permit system tends to eliminate the problems of multiple markets and prices per polluter. In this system permits are defined in terms of emission levels at the source as opposed to the effect of emission on the ambient quality in all reception areas. Discharges within a particular area are treated as equivalent regardless of where they drift. The main problem with this system is that it does not discriminate between sources on the basis of the damage done.

In order to overcome the problems associated with the above systems the off-set system is proposed (Krupnick *et al.*, 1983) in which rights are defined in terms of emissions and trade takes place within a defined zone, but not on a one-for-one basis. No transfer will be allowed unless the air quality is preserved at any receptor point. The buyer must obtain enough to satisfy the standard at all points within the area. The offset system, therefore, tries to combine the characteristics of ambient and emission-based systems.

6.3.10 Public ownership

Advocates of the virtues of the socialist market tend to argue that when industries are owned collectively the outcome will be superior to that resulting from the free market mechanism. Public ownership may become essential when controlling the activities of private firms becomes difficult. Private companies would not have much interest in a clean environment when it conflicts with their profit motives. In theory, public ownership looks attractive as the government is the owner of the industry as well as the controller of pollution. However, it is difficult to find hard evidence for this argument. Today, Eastern Europe is one of the most environmentally deprived regions of Europe, with Poland, the Baltic States, Russia and Czechoslovakia in particular, suffering from serious pollution problems.

6.4   PUBLIC POLICY IN THE UK

In the UK, citizens have no clearly defined rights to a clean environment or to peace and quiet. Nor do they have legal access to the courts to pursue such a right. However, if an ordinary citizen legally occupies property, e.g. a house, farm, factory, etc., the situation is quite different. If a person suffers as a result of pollution whilst inside their property, then a case can be brought under common law against the culprit.

Common law is the body of case law defined by court rulings based on evidence from particular cases, as opposed to statutory law which is determined by parliamentary legislation.

One problem with environmental disputes is that the victim must not only identify clearly the culprit but must also prove unreasonable suffering. Such action can proceed even against bodies meeting their statutory obligations to the community. Water or electricity boards, for example, may be taken to court if they cause damage to ordinary citizens, even during the course of their duties.

If the individual suffered harm not while on their property, but while on common property, for example in the street or in a wilderness area, then there would be no legal case unless the culprit behaved unreasonably. However, instead of taking the case to the law the victim can always place a complaint to the District Health Authorities, a Member of Parliament or a local Councillor.

The basis on which pollution standards are established in the UK is known as the 'best practicable means' (BPM). This expression was incorporated in the Alkali Act of 1874 which enabled the Alkali Inspectorate to determine the best way of controlling pollution. The statutory definition of the BPM is spelled out clearly in the Clean Air Act of 1956, which states that 'practicable means reasonably practicable having regard amongst other things to local conditions and circumstances, to the financial implications and the current state of technology'. The Clear Air Act aims to balance three things:

1. the natural desire of the public to enjoy a clean environment;
2. the desire of manufacturers to avoid unremunerative expenses which would reduce competitiveness; and
3. overriding national interests.

The most important administrative bodies which set environmental standards in the UK are Her Majesty's Inspectorate of Pollution and the Local Public Health Authorities. Limits for pollution emissions are laid down for established industrial units on the basis of the best practicable means by these bodies. However, new plants may be required to keep emissions below the presumptive limits. These presumptive limits and the failure to obey may be used as evidence in legal proceedings. In some cases universal standards are enforced instead of those best suited to meeting the needs of local demand.

Standards are set on the basis of negotiations between polluter, regulator and other interested parties such as the representatives of the local public. These negotiations are mostly confidential and conducted on an amicable basis. The authorities in the UK believe that only an amicable negotiation can give the best result. An antagonized polluter is likely to respond grudgingly and may try to exceed the established limit whenever possible.

Her Majesty's Inspectorate of Pollution is responsible for approving plant design and for the regular testing and monitoring of pollution from scheduled plants. Sandbach (1979) points out that these plants are responsible for 75% of the UK's fuel consumption and represent the most persistent source of air pollution. They include power stations, oil refineries, cement, lead, iron and steel works, aluminium smelters and ammonia plants.

Most pollution control officers do not regard themselves as policemen but more as technical specialists whose job is to help and advise the polluter, not to threaten him/her. The Inspectorate of Pollution works closely with the scheduled industries to encourage innovation in curbing pollution. Many inspectors operate as part of a team of troubleshooters dealing with technically difficult problems of pollution, and trying to find practicable solutions to these problems. Sometimes it is difficult to set an emission standard given the highly diverse and complex nature of modern industry. In such cases it is the policy of the Inspectorate to draft codes of practice, embracing good housekeeping and commonsense to deal with potential problems.

The best practicable means aims first to prevent emission of harmful or offensive substances and, second, to render harmless and inoffensive those necessarily discharged. The latter is usually effected by suitable dispersion from tall chimneys. The alternatives of prevention or emission are not offered to industry. Dispersion is offered only when the best practicable means of prevention has been fully explored. There are many cases where a suitable means of prevention has yet to be worked out. The Inspectorate is interested not only in what goes out of a chimney, it has to ensure that the whole operation associated with pollution and disposal of waste is properly conducted.

The word 'practicable' is not defined in the 1874 Alkali Act, but over the last century it has come to include both technology and economics. Ireland (1983) argues that we now have the technical knowledge to absorb gases, arrest grit, dust and fumes, and prevent smoke formation so that if money was no object then there would be little or no pollution problem. The main reason why we still permit the escape of these pollutants is because economics are an important part of the word 'practicable'. In other words, a lot of our problems are cheque book orientated rather than being technical. The pollution control authorities strive for perfection in prevention, but the further one goes along this path the more difficult it becomes to achieve significant improvements at practicable cost because we hit the zone of diminishing returns.

Pollution control inspectors make routine but unannounced visits to inspect registered units for the purposes of monitoring. For smaller units and the least harmful pollution emitters at least two visits per year is the norm. For the bigger and more serious emitters at least eight visits take place per year. Most tests take place in the workplace. However, sometimes tests require expensive and non-transportable equipment and for this inspectors collect samples and send them away to specialist laboratories. As industries develop they may present new environmental problems. Constant development of new and more cost-effective processes and products result in a constant battle for the Inspectorate to keep pace with environmental problems. The Inspectorate's job is further complicated by the fact that officers have to take a broad and balanced view on each case while pressure for action and criticism often spring from narrow interest groups. There is also the 'media factor' which tends to sensationalize environmental issues.

Limits for emission are reviewed periodically by the Inspectorate to take into account changes in technology. For example, the emission standard for the cement industry was reduced from 0.4 grains per cubic foot in 1950 to a sliding scale of 0.2–0.1 grains per cubic foot in 1975. There are considerable advantages in changing standards without resorting to preliminary legislation. For major sources and more serious pollutants such as sulphuric or hydrochloric acid, unit standards have not changed since 1906. Almost every pollution control technology has a limited life. Extension of deadlines and upgrading old equipment have also been allowed to extend the economic life of pollution control technology.

The government believes that prosecution of offenders must be the last resort, mainly because it indicates an admission of failure of the pollution control system. Under the 1906 Alkali Act, prosecutions could only be brought by an inspector with the consent of the Secretary of State. This power is seldom used, and inspectors always prefer persuasion to prosecution, leaving the latter for extreme and unreasonable offences. For example, in 1976 in England and Wales about 500 complaints were investigated by the Alkali and Clean Air Inspectorate, the relevant government department at that time, but only a couple of prosecutions took place. The situation is very much the same now: for example in Northern Ireland the Alkali and Radiochemical Inspectorate, the regional government department for pollution control, has not taken any offender to the court during the last few years. The most numerous complaints were against mineral works, iron and steel companies and gas and coke works. The Inspectorate's view on so few prosecutions was that legal action will not help to solve the environmental problems. Most complaints refer to nuisances that have a direct impact, others could remain undetected by the general public. Most local public authorities do not have the money for routine pollution monitoring, preferring to rely on complaints. In other countries, such as Holland, the authorities provide a free phone number to allow the public to register complaints. This arrangement is helpful especially when there is accidental discharge and the authorities have to respond quickly.

The nature of water pollution is different from air pollution. Many different substances pollute water; some of them are directly toxic and some encourage the growth of algae and bacteria, causing the oxygen levels in the water to drop and suppressing the natural flora and fauna. Unlike air pollution people do not consume water pollutants directly on discharge. The public are concerned about the quality of drinking and river water, the presence of poisons in seafood, and amenity and recreational aspects of water.

In the UK the 1951 and 1961 Rivers Acts are the principal legislation for the control of river pollution. The conditions which exist in these Acts take into account local environmental conditions, technical factors determining the cost of control, and a multitude of different uses of water such as drinking, agriculture, industry, fishing and amenity. In establishing standards, bargaining takes place between the industry and the relevant water authority which, to a certain extent, represents the interests of consumers. The 1976 National Water Council consultation paper set the general standards for discharge, which stemmed from the Royal Commission Report on Sewage Disposal which took place in 1919. The standard for discharge is 30 mg/litre suspended solids and a biochemical oxygen demand (BOD) of 20 mg/litre. This standard has been criticized by Sandbach (1979) on the grounds that conditions differ from one part of the country to another. Some rivers are highly sewage effluent whereas others are heavily trade effluent. Furthermore, the nature of water demand varies quite considerably from region to region.

In 1978 the National Water Council recommended that the water authorities should establish environmental quality objectives for rivers in England and

Wales, and then readjust existing consent conditions to meet these objectives. This meant that some water authorities could demand much more stringent consent conditions for discharges than others, depending upon local circumstances. Great variations have always existed among the water authorities regarding their vigour and vigilance in prosecuting offenders.

In 1989 the government created the National Rivers Authority to control pollution of rivers, estuaries and bathing waters. In the same year the water industry was privatized. The government believes that with access to private capital the newly privatized water industries will carry out a comprehensive programme to improve the quality of water in Britain.

During the last few decades the application of chemical fertilizers to crops has created serious water as well as land contamination throughout the world. Nitrates, in particular, have been allowed to leak into streams and rivers where they cause oxygen depletion and pollute drinking water supplies. For many years to come there will be an unnaturally high level of dissolved nitrogen in soil and water in many areas. Hanley (1990) reports on this problem and demonstrates that individual decisions by farmers cannot create a socially acceptable situation. Others have investigated to what extent a reduction in farming levels can create enhanced environmental benefits (Russell, 1990, 1993).

6.4.1 1990 White Paper on the environment

On 25 September 1990 the British government published its long awaited White Paper, *This Common Inheritance,* on the environment, covering issues from the street corner to the stratosphere, from human health to endangered species. The paper states that the government will aim:

1. to preserve and enhance Britain's natural and cultural inheritance;
2. to encourage more efficient use of energy and other resources;
3. to ensure that Britain meets its commitments for reducing global warming, acid rain and ozone depletion;
4. to ensure that water and air in Britain are clean and safe and that controls over wastes are maintained and strengthened where necessary; and
5. to maintain Britain's contribution to environmental research and encourage a better understanding of the environment and a greater sense of responsibility for it.

As for the instruments to achieve these objectives the White Paper mentions two methods: regulations and fiscal measures with an emphasis on the polluter pays principle. The regulatory approach has served well in the past and will continue to be the foundation of pollution control. But this method can be expensive to operate and may not deliver the most cost-effective solution. The fiscal system, in particular the recent reduction in tax on unleaded petrol, has been a success and thus the government will look at ways of using the market to encourage producers as well as consumers to act in a more environmentally friendly manner.

On air pollution the White Paper commits the government to new targets for air quality. It will (1) base action increasingly on air quality standards, with advice from a new expert group; (2) press the European Community for much tighter emission standards; (3) add an emission check to the MOT test; (4) press the European Community for new directives on waste inciner-ation and help to develop specifications for less-polluting building products; (5) provide new guidance on random gas emission; and (6) issue new advice on avoiding passive smoking.

On noise levels, the government will (1) seek to tighten international noise regulations for vehicles and aircraft; (2) improve standards for noisy products like burglar and car alarms; (3) consider covering noise in MOT tests; (4) improve regulations for house insulation; (5) extend noise insulations to new rail lines; (6) research into ways of reducing aircraft and helicopter noise; and (7) help local authorities to establish noise control zones.

The White Paper states that the newly privatized water industries in England and Wales will invest £28 billion by 2000 to improve the water quality. To reduce the pollution from agriculture the government will (1) introduce new regulations for the construction of silage and slurry stores; (2) work for a European Community directive on nitrate control and produce a revised code of good agricultural practice including livestock waste.

On dangerous substances the government will (1) participate with international bodies to understand the effects of existing chemicals; (2) increase vigilance on pesticides; (3) reduce discharges into the North Sea of cadmium and lead by 70% between 1985 and 1995; (4) phase out uses of polychlorinated biphenyls (PCBs) by 1999; (5) keep the release of genetically modified organisms under control; and (6) ensure that radiation limits continue to meet international standards.

To meet its target of recycling 50% of the domestic waste that can be recycled the government will (1) work with local authorities to assess the effectiveness of experimental recycling projects; (2) develop a scheme for labelling recyclable products; (3) encourage more plastic/bottle banks; and (4) encourage companies to recycle from domestic waste, wastes from mining and from building material.

On the international level the government will press for European Community action to reduce global warming, impose stricter vehicle emission standards, develop the integration of agricultural and environmental policies and introduce an effective labelling scheme to help consumers play their part. The government believes that on global environmental problems all countries must act together. For example, if other countries take similar action, Britain is prepared to set itself the task of reducing carbon dioxide emissions to their 1990 levels by 2005.

The government believes that the 1990 White Paper has changed forever the way that environment policy is made in Britain. It has firmly brought green policy into every Whitehall department. Furthermore, the Cabinet Committee on the environment, chaired by the Prime Minister, will continue to exist and each government department will also have its own minister responsible for considering the environmental impact of all the department's policies and spending plans. New guidelines are also being drawn up to help departments assess how to take the environment into account in their planning.

Nevertheless, the White Paper has been heavily criticized by all opposition parties as well as environmental pressure groups as being a feeble attempt, a woolly wish list of what the government might like to see happening towards the end of the century. It is long on words such as explore, consider, study, work towards and urge, but very short in action. Indeed, the White Paper's central failing is the absence of any firm proposals to harness market forces to environmental ends. The idea of using the price mechanism to save the environment has not been exploited.

In response to these criticisms Chris Patten, the then Secretary of State for the Environment, defended the White Paper as a foundation for environmental protection for the coming years. He stated that it certainly would not be the government's last word on the environment.

6.4.2 Environmental Protection Act 1990

In 1990 the UK government introduced the Environmental Protection Act, covering a wide area of activity. The Act is in nine parts. Part 1 makes provision for the improved control of pollution in air, water and land arising from various industrial and other processes. Part 2 re-enacts the provisions of Control of Pollution Act of 1974 relating to waste on land. Part 3 restates the law defining statutory nuisances and improves the procedures for dealing with them, to provide for the termination of the existing controls over offensive trades and businesses and to provide for the extension of the Clean Air Acts to prescribed gases. It excludes smoke emitted by military installations, private dwellings, trains and industrial units.

Part 4 amends the law relating to litter and makes further provision imposing or conferring powers to impose duties to keep public places clear of litter and clean. Part 5 amends the Radioactive Substances Act 1960. This section gives the Secretary of State authority to appoint inspectors to assist him in the execution of his duties, to levy fees and charges in respect of applications for registration and to issue enforcement notices for compliance with the law. Part 6 deals with genetically modified organisms. Part 7 makes provision for the abolition of the Nature Conservancy Council and for the creation of councils to replace it and discharge the functions of that council and, in respect of Wales, the Countryside Commission. Part 8 makes amendments to the Water Act 1989, the Control of Pollution Act 1974, the Food and Environmental Protection Act 1985 regarding the dumping of waste at sea. Also it makes further provision with regard to the prevention of oil pollution from ships, burning of crop residues, control of dogs, and financial assistance for purposes connected with the environment.

6.5   PUBLIC POLICY IN THE UNITED STATES

The main reason for the need for public intervention on matters regarding the environment in the United States, and indeed in other countries, is that most environmental problems come about as a result of the failure of the private market system. The United States government uses a variety of means such as regulation, prohibition, fiscal incentives and fiscal disincentives to achieve its environmental objectives with the aim of internalizing for society those environmental effects that are external to individuals.

There are various regulatory commissions created by the legislative branch of the government and their activities are subject to judicial review. These commissions operate at both state and federal levels. In view of the fact that most serious environmental problems in the United States can be traced directly to the various effluents produced by public utilities such as transport, gas and electricity generating units, regulatory commissions deal extensively with these utilities. For example, the federal government estimated that about 50% of the total sulphur dioxide pollutants and 25% of the nitrogen oxide emissions may be attributed to the discharge of coal, gas and oil-driven power points (Seneca and Taussig, 1979). Most of these public utilities are natural monopolies. The most important regulatory federal commissions are: the Environmental Protection Agency, the Federal Power Commission, the Inter-State Commerce Commission, the Federal Aviation Administration, the Nuclear Regulatory Commission and the Department of Energy.

The Environmental Protection Agency (EPA) sets national air and water quality standards. Prior to the 1970 Clean Air Act Amendments air quality and improvement programmes had been largely regulated by individual states. States still issue permits and regulate emissions but the EPA can override these standards and introduce stringent pollution control measures in urban areas or elsewhere. Alongside the individual state authorities, the EPA sets water quality standards and issues permits for discharge. The nature of permits and the responsibility for issuing guidelines which delineate the minimum standards for permit application rest with the Agency. The Federal Power Commission deals with power projects and interstate transmission of electricity and natural gas. The Inter-State Commerce Commission has jurisdiction over inter-state transportation and oil pipelines. The Federal Aviation Administration oversees matters on civil aviation. The Nuclear Regulatory Commission deals with nuclear energy. The Department of Energy is not a regulatory agency, but its activities have important environmental implications.

All these regulatory bodies are involved in national environmental policy-making. From time to time they come under fire from public protest over the location of new power plants, siting of airports, environmental accidents, breaching of regulation by the industry and so on. Seneca and Taussig (1979) argue that there would be considerable advantages in increasing the power of these commissions. However, not everybody would welcome the idea of giving greater power to the regulatory commissions for various reasons. First, greater power may increase bureaucracy and compound inefficiencies which already exist in some government departments. Second, increased powers will no doubt create additional costs which are likely to be passed on to firms and then eventually to consumers, which will deteriorate the competitive strategies of firms. Most regulatory agencies were, in the past, extremely sympathetic to the interests of the firms they are supposed to regulate, but this situation has been changing.

Prohibition means outright legislative bans on activities of firms by Congress, individual states and local legislatures. These laws are then enforced by the judicial system of the state. Prohibition can be temporary or permanent. In emergencies the relevant authority can be given powers to close down industrial plants. Finally, various fiscal incentives (subsidies) and disincentives (taxes) are also available.

Apart from setting standards for air and water purity the EPA also deals with issues such as noise, solid wastes, ocean dumping, pesticides, toxic substances, land use and recycling. These duties have all been vested in the EPA by legislative acts such as the Clean Air Act Amendments (1970), the Federal Water Pollution Control Act Amendments (1972), the National Environmental Policy Act (1969), the Marine Protection, Research and Sanctuaries Act (1972), the Federal Insecticide, Fungicide and Rodenticide Act Amendments (1975), the Noise Control Act (1972), the Resource Conservation Recovery Act (1976) and the Solid Waste Disposal Act Amendments (1976). Normally Congress sets broad environmental objectives and then instructs the EPA to develop and enforce policies necessary to achieve these objectives.

The EPA by and large adopts a regulatory approach in designing environmental policy. Although there is a wide range in the precise nature of these policies the Agency normally specifies what each individual discharger must do with respect to waste emissions. For example, the Agency may set a ceiling on the physical amount of a particular pollutant emitted into the air or water. Needless to say, this ceiling may vary with the residual in question and the type of production process which creates it. Other regulations require a specific waste treatment technology or a change in the existing production process so as to create less waste. Regulations of this type allow for a phase-in period over a number of years. In other cases, regulations specify the nature of inputs, e.g. lead-free petrol, low-sulphur coal, etc.

The standards set by the EPA can be placed in two broad categories: National Primary Standards to protect public health and National Secondary Standards to protect the public from any unknown or unanticipated adverse effects. O'Riordan and Turner (1983) argue that despite the great efforts by the Agency many states still fail to meet at least one of these standards, the legal implication being that no new industrial growth could be permitted in these 'non-attained' fields. The Agency aims to introduce an 'emission offset policy' which allows new industrial units to locate in a non-attainment area provided that their emissions are more than offset by concurrent emissions reduced from other units in the same airshed.

One of the most cumbersome problems in standard setting has been the reluctance of large-scale industries to implement them. In most cases it has been in the industries' interest to delay controlling pollution as long as possible. In this, regulations are sometimes challenged in lengthy court battles or the industries threaten to close down their operations and put emphasis on the diminished employment opportunities in the area. Another tactic which reluctant industries use is to employ the best lawyers and industrial experts armed with research showing that standards set by the EPA are unduly harsh for firms and thus potentially harmful to the economy. Sometimes the officers of the Agency lack industrial expertise and experience and thus find themselves in a weak position in negotiations.

A good proportion of current air pollution control stems from the 1970 Clean Air Act Amendments which require the EPA to identify each air pollutant which has an adverse effect on public health and welfare. Prior to this Act, air quality objectives and improvement programmes had been regulated by individual states. The EPA has identified and set standards for a number of air pollutants such as sulphur dioxide, particulates, carbon monoxide, nitrogen oxides, hydrocarbons and photochemical oxidants. Once standards are determined, each state is required to submit to the EPA an air control plan for existing pollution sources. All plans are subject to the Agency's approval and if they are insufficient the Agency has the authority to develop and implement its own plan for any state.

In 1977 the Congress amended the Clean Air Act making sure that parts of the country which enjoyed clean air well above the previously established standards would be required to maintain their positions. In 1990 the Congress made further amendments in which it required the EPA to establish, with a strict deadline, technology based standards for sources of about 200 specific pollutants. Also the Congress wrote into law a requirement that emission of sulphur dioxide from coal-fired power stations be reduced by about 10 million US tons per annum with a view to moderating problems created by acid rain.

The area of water pollution has the longest history of federal involvement in environmental matters. As far back as 1899, the Rivers and Harbours Act represented the federal concern with water discharges into the nation's rivers, lakes and seas. The current basis for the EPA's water quality regulations stems from the Federal Water Pollution Act of 1972 (see Marcus, 1980). This Act established a permit system of discharge requirements for both municipal treatment plants and industry. These permits are normally issued by the EPA. The Agency also grants subsidies to municipal plants and industrial units so that they can construct new treatment works or improve the performance of old ones. These grants can cover up to 75% of the establishment as improvement costs. The Act enables heavy fines to be imposed upon offenders. The penalty for the first offence is moderate but subsequent convictions carry very heavy fines.

The Water Pollution Control Act 1972 required the Environmental Protection Agency to meet six deadlines.

1. By 1973, it was to issue effluent guidelines for major industrial categories of water pollution.
2. By 1974, it was to grant permits to all sources of water pollution.
3. By 1977, all sources were to install the best practicable technology for pollution emission.

4. By 1981, all major United States waterways were to be bathable and fishable.
5. By 1983, all sources were to install the best available technology to control pollution.
6. By 1985, all discharges were to be eliminated.

Hartwick and Olewiler (1986) argue that these targets were unrealistic given the limited resources available to the EPA. The Agency was faced with an enormous task of carrying out the agenda with insufficient human resources. With reference to the first two deadlines, the Agency had to examine over 200 000 industrial polluters emitting 30 major categories of pollution. Information on production technology for each single unit had to be collected and then converted into guidelines within one year. Consequently, the deadline was not met.

Finally, a few words on the rights of the United States citizens to a clean environment. Since United States law, by and large, is based upon the Anglo-Saxon Common Law principles explained earlier, the legal situation is not very different from that existing in the UK. However, some lawyers have recently argued that there are definable rights to certain democratically determined levels of environmental quality in the United States. Individuals should be allowed to prosecute environmental culprits including the official regulators who set the environmental standards which may cause injury. However, there are two dangers in their arguments. First, courts may become the arbitration body for public health and safety and get overburdened by sheer numbers of environmental cases. Second, if victims win their cases, they may financially cripple the industry. When there are a large number of claimants the sums involved could be enormous and many industries may be driven to bankruptcy. For a contrast between the United States and the British policy approaches see Vogel (1986).

6.6   THE EUROPEAN COMMUNITY AND THE ENVIRONMENT

The relationship between economic growth and the protection of the environment is at the heart of the European Community's environmental policy which was established in October 1972 at the European summit meeting in Paris. These twin objectives might at first sight appear to be contradictory but in fact they need not be. The Community believes that a lasting economic growth can only be achieved within the framework of a protected environment, since the natural resources of the environment constitute not only the basis but also the limits of economic expansion.

In the early 1970s, environment and energy matters reached the top of the political agenda in the Western world. The United Nations' Conference on the Environment in Stockholm, 1972, gave a worldwide platform to the necessity of arresting environmental decline. In the 1972 summit, the heads of states laid down a number of principles on the European Community's environmental policy.

1. The best environmental policy is to prevent pollution at source rather than to treat the symptoms.

2. Environmental policy must be compatible with economic and social policies.
3. Decisions on environmental protection should be taken at the earliest possible stage in all technical planning.
4. Exploitation of natural resources in a way which creates significant damage to the ecological balance should be avoided.
5. Research in the field of conserving and improving the environment should be encouraged.
6. The 'polluter pays' principle should be employed on environmental protection programmes.
7. Each member state must take care not to create environmental problems for other states.
8. The interests of the developing countries must be taken into consideration in environmental policies.
9. The Community must make its voice heard in international organizations dealing with environmental issues.
10. All member states are responsible for the protection of the environment, as opposed to a few selected countries.
11. For each different kind of pollution there should be different emission levels and policies.
12. Environmental problems should no longer be treated in isolation by individual countries.
13. The Community's environmental policy is aimed at the co-ordinated and harmonized progress of national policies, without hampering the actual progress at a national level.

One year later the Community adopted its first environmental action programme which fell into three broad categories:

1. Reduction and prevention of pollution and other nuisances.
2. Improvement to the environment and quality of life.
3. Joint action within the community and in international organizations dealing with the environment.

The first four-year action programme was followed by second and third programmes, and a fourth, to run for six years from 1987 to 1992. In amendments to the Treaty of Rome in 1986, the objectives of the Community's environmental policy endorsed the prime principles that preventive action should be taken, that environmental damage should be rectified at source and that the polluter should pay. These spell out that 'environmental protection requirements shall be a component of the Community's other policies'.

Given the limited resources at the Community's disposal, the European Community is forced to prioritize environmental issues. In this respect the Community outlined a series of priorities for action under the fourth environmental action programme. These are as follows.

Air pollution   A long-term strategy to reduce air pollution both within the Community and outside is currently being worked out (European Community, 1987). This aims to reduce the emission of pollution from all sources and also to reduce ambient air concentrations of the most important pollutants to acceptable levels. It is intended that the long-term strategy should cover pollution emanating from transport, industrial plants, nuclear and conventional power generating units.

Water pollution   An increasing priority will be given to curb marine pollution. Particular attention will be given to the protection of the Mediterranean, reduction of land-based emissions of pollutants to all Community seas and the implementation of the Community's information system for dealing with harmful substances spilled at sea. New action will be taken to protect the Community's fresh water supplies from discharge of livestock effluents, fertilizers and pesticides.

Chemicals   Improvements in the notification, classification and labelling of new and existing chemicals is another priority area. A major new proposal will be the integrated regulation of dangerous chemicals, requiring a review of the adequacy of existing Community legislation. A directive will also be proposed setting out a comprehensive structure for risk assessment and regulation of chemicals already on the market.

Biotechnology   The Community is monitoring the progress taking place in the area of biotechnology. The proliferation of new industries using new biotechnology techniques will no doubt have environmental consequences. The Community feels that appropriate precautions must be taken so as not to harm the environment as progress continues.

Noise   The Community aims to reduce the noise from individual products, such as motorcycles, cars and aircraft. A regulation may be introduced to test these products in government vehicle inspection departments. However, it has been noted (European Community, 1987) that due to staff shortages it has not been possible to implement the anti-noise policy effectively. In line with the 'polluter pays' principle, noise-related changes especially at the Community's airports may eventually become a reality.

Conservation of nature and natural resources   The Community believes that the time is ripe for taking measures in this area. This should be in line with the three central aims of the World Conservation Strategy, namely the maintenance of essential ecological processes and life support systems, the preservation of genetic diversity and the sustainable use of species and ecosystems. Soil protection in particular is on the agenda. This should be achieved by reinforcing the mechanisms and structures for co-ordination to ensure that soil is more effectively taken into account in development policies. In this, contamination, physical degradation and soil misuse should be avoided. Measures to encourage less intensive livestock production, to reduce the scale of use of agricultural chemicals, to promote the proper management of agricultural waste, to prevent soil erosion, to protect groundwater supplies and to aid the recovery of derelict and contaminated land are of particular importance.

Waste management   The development of clean technologies coupled with the creation of the right market conditions for a more rational approach to waste management is another priority area. This would, hopefully, lead to economic and employment gains and a considerable reduction in import dependence as well as a reduction in pollution. Efforts will also be increased to define criteria for environmentally sound products which give rise to little or no waste.

Urban areas   It has been recognized by the Community that many European cities are now in a worse condition than they were ten years ago. This is due to large-scale migration into towns which has led to pressure on, and consequently the deterioration of, the urban infrastructure. Also there has been a move from inner cities to the suburbs, particularly by well-off individuals, which diminishes the income base which is necessary for improvement. Some initiatives have already taken place within the Community to improve the urban structure in Belfast and Naples. These programmes could be extended to other cities.

Coastal zones   The Community has already undertaken work to identify specific problems in certain coastal areas that require urgent remedies. Currently work is in progress to develop a European Coastal Charter, along with a series of measures to combat degradation of the Community's coastal regions.

The Community's policies towards action at international level are: (1) strengthening the Community's participation in protection of regional seas; (2) co-operation by member or non-member states on the protection of the Mediterranean; (3) taking part in the Council of Europe Convention for the protection of vertebrate animals used for experimental purposes; (4) adopting a regulation requiring member states to ratify the international agreements on the transport of dangerous goods; (5) development of an international code of conduct governing exports of dangerous chemicals and further strengthening co-operation on the environment with non-member countries such as the United States, Japan and the European Free Trade Association (EFTA) nations.

The Community also has a genuine desire to help developing countries with their environmental problems, especially at a time when they are facing serious issues such as deforestation, desertification, soil erosion, loss of wildlife and genetic diversity and urban degradation.

6.7   TRANSITION IN EASTERN EUROPE AND THE FUTURE OF ENVIRONMENTAL POLICIES

Appalling environmental problems in Eastern Europe were created by their former communist rulers as a result of overambitious and ill-considered industrial and agricultural policies designed to achieve a maximum rate of economic growth. As mentioned in Chapter 1, although the constitutions of the communist countries were protective towards the environment, when economic growth targets conflicted with environmental protection, the former often won. Informal environmental assessments were carried out on projects in the context of planning and pollution regulations which were somewhat cosmetic.

The transition to a free market system is proving to be a painful process; both economic activity and employment levels are currently declining, and discontent amongst the population is inevitable. In this climate it is highly unlikely that partisan environmental policies will be given high priority in the near future. Strangely enough, although one would have expected a reduction in industrial activity to lead to reduced pollution levels, this has not occurred. For example, in Russia in 1991 when production levels fell by 15%, the pollution level increased by 10% (Lombardini, 1992). However, all the countries of the former Soviet bloc have already adopted, or are debating, environmental assessment legislation.

6.7.1 Poland

In the past, the 1984 Land Use Legislation in Poland required an environmental assessment from most projects. In 1990 an Independent Environmental Impact Assessment Commission was established with the purpose of providing an evaluation of the assessment reports and to advise the authorities before proposed projects were approved. Central government, Voivodship (provincial) governers, the Ministry of Environmental Protection, the Natural Resources Department and the Forest Service have a key role in approving projects.

Assessments are carried out by independent experts in two stages: the first deals with comparative site evaluation and the second gives a fuller assessment, carried out once the project has begun. Public participation plays a role in large projects decided at a national level, but this is not a well-established practice at provincial level. Currently work is in progress to improve the system, especially to reduce the two stage assessment to one for large projects. Proposals to widen public participation are under consideration.

6.7.2 Hungary

In the past a number of *ad hoc* environmental studies have been prepared in relation to mining and construction of highways in Hungary. There was no explicit legislation on environmental control. At the moment a draft law inspired by the European Commission is being prepared for enactment. According to this there are a number of environmental categories for projects. One group is subject to mandatory assessment, the other group requires assessment only if demanded by the Ministry of the Environment. Independent consultants are not always required to carry out environmental impact assessment: it could be done by the investor. There are 12 regional environmental inspectors who play a key role in ensuring that reports are sufficient. Public participation is an important part of the environmental assessment.

6.7.3 Czechland and Slovakia

In January 1993 the former Czechoslovakia was divided into two countries, the so-called velvet divorce. Before the partition new environmental procedures had been established and a number of laws passed. The 1991 Federal Act on the Environment required a rigorous environmental assessment of projects but the two republics have yet to ratify this. The 1992 Czech Law on Environmental Impact Assessment focuses on construction projects. Investors must submit an environmental assessment report carried out by a reputable firm specializing in environmental consultancy. Public participation is encouraged. The Republic of Slovakia is also moving in the same direction.

6.7.4 Romania

Currently Romanian government bodies carry out environmental assessments on proposed projects. A draft legislation exists which aims to widen the assessment exercise to include non-government bodies. There are no specific provisions for public participation.

6.7.5 Slovenia

A number of draft legislations, mainly modelled on the European Community directive, exist in Slovenia to be enacted in due course. Construction, modification, removal and introduction of new products requires environmental assessment. It is expected that after enactment the central government will require mandatory assessment on many projects. There will also be a substantial role for the regional planning bodies. The investor is mainly responsible for the assessment which must be prepared by a certified firm. For technology-based projects the investor must explain the rationale for using a particular method over alternatives. The assessment is open to public scrutiny and the government must let its decision be known within 60 days.

6.7.6 Bulgaria

Legislation based upon the European Community directive was introduced in 1991 and requires compulsory assessment for environmentally sensitive projects. Assessments must be carried out by reputable independent experts at the investors' expense. In addition, large industrial projects require regular assessment every five years. Public participation in assessment is thought to be desirable.

7

International environmental problems

The fact that a problem will certainly take a long time to solve, and that it
will demand the attention of many minds for several generations, is no
justification for postponing the study.

T.S. Eliot

Pollution does not stop at national boundaries. One country's waste all too easily
becomes its neighbour's problem and this is particularly so for water and air-
borne pollution. Therefore, much public discussion of programmes for the protection of the environment has emphasized their international implications.
Needless to say, international environmental pollution requires an international
solution. Although there is a good deal of talk about international co-operation
in the control of transnational pollution, joint programmes will undoubtedly
prove to be very difficult to develop.[1]

In general terms there are two types of pollution: flow pollution and stock pollution. Flow pollution presents less intractable environmental problems and can
be measured relatively easily. Examples of this kind of pollution are smoke,
dust, noise, chemical spillage and odour, created mostly by local industry which
causes loss of amenity, health and general comfort in the immediate area. Flow
pollution normally degrades and disappears from the affected area in the course
of time. The damage function which appears in most environmental textbooks is
related to flow pollution. Most government departments deal with this type of
pollution by trying to find the socially optimal point for it.

Today, stock pollution has become a much more topical issue. This can be
defined as the type of pollution that accumulates in the environment over time.
Stock pollution remains unnoticed until a critical threshold level is reached.
Existing national and international institutes are totally inadequate to deal with
it. Some conservation economists (e.g. Page, 1983) call for a complete re-
design of pollution management institutions'. The reasons for the inadequacy of

---

1. Some economists, such as Dasgupta (1991), use game theory to illustrate how co-
    operation may be sustained over time by means of conduct.

the *status quo* are as follows. First, there are insufficient data on pollution generation, especially for stock pollution. New restructuring should aim at gathering information on this type of pollution. Second, the potential cost of stock pollution could be catastrophic, and no existing institute is trying to estimate its complete cost. Third, there is insufficient international law regulating the problems caused by transfrontier pollution. For example, it is very difficult for victims in Sweden to claim compensation from the culprits who live in the UK. At present there are three serious cases of global pollution affecting many nations, namely, acid rain, the greenhouse effect and the destruction of the ozone layer. These all result from gaseous emissions into the atmosphere.

7.1   ACID RAIN

The chorus of complaints on this issue is aimed mainly at emissions of sulphur dioxide from coal-fired power stations, although it is recognized that nitrogen oxides, unburnt hydrocarbons from car exhausts and other pollutants are also blameworthy. These fumes have, of course, been with us for more than 100 years, and although tall chimneys built at power stations have helped to end the dreadful polluting city smogs of the 1950s, it has only been at the expense of polluting the upper atmosphere and turning rain into acid, especially throughout Europe and North America.

In the early 1970s the issue of acid rain was first brought to public attention and the culprit was pinpointed as sulphur dioxide. This is emitted largely by industrial burning of coal and heavy oil and is transformed into sulphuric acid in the atmosphere. Many countries generate this pollution but Britain was identified as a particular villain because its pollution is carried across the European frontiers by the westerly winds. Countries like Germany, Holland, Norway and Sweden are affected by British pollution. However, Britain's sulphur emissions have been declining since 1970 and acidification in the affected areas has gone on unabated. Increasingly, concern has focused on the nitrogen oxide emitted by cars, lorries and power stations. Although the consensus appears to be that all these elements make a contribution to contaminating the air and soil, the question is, what contribution? Furthermore, who should pay for its elimination, and when? The answer to these questions lies in national and international policies.

The damage caused by acid rain is widespread and extensive (Commission of the European Communities, 1983; World Conservation Union, 1990). In Scandinavia, trout and salmon stocks have died because the acid rain washed aluminium out of the soil and into the rivers and lakes. The aluminium poisons the gills and starves the fish of salt and oxygen. Other biological processes in the lakes fail and as the lakes die, a thick carpet of algae on the bottom develops as the only sign of life. It was estimated in a report commissioned by the European Community that half the fish stocks had been lost in an area covering 20 000 square kilometres of Norway (Gowers, 1984). In Sweden, fish have been killed in up to 4000 lakes. The same scene greets fishermen in Canada, the United States and Britain. In effect, fish disappeared from many rivers and lakes of Scotland some years ago.

The damage to forests is also extensive and there are a number of theories about the way that acid rain attacks trees. One is that acid rain starves and eventually poisons tree roots by altering the chemistry of the soil. If enough acid accumulates in the soil, aluminium is released which poisons tree roots. This theory, however, is somewhat defective. Although damaged trees are often short of some nutrients, there is not much evidence of unusually high levels of free aluminium in the soil around damaged trees. Furthermore, most damaged trees are coniferous and they are reasonably tolerant of aluminium. Another theory is that the trees are poisoned directly through their leaves.

The characteristic signs of damage due to acid rain on a coniferous tree are limp branches with yellowing or brown needles. Soon the needles begin to drop and eventually the tree stops growing new ones. The tree becomes bald at the top and an excessive number of cones may be produced or shoots sprouted. Finally, insects, drought and frost kill the weakened tree.

The damage to European forests is quite extensive. In France over 10 000 ha of woodland has been seriously damaged and 30 000 ha are showing signs of deterioration. In what was formerly West Germany, where 750 000 jobs depend on the forest industry, some 2 million ha of forest, covering over 10% of the nation's total forest area, are affected.

In what was formerly East Germany the situation is thought to be even worse and government officials are reluctant to provide data on the extent of the disease. In Czechoslovakia, approximately 1 million ha of woodland, more than 20% of the total, is now irreversibly damaged. In Sweden, where forest products constitute 20% of export earnings, the damage to the trees is just as serious, especially in the southern part of the country.

The damage caused by acid rain is not confined to fish stocks and forests. The pollution also clings to stone; if the structure is a form of limestone, such as marble or chalk, it is turned into the soft and soluble powder that builders call gypsum. Many historic monuments throughout Europe have been eaten away by acid rain. For example, it is now well-documented that St Paul's Cathedral in London has lost an inch-thick layer of its stone over some considerable period of time. But on the cathedral's southern face, which looks directly across the River Thames to Bankside power station chimneys, the corrosion has accelerated in the past few decades. Apart from stones, more than 100 000 stained-glass treasures in Europe, some of which are more than 1000 years old, are affected by acid rain. The United Nations Economic Commission for Europe documented that stained-glass objects were generally in good condition up to the turn of this century. It has been argued that in the last 30 years the deterioration process has apparently accelerated to the extent that a total loss is expected within a few decades if no remedial action is taken. Medieval and post-medieval glass is particularly endangered because of the production process. Sulphuric acid etches stained glass, the surface corrodes, and forms a chalky crust which accelerates decomposition, letting paint peel off. The glass substance finally splits and disintegrates into minute particles.

Wildlife in many parts of Europe has also been badly affected by the ravages of acid rain. As rivers and lakes die, in addition to fish many smaller species such as mayflies, caddis flies and their nymphs and larvae disappear. As they go, so do birds and animals that depend on them for food. Wildflowers such as marjoram, the woodland violet, frog and frogroot orchids, mistletoe and many other species are under pressure. Acid rain is also damaging agricultural crops and deteriorating plumbing systems in houses. The annual cost of this damage may amount to billions of pounds in European Community countries alone.

7.2   DESTRUCTION OF THE OZONE LAYER

Ozone is produced in the ultra stratosphere and ionosphere by the effect of ultraviolet light on oxygen molecules. The process that controls the rate of change in ozone concentration is complex and incompletely understood by scientists, but it is known that concentrations vary significantly according to the time of day, geographic location, latitude and altitude. Scientists were first alerted to the fact that levels of ozone in the stratosphere were being seriously reduced because of air pollution in 1985 when a large hole in the ozone layer was discovered. In 1987 it was estimated that the depletion of the total ozone column over Antarctica was reduced to about 60% of its 1975 level. At altitudes of 14–18 km the depletion, on occasions, exceeded 95% (Everest, 1988). The ozone layer protects living organisms from the harmful effects of ultraviolet light. This protective element is crucial for human well-being because ultraviolet light is a major cause of skin cancer.

During the last few years it has been brought to the attention of the public that the ozone layer is a vital but exhaustible resource, which has been depleting quite rapidly. It is generally agreed that chlorofluorocarbon (CFC) gases are the biggest culprits which eat up the ozone layer. Although at least 20 types of chlorofluorocarbons are currently being produced and emitted into the atmosphere, CFC-11 and CFC-12 are the dominant types. They are mostly produced by the aerosol and refrigerator industries. Table 7.1 shows the historic development of the CFC market in the United States, one of the largest producers. Much of the CFC embodied in refrigerators, air conditioners and certain kinds of closed-cell foams remains trapped until these items are finally scrapped years later, and then the gases are released. In other cases, such as aerosols and some open-cell foams, CFCs are released into the atmosphere during manufacture, or within a year of production.

International action on the ozone problem has been co-ordinated by the United Nations Environmental Programme (UNEP). Some agreements on international co-operation about the underlying science and relevant control measures have been established at gatherings in 1985 in Austria (the Vienna

Table 7.1   Historic development of the United States CFC market

Application	Time period	Reason for development	Estimated income elasticity of demand
CFC-11 aerosol	1953–75	Personal care products expands	2.96
CFC-11 non-aerosol	1951–59	Refrigerator market develops	3.30
	1960–82	Foam products market develops	4.39
CFC-12 aerosol	1953–75	Personal care products expands	2.84
CFC-12 non-aerosol	1951–57	Market for refrigerators develops	3.02

Source: Quinn (1986).

Convention) and in 1987 in Canada (the Montreal Protocol). A good deal of attention has been focused on understanding and modelling the complex chemical reactions which underlie ozone depletion. At early stages of the study these models indicated substantial depletion (15–20%) over the next 50 years. More recent models, however, have been predicting a much lower depletion rate of 3–5%. Nevertheless, it is still thought that depletion rates over the Antarctic are over 50%, which gives rise to serious concern. Some scientists (e.g. Everest, 1988) argue that conditions over Antarctica are very different from elsewhere so that it is possible that the Antarctic ozone hole is only a local phenomenon. Recent satellite data have indicated only a 2–8% fall in global ozone levels over the past seven years.

The Montreal Protocol, which took place in 1987, stated that by June 1990 participating countries must freeze their domestic consumption of CFCs at 1986 levels and limit production to 110% of 1986 levels. These limits fall to 80 and 50% for production by 1994 and 1999, respectively. Although these figures are determined mainly on the basis of political considerations, they do still have an important scientific component. The overall production of CFCs is determined by multiplying the annual output by its 'ozone depleting ability', which is based on the models of stratospheric chemistry. Mintzer (1987) modelled predicted global production of CFCs and the resulting concentrations give four differing scenarios: business-as-usual; high emission (in which accelerated growth in production and consumption of related commodities is permitted without any policies to slow down growth); modest policy (in which production of all related commodities is restricted eventually to maintain constant levels); and slow

Table 7.2   Projected total production of CFC-11 and CFC-12 under four different scenarios

| Year | Thousands of tonnes per year | | | |
	Business as usual	High emission	Modest policy	Slow build-up
1990	1100	1100	1100	750
2000	1200	1600	1200	750
2010	1300	2100	1200	750
2020	1400	2600	1200	750
2030	1500	3200	1200	750
2040	1600	4000	1200	750
2050	1800	5400	1200	750
2075	2100	9100	1200	750
Annual growth rate, 1985–2075	0.95%	2.6%	0.35%	0.0%

Source: Mintzer (1987).

build-up (in which reasonably rigorous policies are adopted to protect the atmosphere). His results are shown in Tables 7.2 and 7.3.

In March 1989 a delegation from more than 120 countries met in London at the Saving the Ozone Layer Conference to discuss the serious problem of ozone depletion. The conference was opened by the then British Prime Minister, Margaret Thatcher, who urged consumers not to buy a new fridge until 'ozone-friendly' models were made available by manufacturers. She praised 'the intense efforts of chemical companies to find safe alternatives to the CFCs. However the chairman of the Refrigeration Industry Board, who represents the UK manufacturers and contractors, pointed out that it could be some years before fridge manufacturers found viable alternatives to ozone-depleting CFCs. On the second day of the conference the Prince of Wales urged developed nations to accelerate dramatically programmes to phase out ozone-depleting CFCs, and to help the Third World to find alternatives in their emerging technologies. He also warned that legislation forcing industries to adopt more environmentally sensitive manufacturing processes was essential to avoid global disaster. Furthermore, the Prince also argued that millions of people were now looking to their governments to act with the urgency demanded by environmental threats. He said there was an overwhelming scientific case to change the terms of the Montreal Protocol and eliminate completely the use of CFCs as soon as possible.

Delegates from the Third World, especially China and India, made determined pleas for a new fund to bring a degree of equity into the saving of the

Table 7.3 Projected atmospheric concentrations of CFCs under four different scenarios

Year	Business as usual		High emission		Modest policy		Slow build-up	
	CFC-11 (ppbv)[a]	CFC-12 (ppbv)[a]	CFC-11 (ppbv)[a]	CFC-12 (ppbv)[a]	CFC-11 (ppbv)[a]	CFC-12 (ppbv)[a]	CFC-11 (ppbv)[a]	CFC-12 (ppbv)[a]
1980	0.170	0.285	0.170	0.285	0.170	0.285	0.170	0.285
1990	0.303	0.521	0.303	0.521	0.302	0.517	0.282	0.500
2000	0.471	0.793	0.512	0.827	0.464	0.748	0.381	0.693
2010	0.648	1.080	0.815	1.225	0.635	0.987	0.470	0.875
2020	0.826	1.377	1.196	1.714	0.796	1.232	0.547	1.046
2030	1.005	1.687	1.651	2.295	0.959	1.488	0.615	1.206
2040	1.190	2.014	2.198	2.983	1.127	1.756	0.674	1.355
2050	1.379	2.359	2.897	3.828	1.299	2.039	0.726	1.494
2060	1.573	2.720	3.859	4.904	1.474	2.333	0.771	1.625
2075	1.873	3.291	5.749	6.906	–	–	0.829	1.805

[a]ppbv, parts per billion by volume.
Source: Mintzer (1987)

ozone layer. The leader of the Chinese delegation pointed out that his country's 1.1 billion population would probably suffer more than those in the developed world from the results of damage to the ozone layer and change in climate. The present problems of CFC pollution which have yet to unfold are overwhelmingly the product of 30 years of environmental abuse in the West. The Third World will need to be compensated for the damage it will sustain, and the development it will have to forgo, as a result of CFCs and future restrictions on their use.

The Indian delegate argued that the West had a moral duty to consider the way developing countries were being invited to keep their consumption of CFCs at a level 100 times lower than those which are enjoyed in the West. He pointed out that there was no wish in the Third World that chemical multinationals should suffer, but the entire world should be free from ozone depletion. At the same time, according to perceptions in the developed countries, the chemical giants should be safe from profit depletion. Endorsing the idea of a fund to help developing nations, he concluded that this should not be thought of as charity; there is an excellent principle established in the West – the polluter pays.

In 1979 NASA's Nimbus-7 satellite began recording daily ozone levels in both northern and southern hemispheres. Since then its total ozone mapping spectrometer has charted a steady decline. Other ground-based research institutes such as the United Nations' World Meteorological Organization confirm NASA's findings. According to these findings ozone levels in 1992 were the lowest on record and thinning took place simultaneously in the northern and southern hemispheres. The satellite also revealed that ozone thinning accelerated in the northern hemisphere in the first three months of 1993. Over latitudes that include London, New York and San Francisco, ultraviolet radiation from the sun as a result of ozone thinning has been increasing steadily (Connor, 1993).

NASA scientists could not offer a definite explanation for the rapid decline in the ozone layer. One influential factor may have been unusual atmospheric effects that have lingered on from the 1991 Mount Pinatubo volcanic explosion in the Philippines. Another factor may be the persistence of CFCs in the atmosphere. Ozone will continue to decline before it begins to level out; clearly we have not seen the worst yet.

Connor (1993) reports that scientists at the UN estimate that for every 1% decrease in ozone there could be a 3% increase in non-malignant skin cancers and a smaller increase in potentially fatal melanomas. Furthermore, the increase in ultraviolet light that results from a dwindling ozone layer can lower productivity of crops substantially. This could be a particularly serious problem in the next century when the full effects of ozone depletion on crop production will be felt and when food demand for a growing population will be greater than ever.

One of the most important areas for concern is the effect of ozone depletion on marine plankton which have a vital role in the absorption of carbon dioxide globally. It is estimated that plankton account for over half the global amount of CO_2 absorbed each year, so clearly any harm to plankton could have a knock-on effect on the greenhouse effect (see below), exacerbating global warming.

Furthermore, anything that kills the plankton on which shrimps and fish depend will have a significant effect on the human food supply. Another adverse effect of increasing levels of ultraviolet light could be greater degradation of materials such as plastics that are sensitive to this type of radiation.

A number of studies have shown that the benefits to be gained from reducing the emission of ozone-depleting CFCs are very large indeed. In 1987 the United States Environmental Protection Agency published the results of research showing that the benefits of CFC control were far higher than the costs (EPA, 1987). In this study benefits were identified to be values of lives saved, benefits in the form of avoided medical treatment and avoided reduction in agricultural yields. The cost of control was largely based on the assumption of additional expenditure in switching to more expensive substances that are not ozone depleting. Table 7.4 summarizes the results of this research. The study covers the period 1987–2075 and assumes costs to apply from 1986 levels. All figures are discounted to the base year, which is 1987.

This study makes a strong economic case for reducing CFC emissions and shows that benefits overwhelmingly outweigh the costs. Many CFC-producing companies such as ICI and Dupont have already begun to invest in research into alternatives, anticipating some regulatory control.

7.3   THE GREENHOUSE EFFECT

Many economically important activities emit gaseous pollutants such as carbon dioxide (CO_2), nitrous oxide (N_2O), chlorofluorocarbons (CFCs) and methane (CH_4) into the air. These emissions have infrared radiation absorption bands in the range of 8–13 μm and alter the heating rates in the atmosphere, causing the lower atmosphere to heat and the stratosphere to cool. Radiation from the sun heats the earth's surface and this heating is balanced by the emission of long-wave thermal radiation. Gases like CO_2, N_2O, CFCs and CH_4 strongly absorb the long wavelength radiation and return part of it back to the surface, causing an increase in the surface temperature. This process, known as the greenhouse

Table 7.4   Costs and benefits of CFC control in the United States

Level of control	Discounted benefits ($ billion)	Discounted costs ($ billion)
80% cut	3533	22
50% cut	3488	13
20% cut	3396	12
Freeze	3314	7

Source: EPA (1987).

effect, has been happening over at least the last two hundred years, and raises serious questions about the extent of future global warming and its impact on the future climate.

Everest (1988) argues that even a limited temperature rise of 0.6°C would be comparable in magnitude, although opposite in sign, to that which occurred in the 'Little Ice Age' of western Europe between 1400 and 1800. Given the worst possible scenario, a warming at the top of the possible range would be comparable to, or even greater than, the difference between present conditions and those in the last ice age and could result in profound social changes.

Over the last ten million years the naturally occurring concentration of greenhouse gases, especially carbon dioxide, has fluctuated quite substantially. Throughout this period CO_2 has warmed the planet's surface. Mintzer (1987) argues that CO_2, together with water vapour and clouds, warms the earth's surface by approximately 33°C from an estimated average temperature of $-18°C$. Without this natural process the earth would be a comparatively cold and lifeless planet. Ironically, the enhanced greenhouse effect now threatens to disrupt human society and the natural ecosystems. Clark (1982) suggests that the combined atmospheric build-up of CO_2 and other gases since 1860 has already increased the earth's surface temperature by approximately 0.5–1.5°C above the average global temperature of the pre-industrial period.

Many important social and economic decisions are being made on major irrigation, hydropower and other water projects, on drought and agricultural land use, and on coastal engineering projects, based on assumptions about the climate a number of decades into the future. In the light of the greenhouse effect these predictions are no longer valid since increased gases are expected to cause a significant warming of the global climate. The consequences of climate change on energy and agricultural projects and consequently on human settlements are not known. The impact is likely to vary between regions with perceived potential for winners and losers, although such perceptions tend to be oversimplified. For instance, will the output in the United States grain belt be adversely affected? Or, will increased rainfall in currently dry regions lead to improved agricultural productivity? These questions are as yet unanswered.

Perhaps the most important problem associated with the greenhouse effect is the anticipated melting of the polar ice sheet, leading to a rise in the sea level. Many scientists agree that the global average sea level has risen by some 10–20 cm during the twentieth century. If this trend accelerates, it would threaten low-lying regions in both developed and developing countries. At the very least it would require the erection of new and large coastal defences, a serious problem for developing countries. It has been estimated, on the basis of observed changes since the beginning of this century, that global warming of 1.5–5°C would lead to a sea-level rise of 20–170 cm towards the end of next century.

The underlying consequences of a rise in sea level, which are complex and still incompletely understood, are a combination of local and regional factors superimposed on average global trends. The latter appear to be controlled by the thermal expansion of the oceans' water. Although some progress has been made in predicting thermal expansion, there is as yet little quantitative understanding of the rate of melting of land-based ice. Both these processes are likely to lag behind any transient heating associated with an increase in the atmospheric concentration of greenhouse gases. A further problem is that the damage caused by higher water levels is linked to the occurrence and severity of storms, which are also influenced by the change in climate. This means that when designing current coastal projects account must be taken of any increase in the sea level and storm damage.

There are numerous models which try to ascertain projected levels of carbon dioxide and other greenhouse gases in the atmosphere and the warming effect which would result from them. Using the scenarios outlined on pp. 216–17), Mintzer (1987), for example, estimated CO_2 and N_2O levels and consequent global warming for the years up to 2075 (Figures 7.1 and 7.2, Table 7.5). However, it is worth noting that significant uncertainties exist in the data used

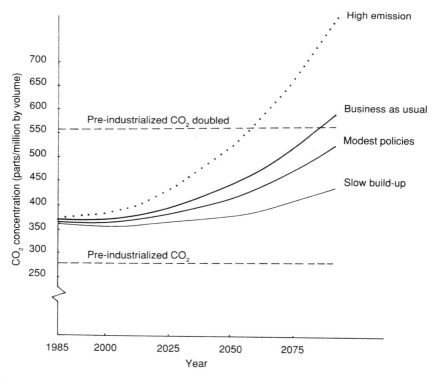

Figure 7.1 Atmospheric CO_2 concentrations, 1985–2075, given four different scenarios. (Source: Mintzer, 1987.)

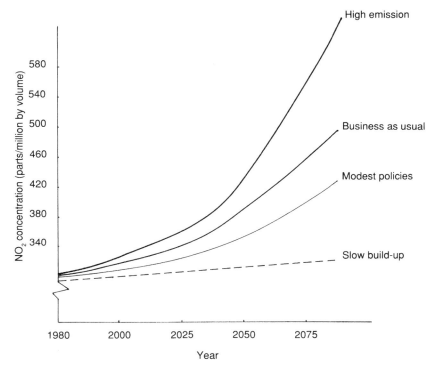

Figure 7.2   Atmospheric N_2O concentrations, 1980–2075, given four different scenarios. (Source: Mintzer, 1987.)

by Mintzer to predict levels of economic activity, energy use and hence emissions of greenhouse gases. For example, although Mintzer used population estimates in line with those produced by the World Bank and the UN – a global population of 6 billion in 2000 and 10 billion in 2100 – he fails to take other demographic factors, such as urbanization, into account. These factors can have a significant effect on economic activity and hence energy use. In developing countries urbanization can triple or quadruple per capita demand for primary energy as charcoal and oil are substituted for firewood. Various technological factors could also affect the rate of growth of future emissions. How fast the efficiency of energy supply and use will improve is uncertain. There are even more complicated feedback effects between greenhouse warming and emission of gases. Global warming could decrease the heating season and increase the cooling season significantly. None of these factors are considered in the above-mentioned models.

The greenhouse effect is indeed an international problem. All countries of the world are potentially affected by greenhouse-related environmental changes and

Table 7.5   Warming of the atmosphere on the basis of the combined effect of the greenhouse gases (CO_2 + N_2O + CH_4 + CFC + ozone)

| Year | Temperature relative to 1980 atmosphere (°C) | | | |
	Business as usual	High emission	Modest policy	Slow build-up
1980	0.0–0.0	0.0–0.0	0.0–0.0	0.0–0.0
1990	0.2–0.5	0.2–0.5	0.1–0.4	0.1–0.4
2000	0.4–1.1	0.5–1.4	0.3–1.0	0.3–0.8
2010	0.6–1.7	0.9–2.6	0.5–1.5	0.4–1.1
2020	0.8–2.4	1.3–3.9	0.7–2.1	0.5–1.4
2030	1.1–3.2	1.8–5.5	0.9–2.7	0.6–1.7
2040	1.3–4.0	2.4–7.1	1.1–3.3	0.7–2.0
2050	1.6–4.8	3.0–8.9	1.3–3.9	0.8–2.3
2060	1.9–5.7	3.7–11.0	1.5–4.5	0.8–2.5
2075	2.4–7.1	4.8–14.5	1.8–5.5	0.9–2.7

Source: World Resources Institute (1987).

no country can do much on its own to combat those changes. As mentioned above, the severity of the greenhouse-led climate change is likely to differ from region to region. Low-lying countries and those who have established communities around irrigation and hydro-energy projects are likely to be the greatest losers. As the environmental changes related to greenhouse gases are likely to take place over many decades or even centuries, the change in social structures and volumes will be most disruptive to future generations.

What are the events that are likely to cause the public policy-makers to take action? One event would be a dramatic scientific discovery, similar to the Antarctic ozone hole, which could indicate substantial future greenhouse heating. Another event may be a rise in public consciousness, as human beings see themselves as victims of a climatic change. Furthermore, some modern illnesses may eventually be correlated with the greenhouse gas pollution. In the Western world where the media are extremely powerful and sophisticated, they may prove to be a catalyst which increases public pressure on policy-makers to come up with solutions. In the Third World, natural disasters caused by drought and floods may be linked to the greenhouse warming effect and this is likely to raise public consciousness there, as well as in the West. Some scientists argue that over the last 40 years there has been a decrease in precipitation at latitudes 5–35°N but an increase in the region 35–70°N (Everest, 1988). However, not everybody accepts that this trend can be firmly linked with the greenhouse effect.

International pressure on industrialized countries is bound to grow as they consume about 80% of the world's fossil fuels with the associated emissions of greenhouse gases. Such arguments could well be used effectively by developing countries to obtain increased aid and establish funds to combat pollution. In effect this has already happened, at the Saving the Ozone Layer Conference in London, March 1989. As developing countries are likely to resist any limitations on the future growth of their own use of fossil fuel, they are also likely to claim that any such reduction and control should fall primarily on developed nations.

What is the British government's policy, towards the greenhouse effect? Everest (1988) summarized this as 'do nothing and await the results of further research'. The assumption is that there is time for better scientific understanding of the greenhouse warming to emerge before action is taken which might close options and thus reduce future flexibility. This policy may be plausible if it is considered that any greenhouse warming is likely to be very slow and that the counter-effects, such as a reduction in the heating season, may slow down the process even further. Furthermore, our greater capacity to tap renewable energy sources such as solar, wind, wave and tidal power, and the promotion of more effective use of energy may lead to a reduction in the emission of greenhouse gases. These are very optimistic assumptions which will not be shared by everyone.

7.4 DESTRUCTION OF BIOLOGICAL DIVERSITY

The earth contains a staggering number of different plant, animal and insect species, each equipped with a different genetic make-up. Up to now, biologists have classified 1 700 000 species but some believe that the final total could reach 40 million, insects alone accounting for 30 million species. The number of genes per species ranges between 1000 in bacteria to 10 000 in some fungi, and averages about 100 000 in mammals. Flowering plants contain the greatest number of genes, often in excess of 400 000. Genes determine the particular characteristics and capabilities of a given organism. Such genetic diversity is vital to the maintenance of ecological stability, enabling different species to fulfil different functions within the biosphere. Scientists believe that reduction of diversity greatly increases the vulnerability of our ecosystem.

Today it has become common knowledge that the genetic diversity of the biosphere is being rapidly eroded; 1000 species become extinct every year. Goldsmith and Hilyard (1988) estimate that by the end of the century that figure will rise to 10 000 per year. An estimated 25 000 plants are now threatened with extinction and given the present trends 60 000 plant species may be lost by 2050. Although nature has considerable resilience, ensured in part by the sheer volume of species, there is a limit to how far that resilience can be stretched. It has been estimated that about one half of the world's species are found in what is left of the tropical forests. The Worldwide Fund for Nature has identified Mexico, Colombia, Brazil, Zaire, Madagascar and Indonesia as megadiversity states and it is thought that 50–80% of the world's biological diversity is to be found in just 6–12 tropical countries (Mittermeier, 1986; Swanson, 1991). Forest destruction in the tropics is the major reason for the increasing rate of species destruction. Of course, extinction of species is a natural process. Some scientific research shows that the longevity of many species lies in the range of 1–10 million years. As some species are lost, new ones are created. The problem of biodiversity arises when the rate of extinction exceeds the rate of creation. Destruction in megadiversity states creates a potential threat to the entire global biology.

In the distant past there have been a number of mass extinctions; probably on five occasions over 50% of the then-existing animal species became extinct (Raup, 1986). Averaging over three periods, the natural rate of extinction is in the neighbourhood of 0.000 009% per year. Therefore, the present stock of biodiversity is the result of several billion years of mostly low-frequency change, making current stock ancient natural capital alongside fossil fuel, freshwater and soil fertility which cannot be replaced once destroyed. In other words, the stock of biodiversity is an exhaustible resource (Swanson, 1991). Today many species are becoming extinct at an alarming rate without ever having been utilized. The human population is expanding and natural habitats declining most rapidly in areas of rich genetic diversity.

Apart from the ecological importance of biodiversity, its conservation is important for many areas of business. For example, it has recently been discovered that the red liquid which comes from *Carina domestica* in the Brazilian forest of Rondonia contains a powerful blood-thinning substance that may someday save the lives of heart attack victims. Cures for many of our illnesses, including cancer, could one day be found in species in these tropical forests. Xenoud, an English company, is now screening organisms, many of them from tropical rainforests, for the ability to fight cancer and heart disease. Vinblastine, a derivative of the Madagascar periwinkle, is now used to treat childhood leukaemia.

The whole field of natural plant products is expanding rapidly. Science and industry have already discovered more than 120 drugs from natural sources. According to a drug company, Missouri Botanical Garden, every one of their top-twenty selling prescription drugs comes from natural sources or is based on them. Vinblastine alone, mentioned above, generates about $50 million a year. A potential anti-AIDS drug, castanoupermine, comes from the seeds of the tropical Alexa tree found in South America. Drug companies have long realized that conservation of biodiversity makes good business sense.

Apart from its role in medicine, diversity is important in many areas of biotechnology. The term biotechnology embraces a collection of scientific techniques that use living organisms to make or modify products, to improve plants and animals or to develop microorganisms for specific tasks. It includes the old-fashioned methods of cheese-making, baking, brewing and plant breeding. Biotechnology can help us to produce disease-resistent crops or plants that thrive on poor soils.

The tropical forests that contain a good proportion of the world's biodiversity are largely unprotected and open to the forces of land conversion to produce timber and a few species of cash output such as beef, rice, maize, etc. Specialized production of these commodities is taking its toll on biodiversity.

Swanson (1991) believes that although conversions from heterogeneous and diverse means of production to homogeneous and single-species production have been seen as a necessary step along the path to development, this is not necessarily true today. There is a good deal of evidence to show that the economic benefits stemming from the classical path to development are declining rapidly. One reason for this is the absence of significant demand for the additional quantities of the primary commodities. On the other hand, the relative scarcity value of natural habitat and diverse products that stem from it is increasing in relation to primary goods.

Swanson (1991) reports on the conservation value of natural habitats and contends that on occasions it exceeds the value of commodities currently produced. For example, take the value of the Great North American Lakes as a source of recreation – fishing was valued at $0.5 billion in 1984, while the value of the fish taken was only $6 million. Many indigenous peoples in the tropics depend for their survival on food that comes from the forest. For example, regional reliance on game meat exceeds 80% in Peru and 70% in rural Ghana. Wood is a primary source of energy in the tropics as well as in much of the developing world. Building material comes from forests. In addition, the tourism value of tropical forests is appreciating. Some natural habitats generate food, energy, building materials, water supply, recreational material as well as raw materials for the multibillion pound biotechnology industry. A good proportion of the benefits captured by the indigenous communities in the tropics do not go through the market mechanism and in this way they may be undervalued.

Loss of biological diversity was recognized in the publication of the World Conservation Strategy in 1980 (Swanson, 1991). Some solutions to the problem were suggested such as establishment of wilderness areas and national parks, debt-for-nature agreements and the development of an international fund for conservation. Such measures achieved a limited amount of success. For instance, about 5% of the remaining tropical forests have received some protection. Some private companies initiated protection measures. In one instance, Merck Co. Inc. signed an agreement in 1991 with Costa Rican authorities and provided $1 million to help fund bioprospecting and conservation efforts until the end of 1993 in exchange for the opportunity to screen collected samples for leads in developing new drugs.

At the Rio Conference in 1992 a biodiversity treaty was signed. This recognizes that nations should conserve the diversity of plant and animal species on their territories. However, there is nothing in the treaty that compels nations to protect their biodiversity. There is a provision to help developing countries via the Global Environmental Facility Fund. The United States refused to sign the biodiversity treaty on the grounds that it compromised its biotechnology industry, which is based upon manipulating genes that may come from plants and animals in the developing world.

Although biodiversity is an exhaustible, or even non-renewable, resource, it would be a mistake to treat it in the same way as other exhaustible resources such as fossil fuel. To start with, biodiversity is composed of living creatures whereas fossil fuel and metal deposits are non-living. Most people would be very uncomfortable with a sustained depletion of the world's biological ancient capital on moral grounds. Economic theories that were developed for metal deposits and fossil fuel, explained in Chapter 5, cannot possibly be used on biodiversity. There are some similarities between stock of genetic diversity and stock of natural wonders which will be explained in Chapter 9. It would indeed be quite reasonable to look upon the whole spectrum of biodiversity as the ultimate wonder of nature. Thus far I have not come across any economic theory of biodiversity but I feel sure that a few theories will be developed in due course.

7.5   INTERNATIONAL GATHERINGS AND INSTITUTES ON THE ENVIRONMENT

In the light of growing concern in environmental matters, the 1970s marked the beginning of a range of international gatherings, conferences and declarations. The first serious conference on the human environment took place in 1972 in Stockholm under the auspices of the United Nations. This conference declared that 'to defend and improve the environment for present and future generations has become an imperative goal for mankind'. It was also emphasized that solidarity and equity in the relations between nations to achieve these objectives should constitute the basis for international co-operation. Environmental education should play a leading role in the defence and improvement of the human environment, which should be provided for all ages and at all levels. Four years later another important event, the Belgrade Charter, took place in former Yugoslavia. This too emphasized the importance of environmental education and international solidarity on matters concerning humans and nature.

The first official intergovernmental conference on the environment took place in Tblisi, Georgia, on 14–22 October 1977. The conference concluded with a number of declarations. It appealed to governments to include environmental issues in their education systems, invited education authorities to promote and intensify efforts in thinking, research and innovation on environmental matters. It urged governments to collaborate in this field especially by exchanging experiences, research findings and by making their training facilities widely available to teachers and specialists from all countries. It also appealed to the international community to give generously of its aids for the promotion of international understanding on the environment. A great deal of emphasis was given in the Tblisi Declaration to environmental education. In particular it was stated that environmental education should be a lifelong process, beginning at the pre-school level and continuing throughout life. Environmental education and issues should form part of development plans, especially in the Third World countries. There should be emphasis on the complexity of environmental problems and thus the need to develop critical thinking and problem-solving skills.

Towards the end of the 1970s there was a slow down in the number of international gatherings, and a reduction in the size of government departments dealing with environmental matters. In order to combat inflation many Western governments were cutting back on public spending, and many environmental institutions suffered quite heavily. For example, in the UK the government closed down two important environmental departments – the Waste Management Advisory Committee and the Noise Advisory Council – and greatly reduced recruitment to its environmental inspectorate. Many departments, in particular the environmental ones, began to operate with a skeleton staff. Towards the end of the 1980s the size of the civil service was the smallest in post-war Britain, and many leading Conservative politicians expressed pride in this.

However, there were some notable conferences during this period. One was the 1979 Geneva Convention on the prevention of long-range transfrontier air pollution in Europe. Then there was the 1982 Stockholm conference on the environment where British Minister Giles Shaw told the conference, with regard to the acid rain problem, that it would be wrong to spend public money on a clean-up unless it was proved necessary. At an international conference on the environment in Munich in 1984, William Waldegrave, the then Parliament Under Secretary of State at the Department of the Environment, said that reducing sulphur dioxide pollution was the wrong solution to the acid rain problem; he added that for Britain to install the necessary 'scrubbers' to reduce power station pollution would cost £1 billion (1984 value) in capital investment with no guarantee that it was the environmental answer. He pointed out that there were other contributory factors to the problem such as nitrogen oxide. Some delegates accused Britain of taking a 'do nothing' attitude.

In the second half of the 1980s the international community was becoming increasingly concerned about environmental issues. In 1985, a conference sponsored by the United Nations took place in Villach, Austria, on the greenhouse effect, and this was followed by similar conferences in Laxenburg, Germany, in 1986 and Bellagio, Italy, in 1987. In 1988 there was the United States Congressional Staff Retreat on Climatic Change, the Commonwealth Study Group on Climatic Change and Sea Level, and the Toronto Conference in June on the Changing Atmosphere which proposed a 20% cut in carbon dioxide emissions by industrialized countries by 2005 and a levy on their fossil fuel consumption. This conference was attended by more than 300 scientists and policy-makers from 48 countries, United Nations organizations, other international bodies and non-governmental organizations. Perhaps one of the most important international agreements drawn up in the 1980s was the 1987 Montreal Protocol on the control of CFC production.

The first years of the 1990s have seen further conferences and agreements but none as high-profile as the Conference on the Environment and Development held in Rio de Janeiro in 1992 (see Section 7.5.1).

There are a number of international organizations set up to deal with matters relating to the environment. One of the most important international institutes is the United National Environmental Programme (UNEP) which was instrumental in a number of recent conferences on the greenhouse effect and ozone depletion. The headquarters of UNEP is in Nairobi, Kenya, and many countries which belong to the United Nations are members. The United Nations also has a regional body in Europe called the Economic Commission for Europe (ECE) located in Geneva, Switzerland. One of its original objectives was the exchange of information and experience on environmental issues between the countries of Europe. The Organization for Economic Co-operation and Development Environmental Commission (OECDEC) is another notable international body. It was established by the OECD countries in 1970 and includes a number of committees, each dealing with specific aspects of environmental issues such as noise, air and water pollution, recycling, urban development and energy. There are also various departments which deal with the legal consequences of international environmental problems. The main objectives of the OECDEC are to promote dissemination of knowledge regarding the environment, harmonization of environmental policies and further international solidarity in order to solve transfrontier pollution problems.

7.5.1 Rio 1992

Twenty years after the United Nations Conference on the Human Environment which was held in Stockholm in June 1972, the United Nations Conference on Environment and Development took place amid great publicity in Rio de Janeiro, Brazil. The Conference was given a very high profile by the world's media. For example, on the opening day, 3 June 1992, the prestigious London Newspaper, *The Independent*, devoted its entire front page to the Rio Conference and gave data illustrating the worsening environmental problems during the last 20 years (Table 7.6). It was noted, however, that there were a number of positive developments giving grounds for optimism. There has been progress: global wealth production has more than doubled over those two decades and a greater proportion of the world's population is adequately fed, clothed and sheltered than in 1972. None the less, the importance of the environment and development problems debated in Rio cannot be overstated.

Many felt that it would be much better for us, and certainly for our children and grandchildren, if the leaders gathering in Brazil showed more courage and wisdom, if they had gone further and faster than they seemed inclined to do. In delaying we would only be storing up greater hardship for the next generation. However, at the start of the Conference many media articles drew an analogy between Rio and Cancun 1981, the last occasion at which the rich and poor worlds sat together to discuss the future of the planet. A number of commentators pointed out that nothing of value had emerged from Cancun, and it would have been better if it had not taken place. Despite impressive rhetoric and participants such as Margaret Thatcher, Ronald Reagan, Pierre Trudeau, Ferdinand Marcos and Indira Ghandi, Cancun produced almost nothing tangible.

Table 7.6 Compounding environmental problems during two decades

Case (global)	1972	1992
Population	3.8 billion, 72% lived in developing countries	5.5 billion, 77% live in developing countries
Military spending	$680 billion (1988 prices)	About $800 billion (1988 prices)
War refugees	3 million	15 million
Nuclear power	100 reactors in 15 countries	428 reactors in 31 countries
Major nuclear accidents	None	Three Mile Island and Chernobyl
Transport	250 million motor vehicles	Over 600 million motor vehicles
Global warming	Annual 16 billion tonnes of CO_2 emissions; concentration level 327 parts per million	Annual 23 billion tonnes of CO_2 emissions; concentration 356 parts per million
Ozone layer	Chlorine concentration below 1.4 parts per million	Chlorine concentration more than doubled
Mega cities	Three cities over 10 million, 38% live in cities	13 cities over 10 million, 46% live in cities
Rainforests	About 100 000 km^2 destroyed every year	About 170 000 km^2 destroyed every year
Fisheries	56 million tonnes annual catch	90 million tonnes annual catch
Species	Destruction gathering pace, 2 million African elephants left	Destruction intensifying, 600 000 African elephants left

Source: *The Independent*, 3 June 1992.

The Rio summit turned out to be the greatest gathering of presidents and prime ministers ever. The most reliable tally of leaders was reported to be 110, although the organizers did struggle quite hard with problems of definition in their attempt to reach a correct figure. There were some 'amusing' situations; for example, George Bush and Fidel Castro of Cuba found themselves in the same room for pre-lunch drinks and sat down at the same table for the meal, however, owing to the size of the table the two leaders never came within 10 m of one another! Also there were concerns in some Western countries about the ability of Brazil to manage the smooth running of a conference of this magnitude. Some journalists, photographers, camera crews and broadcast technicians – over 9000 in total – came fearing chaos and incompetence. In fact their fears proved unfounded and almost everyone was impressed by the efficiency of the conference organizers and the host government.

It was agreed by many experts and participants that the Rio Conference was not a place for scientific debate, the underpinning science was accepted and in no way questioned (Thornton, 1992). The issues to be discussed were how policy-makers and politicians would react, particularly in relation to financial obligations and institutional procedures. The Conference drew up a detailed programme of action, *Agenda 21*, and provided indicative costs for it. *Agenda 21* amounts to some 40 chapters, covering a wide range of social, economic and scientific issues, planning for future conservation and management of resources for development and the means of implementation. Some specific topics are: environmentally sound management of toxic chemicals and other hazardous wastes, environmentally agreeable technology, debt relief, population growth, poverty, aid programmes and science for sustainable development. The overall resource transfer for *Agenda 21* is estimated to be $125 billion per year for the period 1993–2000. Markyanda (1992) reports on a number of alternative estimates. Some of these are $71 billion per annum by the Worldwatch Institute; $20–50 billion per annum by the World Resources Institute, Washington; and $60 billion by 2000 by the United Nations University in Finland.

It is not clear where the money will come from but *Agenda 21* identified a number of possible sources without specifying their contribution. These are: official development assistance channelled through multilateral funds, United Nations agencies and bilateral aid programmes; debt relief through the Paris Club; direct foreign assistance and investment by private concerns; and the use of international taxes. Markyanda (1992) argues that a number of steps must be taken to strengthen *Agenda 21*, including estimating precise financial resources required by the poor countries; establishing financial institutions and a monitoring network for transfer of resources; and establishing guidelines regarding actions needed to integrate economic and environmental management at national, regional and sectional levels.

In addition to *Agenda 21* the Rio Conference produced a number of conventions, declarations and principles. The Climate Change Convention and the Biodiversity Convention were signed by more than 150 nations. These conventions recognize the threat of global warming and mass extinction of species but do not compel nations to tackle the problem. They will not come into force until at least 30 nations have ratified them which may not happen for several years. The United States did not even sign the Biodiversity Treaty. It was also agreed that a mechanism for assistance for both conventions should be set up but the size of the fund has not been established. In the case of the Biodiversity Treaty there is no indication that concessional transfers will be made. The only clear source of funding is the Global Environmental Facility which amounts to $3 billion for the first three years, but this is not restricted to poor countries.

Another agreement to come out of the Conference was the Rio Declaration, whose goal is a new and equitable global partnership. This sets out the principles of sustainable development; that is, development that meets present needs without wrecking the environment for future generations. There was also agreement to establish a UN Sustainable Development Commission to supervise the implementation of *Agenda 21*, and a Statement of Forest Principles, a set of non-binding guidelines for conserving forests.

Although a great deal was achieved at the Rio Conference, in his closing speech the Conference Secretary-General, Maurice Strong, a Canadian business tycoon, said that the leaders had not gone far enough. This view is shared by many conservationists and development workers.

8

Valuation methods for environmental costs and benefits

> We need to rely less on the authority of reductive empirical research and more on the habit of aesthetic apprehension.
>
> A.N. Whitehead

Almost every type of economic activity affects the environment to some extent; they are intrinsically linked. Changes in the economy in the form of changes in technology, consumption patterns, investment levels, international trade links, spatial relocation, macroeconomic policy, etc., will all have an impact, sometimes profoundly, upon the natural environment. This in turn will effect human well-being. For example, if the Chinese government decided to make every householder the owner of a CFC-using fridge within, say, five years and thereafter attempted to capture a good proportion of the world market with its product, this would be an environmental disaster. Even the Chinese would not be able to escape from the harmful effects of ozone depletion aggravated by their own policies. It is quite conceivable that decision-makers in their policy to expand the refrigerator industry would think about the global environmental harm, and damage to Chinese population, alongside the benefits of their project. Alternatively, nations may decide to eliminate completely ozone-depleting CFCs by using alternative methods in production to avoid further damage to the ozone layer. What are the costs and benefits involved in schemes like these? Unfortunately, at this point in time economists cannot give a precise answer.

Today everybody has realized that environmental quality does matter, although we may not be able to put a monetary value on it. In well-developed market economies quality of the environment is an important factor in market transactions. For example, if the environmental quality in a residential district deteriorates, people tend to move out of the area and thus house prices begin to move downwards, reflecting the falling environmental attributes. Two identical houses, one in a dirty, noisy and congested neighbourhood and the other in a more agreeable part of the town, are bound to differ substantially in price.

In developing countries the idea that environmental control is a luxury that can wait for another decade is no longer valid. Environmental problems are most acute in the developing world and sometimes the survival of thousands of poor people depends upon the quality of the environment they live in. Once it was fashionable to argue that a high priority must be given to meeting the basic material needs of the population in the developing world without being fussy about dirty industries. We know that clean water in, say, Kenya is not a luxury but an item of survival. Desertification in sub-Saharan African is not a discomfort to the people in the region but a threat to their lives. For example, it has been estimated that in Burkina Faso environmental damage to crops, livestock and fuel-wood due to land degradation amounted to 8.8% of the GNP in 1988. In Ethiopia the adverse effects of deforestation on crops and fuel-wood supplies is estimated to be 6–9% of the GNP (Pearce, 1991).

Placing a value on environmental damage now has an important part to play in environmental economics for a number of reasons. First, the valuation makes it clear that the environment is not an infinite and free resource, even in the absence of well-established markets. Especially when projects are making substantial claims on the environment, the valuation of such claims signals the growing scarcity of the environmental input. Second, development proposals that are in conflict with conservation will be judged from a better perspective when all environmental impacts are considered. In this way the decision-maker will arrive at a better and fairer decision. Third, when restoration of an environmental quality is considered, valuation can help to identify the extent of a project's worth. Fourth, valuation will free the decision-making process from subjective, or even arbitrary, judgement when environmental issues are involved. In the early stages of cost–benefit analysis environmental consequences of projects were not included in calculations on the grounds that they were largely unquantifiable. Valuation will narrow the gap between quantifiable and non-quantifiable costs and benefits. Fifth, valuation can provide a true picture about the economic worth of projects, the performance of a region or the nation as a whole. For example, relocation of some of the United States industries in northern Mexico, to take advantage of competitive labour costs and less rigorous environmental regulations there, will increase incomes there. But, at the same time, the environmental quality could deteriorate due to air and water pollution, noise and congestion. Ignoring the deteriorating effects of the growing industrial activity on environmental assets will no doubt overstate the economic performance of the region. Finally, valuation can help public sector policy-makers in using various environmental regulation tools such as taxation, subsidies, marketable permits, etc. One of the widely accepted views on environmental change is the 'polluter pays' principle, as explained in Chapter 6. How much should the polluter pay? Valuation could help to throw some light on this matter.

Economic valuation methods can, of course, be criticized in a number of ways. First, reducing environmental effects into monetary figures could be regarded as being morally objectionable. The value of life, health, diversity of wildlife and beauty of the landscape should not be measured in terms of money in the opinion of some people. The value of these things is rooted in deeper fields than economics, such as moral philosophy and even religion. Many people would agree that destruction of species is morally wrong, but suppose the construction of a large-scale water project is proposed which would help millions of poor people but would wipe out a few rare species in the area for ever. Should the project go ahead? No doubt some people would argue that no economic benefit could justify the extinction of God's creatures. Even those who are favouring the project would find it impossible to express the cost of extinction of a species in monetary terms.

Valuation requires extensive economic data which are costly and time consuming to acquire. When an environmentally sensitive project is under consideration, normally there is a tight deadline imposed on the final decision. Even if a study is authorized, a tight time schedule could make it difficult, or in cases impossible, to carry out comprehensive surveys. Furthermore, some environmental economists such as Bowers (1990) question the objectivity of the valuation exercise. The political agency which is ultimately charged with deciding whether a project should go ahead can commission a team of researchers sympathetic to their inclination. Many years ago a number of economists warned that cost–benefit analysis could become a bogus and corrupt discipline (Leopold and Maddock, 1954). Government agencies may inflate benefits or underestimate environmental costs in order to justify projects and thus perpetuate their own growth. The economic test may turn out to be not a proof of merit, but a number of concocted studies designed to gain more widespread acceptance of a decision already made on some other basis.

A number of criteria have been used over the years for valuation of environmental benefits and costs, some of these are detailed below.

8.1   COST–BENEFIT ANALYSIS

Cost–benefit analysis is the oldest project appraisal method, in which costs and benefits of communal projects are scrutinized to determine whether they are worthwhile proposals. The theory of cost–benefit analysis does not impose any restriction on the nature of the costs and benefits involved. That is, all costs and benefits, including environmental ones should, in principle, be part of a cost–benefit study.

The origins of cost–benefit analysis go back to 1844 when Jules Dupuit began to puzzle about the costs and benefits of constructing a bridge. Dupuit (1844) contended that all costs and benefits, including those which are not actually paid, should be assessed before the construction of a public work. Many economists relate the concept of unpaid price to the consumer surplus and mark Dupuit as the first person to identify it. Almost a century later in the USA water projects played an important role in the development of cost–benefit analysis. For

instance, the United States Bureau of Reclamation has been concerned with economic justification of its projects since the advent of its programme in 1902. In
1936 the Flood Control Act established a principle that a project should be
declared feasible if the benefits, to whomsoever they may accrue, are in excess
of the estimated costs. After the Second World War two official reports, the
Federal Inter-Agency River Basin Committee's Sub-Committee on Benefits and
Costs, 1950, and the Bureau of Budget's Budget Circular, 1952, produced form-
alization procedures for costs and benefits of water development and other projects. Over the following years a number of academics attempted to lay down
cost–benefit criteria in relation to water development projects (Eckstein, 1958;
McKean, 1958; Krutilla and Eckstein, 1958; Maass, 1962; Lee and Douglas,
1971).

In the UK transport projects played a crucial role in the development of
cost–benefit techniques. Towards the end of the 1950s a substantial number of
economists were recruited into the civil service who were familiar with the rudi-
mentary appraisal techniques that were developing in the United States. They
were quite keen to use, and even develop, early cost–benefit methods on the
UK's expanding transport projects. The London–Birmingham motorway project,
the M1, was the first major public sector investment which was subjected to a
rigorous cost–benefit analysis. This work, in the main, was a methodological
study, but since then the principles established in it have been widely used on
many other transport studies. After the M1, other sizeable transport projects
such as the Tay Bridge, the Severn Bridge, London's Victoria underground line
and the Cumbrian coast railway line, received the scrutiny of cost–benefit analysts. Apart from being vital for the nation's economic development, transport
projects exhibit a number of interesting environmental characteristics. Most projects, especially the large ones, are creators of substantial environmental
problems such as noise, pollution, loss of habitat and visual amenity.

Some economists, such as Winpenny (1991), contend that cost–benefit analysis has made relatively little headway in capturing the environmental effects of
communal projects. A number of reasons can be given for this. First, communi-
cations between economists and scientists are not always easy. Economists are
not always able to identify the full environmental consequences of projects
where a substantial input from scientists is required, and they often use jargon
that is unintelligible to others. In this way, to some degree, they have isolated
their subject from other disciplines (Cooper, 1981). Second, difficulties in, and
in some cases the impossibility of, quantifying environmental costs and benefits
has led economists to sidestep these issues. In the 1960s and 1970s when
cost–benefit analysis became a widely used investment appraisal method there
was little concern about environmental matters. A typical cost–benefit study
would make some reference to environmental aspects of the project on focus but
not much effort was put on quantifying and valuing them.

In the 1980s, as concern about the environment was growing, the cost–benefit
analysts were required to take environmental issues more seriously. For

example, in the 1984 British Manual for Investment Appraisal in the Public Sector, HM Treasury (1984) stated

Many costs and benefits are measured directly in money terms; for example, savings in expenditure on resources, and sales revenues. Where they are not (examples are travel time saved, noise and other forms of pollution, and broad managerial or political factors) costs and benefits can sometimes still sensibly be given money values, often by analysing people's actual behaviour and declared or revealed preferences. These imputed money values can be used in the appraisal as if they were actual cash flows. Other factors which cannot be valued should be listed, and quantified so far as practicable, making it clear that they are additional factors to be taken into account . . . Account sometimes needs to be taken of the value which individuals may place now on the possibility of using a service or visiting an attractive area even when they do not currently use the service or make visits.

Indeed, if properly applied in a comprehensive manner, cost–benefit analysis can be a very useful discipline to incorporate environmental values associated with communal projects. Dixon (1991) contends that the simplest approach in valuing environmental effects is usually the best. Those that require extensive data or complicated experimental techniques have very limited applications, especially in the developing countries where environmental problems are most acute and sophisticated data are not available.

8.2 COST-EFFECTIVENESS ANALYSIS

Cost-effectiveness analysis is a common form of appraisal in health, crime prevention and environmental projects, where benefits cannot be measured easily. It is essentially a cost minimization exercise in achieving a particular objective such as clean air, water, noise reduction, protection of species, etc. Costs could be broadened to include forgone environmental benefits, when different alternative goals are involved. Cost-effectiveness analysis suffers from a number of defects. It cannot be used to determine whether or not a particular objective is worth pursuing. Once a decision is made to go ahead with a project, say, on political grounds, then the agency may think about achieving it in the most cost-effective manner. Furthermore, it is difficult to use cost-effectiveness analysis for decisions concerning marginal changes in objectives.

Where attainment of objectives cannot be exactly measured, or where project alternatives cannot achieve an identical result, cost-effective analysis becomes ambiguous. In particular, when a project has a number of objectives and a number of alternatives it is difficult to obtain a satisfactory answer by using cost-effectiveness analysis. However, despite all these problems cost-effectiveness analysis is widely used in developed as well as in developing countries. For example, in the UK the 1968 Roskill Commission's study on the siting of the Third London Airport was, essentially, a cost-effectiveness study. The

Commission was required to identify where the Third London Airport should be sited by taking into account the costs involved. Many examples of cost-effectiveness analysis can be found in health studies (Drummond, 1980; Drummond *et al.*, 1986). There are also many examples of case studies in developing countries. One case study was in Indonesia where cost-effectiveness analysis of an immunization programme was evaluated (Barnum *et al.*, 1980). Due to poor environmental quality and sanitation methods, diptheria, tetanus, cholera, tuberculosis and pertussis are common in many areas of Indonesia. A study was set up to determine whether it would be cheaper to treat certain infections or to prevent infections through the extension of the existing immunization programme. The work considered the direct health service costs including both capital and operation components. These costs were valued by using market prices. For example, hospital costs were estimated by multiplying length of stay by daily charge. Deaths averted were the chosen criterion and the assessment of effectiveness of the programme was made by extrapolation from current knowledge and by technical judgement. It was concluded that the immunization project would be highly cost-effective in death prevention compared with treatment.

There are some projects in which it is best to combine both cost–benefit and cost-effectiveness analyses, when some benefits are measurable, and others are not. Beijing Water Project is a good example (Hufschmidt *et al.*, 1983). This project compares various ways of dealing with water pollution in a southern suburb of Beijing where household, industrial and agricultural pollution heavily affect the surface water in rivers and canals. A clean-up is required to improve the quality of drinking water and water used by industry. A number of control measures in different combinations were considered for the cleaning-up operation such as eliminating sewage overflow from the treatment plant, separating part of the clean water from the sewage, primary treatment of sewage for irrigation, recycling of clean water used in industry, etc.

Each measure was appraised in two ways: on the basis of net present value results after comparing costs with measurable benefits and on the basis of water quality measured by oxygen demand and dissolved oxygen. Options were ranked in each method, and the alternative giving the highest score was chosen.

8.3 HEDONISTIC PRICE

In the absence of a market the hedonistic price approach measures the quality of the environmental services by observing the prices of surrogate goods affected by different environmental conditions. The origins of the use of hedonistic methods to measure the value of the environment can be traced to writings on the pure theory of public expenditure by Samuelson (1954). After that, various economists, using mainly housing and labour markets, measured various environmental factors such as air pollution (Anderson and Crocker, 1971; Harrison and Rubinfeld, 1978); noise (Nelson, 1979); social infrastructure (Cummings *et al.*, 1978); climate (Hoch and Drake, 1974); neighbourhood ethnic composition (Schnare, 1976); and earthquake risk and property prices (Brookshire *et al.*, 1984).

Property market studies which were elaborated by Rosen (1974) start from the assumption that property prices capture the quality of the environment. For example, other things being equal, one would expect properties in areas with clean air to command higher prices than houses in areas with polluted air. House prices differ for a number of reasons, but if the quality of the air is the only different factor between two houses it would then be reasonable to argue that the higher price for the house located in the clean air area explains a good proportion of the value placed on clean air.

Rosen's analysis sees houses as individual commodities that are differentiated by various characteristics, internal as well as external. Although the main body of the value for a house stems from the fact that it is essentially a living space, the unique characteristics that it contains affect the value quite substantially. Residents derive utility from the internal as well as external attributes of a house. Now the question is how much do the environmental factors affect the value of a house? In order to be able to answer this question we need empirical research involving econometrics. A multiple regression analysis to capture environmental effects on house prices could be based upon cross-section data, time series data or a mixture of both, known as the pooling technique.

A property value function is likely to have many variables such as living space, quality of accommodation, income levels of residents (actual as well as potential), availability factor, accessibility, level and quality of public facilities in the area, level and quality of other facilities, e.g. shops, churches, etc., rates (property taxes), air quality, noise levels, congestion factor including housing density as well as traffic, access to recreation areas, general view, and some other factors. In order to measure the impact of any of these variables on property prices they all must be included in the regression equations, i.e. the property value function must be comprehensively defined. Furthermore, the functional relationship in the regression model must be correct.

Some studies rely upon a few key explanatory variables in aggregate forms. Let us say that to start with the analyst used four explanatory variables:

$$PV = f(H, A, N, E), \qquad (8.1)$$

where PV is the property value, H is the housing variable, A is the accessibility factor, N is the neighbourhood factor, and E is the environmental factor.

An index number for each one of these variables can be calculated to carry out a regression analysis, say, on a cross-section basis. Butler (1982) discovered that a more comprehensive specification than the one containing a few variables would have a significant effect on the results. Furthermore, leaving out the less important variables would have little effect on the principle items.

Palmquist (1991) argues that the temporal stability of the hedonistic value equations must also be considered. If there is an active forward-looking market then observed prices will be safe to use unless new information becomes available. If forward-looking markets do not exist, temporal stability cannot be guaranteed. If there was a major change announced in an area, no doubt property prices would be affected. For example, the announcement of a motorway construction or the location of a waste incineration plant would be likely to change property prices in affected areas. If the study is based on time series data a lagged model, similar to the one used for forestry planting function (pp. 101–2), could be used as the full effect of the announcement is bound to take time. When a hedonistic regression function contains a large number of variables multicollinearity may undermine the results.

Hedonistic price analysis measures indirectly people's willingness to pay for/accept the changing environmental quality. If improvement was the issue then willingness to pay by residents or owners would be relevant. If an environmental deterioration is contemplated, willingness to accept a nuisance may be considered. For example, when granting a permit to build a waste incinerator the government may consider ways of compensating the households who expect a loss in the value of their property. Figure 8.1 illustrates the relationship between property values and environmental quality. In the case of improving environmental attributes, moving right along the horizontal axis, property prices increase. In the event of deterioration, moving left say from point E_1 to E_2, the property prices fall.

Extensive research has been carried out to find the effect of changing environmental quality on property prices. Pearce and Markyanda (1989), for example, report the effects of air pollution on property values in a number of North

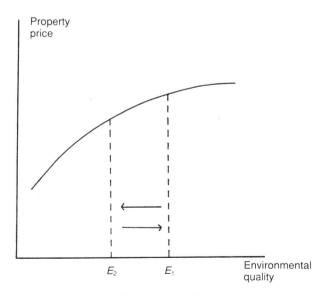

Figure 8.1   Property prices versus environmental quality.

Table 8.1 Air pollution and property prices

City	Pollution type	% Fall in property prices for % increase in pollution (pollution elasticity of hedonistic value function)
Chicago	Particulates and sulphation	−0.26 to −0.5
Los Angeles	Particulates and oxidants	−0.22

American cities. Table 8.1 summarizes their results in two large cities, Chicago and Los Angeles. According to this a 1% increase in air pollution in Los Angeles *ceteris paribus* would lead to a 0.22% fall in property prices.

Noise levels from aircraft, motorways, industrial activity and entertainment facilities also affect property prices. At 70 decibels noise becomes intrusive to the human ear, at 85 decibels it can damage hearing if exposure continues for 9 hours or more. Even at lower levels, annoyance and interruption of economic activities can take place, affecting production levels, which may lead victims to take defensive actions such as installing soundproof windows, walls, etc. Most of the economic literature focuses on aircraft noise, possibly because it is clearly identified within a geographical boundary. O'Byrne *et al.* (1985) make Atlanta Airport and the affected areas a case study. The airport is one of the busiest in the United States, handling over 1500 movements a day. The research project investigates the relationship between aircraft noise and discounts on house prices by using cross-section data at two different time points, 1970 and 1979/80, a sort of pooling technique. The result is that for each decibel increase in aircraft noise there is a 0.67% decline in house price in the affected area.

There are certain limitations to the hedonistic approach in both labour and housing markets. First, the method requires extensive and refined data which are not always available. A good proportion of the data are collected by a number of private and public bodies such as various environmental regulation bodies, trade unions, environment research laboratories, local authorities, private construction companies, and so on. Compiling a suitable data set can be time consuming as it usually involves obtaining information from various sources. This requirement restricts the application of hedonistic methods to developed market economics. In the early stages of housing studies in the United States researchers used census data on average owner-estimated house values which were easily available, but nowadays disaggregated data are required with more reliable value estimates such as actual rental prices. Second, as mentioned above, estimating the values of hedonistic parameters is highly sensitive to model specification and level of disaggregation. Using the same data set, different specifications such as linear, log-linear, translog, distributed lags, etc., can give different but equally acceptable results.

A third problem is that although some environmental attributes such as air quality can be measured fairly reliably, this cannot be said to be true for visibility, smell, turbidity, likelihood of industrial and environmental accidents, premature deaths, class factor in a neighbourhood, etc. Even when these problems are overcome, it is still unclear whether impartial or scientific measures are good proxies for individual perceptions. Furthermore, there is the so-called awareness factor. Although scientific instruments can pick up that the air quality in a district has been slowly deteriorating over a number of years, people in the area may not be aware of it or even if they are aware they may not take it to be too serious a problem. Especially in the housing sector when the effect of pollution is unclear, hedonistic research can become ambiguous.

Distortions in the housing market exist in most countries. For example, in the UK the public sector owns a substantial part of the housing stock and rents, essentially, are politically determined. Some regions are segmented by class, racial origin, occupation and religion. For example, in Northern Ireland religious division is forceful in certain districts and thus people live in segmented housing areas consisting of almost entirely one faith. Therefore, some people have little choice regarding the location they wish to live in. Sometimes people buy houses for speculative purposes, as in the UK in the 1980s when 'moving up the ladder' became a fashion. In normal times the transaction cost of moving house, i.e. fees for the estate agent, surveyor, solicitor, financial advisor, removal companies, etc., is a deterrent. Finally, in hedonistic studies on property it is assumed that the change in the environmental quality is fully absorbed into the value of the houses. Braden and Kolstad (1991) contend that this is a very strong assumption; essentially there is a weak relationship between the value of the house and the environment.

Similar points can be made for hedonistic studies on the labour market. It is well known that due to factors, such as sexual, racial and religious discrimination, restrictive trade union practices, monopsony, etc., labour markets function imperfectly. Work-related risks may be difficult to measure and workers may not be fully informed about risks in all cases. Some individuals, by their nature, are risk takers, others are risk averse. All these factors complicate the picture.

Despite all its problems, the hedonistic approach could be useful in many instances in providing information about improving the quality of the environment and conditions of work.

8.4 CONTINGENT VALUATION

An alternative approach to hedonistic price is to ask individuals, using survey methods, to state their willingness to pay for a certain environmental quality. For example, consider that a sparsely populated but valuable scenic area is threatened by smoke from a proposed coal-fired power station. The smoke would reduce the number of visitors following the construction of the power plant. In a case like this the hedonistic approach is unavailable because of the scarcity of local population which makes wage and property value data insufficient. Davis (1963), Bohm (1972), Hammack and Brown (1974), Randal *et al.* (1974) and Brookshire *et al.* (1976) were the earliest economists to use the survey approach to value environmental attributes.

The contingent valuation approach tries to identify the value that people derive from consuming an environmental benefit by questioning a sample of individuals in order to obtain their maximum willingness to pay to have the benefit or minimum compensation to go without it, i.e. willingness to accept a loss of benefit. It should be mentioned at this point that willingness to pay and willingness to accept are by no means the same. In effect, the divergence between the two concepts is wide and economic theory does not prepare us to anticipate this. Empirical research reveals that willingness to pay is several times, typically one-third to one-fifth, lower than willingness to accept (Winpenny, 1991). The answer to this lies in human psychology; apparently people value more highly the loss of something they already own, than the gain of something they do not yet have. Bishop and Heberlein (1979) studied the value of wildfowl to hunters in Wisconsin on the basis of day value criterion. In this, the willingness to pay was found to be $21 and the willingness to accept $101. In a recent survey it has been argued that marked differences between willingness to pay and willingness to accept have a strong theoretical basis stemming directly from neo-classical economics (Bateman and Turner, 1993).

Asking individuals to state their views about environmental attributes and their willingness to pay for improvement is similar to commercial market research. Before a new product is put on the market, researchers enquire about the preferences of potential buyers regarding colour, taste (if applicable), packaging, etc., and their willingness to pay for various qualities. In the case of environmental attributes such as vistas, clean air, noise levels, etc., there is no direct market but nevertheless individuals could be asked about the goods in question as if a market existed.

The questionnaire method normally contains 'bids' which are close to what would be revealed if a market existed. One important requirement is that the environmental attribute in question must be known to the respondent. It is customary to start from an initial bid, the starting price, to find out whether or not the respondent accepts it. If he/she agrees with the starting point then an iterative process begins, in which an increase in the base price takes place step by step until the individual declares that he/she is not willing to pay the extra increment in the bid. The last accepted figure is the respondent's maximum willingness to pay. In the event of willingness to accept, the process is reversed and bids are systematically lowered until the individual's minimum willingness to pay is reached. Enquiries can be conducted by using experimental techniques in laboratory conditions as well as by using survey methods.

Apart from individuals' willingness to pay or willingness to accept, enquiry can be conducted to identify other relevant items such as respondents' income levels, education, age, gender, cultural background, tastes, familiarity level with the attribute in question, etc. On the basis of acquired information a regression analysis can be conducted. A general willingness to pay function for individual i is likely to be:

$$WTP_i = f(Q_i, Y_i, T_i, S_i), \tag{8.2}$$

where WTP_i is willingness to pay, Q_i is quality/quantity of the attribute, Y_i is the income level, T_i the index of tastes and S_i is a vector of relevant socio-economic factors. For a discussion of this methodology see Cummings *et al.* (1986) and Hanley (1988).

There are a number of factors which led some economists to express concern about the contingent valuation method. They are as follows:

Strategic bias This arises when the individual provides a biased answer in order to influence a particular outcome. This problem is known in public sector economics as the free rider problem; that is, if a person has to pay on the basis of his or her stated willingness, he/she may try to conceal the true figure by stating a much smaller one in order to qualify for a lower price. If individuals are told that due to the removal of a subsidy on a recreational facility additional funds are required by users on the basis of their willingness to pay, it is quite possible that each person will understate his/her price. Or, alternatively, if a decision to preserve a lake for recreation purposes depends upon whether or not the survey produces a sufficiently large value, then respondents may be tempted to overstate their price.

Some economists, such as Bohm (1972) and Schulze *et al.* (1981b), contend that strategic bias could be eliminated in a simple way. Instead of telling individuals that they will be charged according to their stated willingness to pay, they should be told that each one will pay the average bid price. It would then be difficult for individuals to tailor a price for their advantage. However, there would still be a problem with this type of survey. Each individual may think that if he or she understates his/her price the average will be smaller than the true figure and thus they, in the end, will pay less.

Information bias This may arise when individuals are asked to value attributes with which they have very little or no experience. For example, if an individual is asked to value a lake which is used for recreational fishing and if that person has no experience of fishing on that particular lake, or perhaps he/she may not be too keen on fishing, the valuation will be based upon an entirely false perception. Furthermore, new information may affect the bidding behaviour of individuals. For example, when individuals are told that the aggregate willingness to pay is not sufficient to keep the service in operation, then they may revise their bids (Rowe *et al.*, 1980; Schulze *et al.*, 1981b).

Instrumental bias It was mentioned above that the researcher starts from a base level and goes up in order to identify the respondent's willingness to pay. It may be possible that the starting bid could be suggestive, leading the respondent to settle at a figure around it. For example, a starting bid of £0.5 may yield quite a different result than £3. Furthermore, the range could also have a similar effect; a range of £0.5–£30 may produce a different valuation from respondents than a range of £0.5–£100.

Hypothetical bias This occurs when individuals are faced with contrived as opposed to actual choices. In the latter, when errors are made there is a cost; regret at having paid more than they should have, i.e. the feeling of having been 'taken for a ride'. In the former, since individuals will not have to pay the estimated value they may treat the experiment or the survey casually. Some experimental work has been done to determine how serious hypothetical bias is (Cummings *et al.*, 1986).

Payment bias The proposed method of payment for the environmental attribute can affect the respondent's answers. Payment by way of entrance fees, taxes, surcharges, rates, higher prices for goods, etc., can have an impact. Some individuals may be sensitive to tax changes. For example, they may prefer a, say, £50 annual user charge to £50 annual tax.

8.4.1 An example using contingent valuation

There are numerous case studies to value environmental attributes, some of which are reported below. One recent study is by Hanley (1989) who estimates non-market recreation benefits derived by visitors to a part of the Queen Elizabeth Forest Park in Central Scotland by using the contingent valuation method. The forest park, which covers 17 000 ha, is in the Loch Lomond area, easily accessible from the population centres of Glasgow, Edinburgh, Sterling and surrounding districts. It is estimated that the park is visited by 145 000 people each year.

For the study, 1148 questionnaires were collected, partly by interviewing visitors and partly by respondents filling them in and putting the completed forms in boxes provided in the area. The main objective was to value wildlife, landscape and recreational facilities by requiring respondents to answer a number of questions:

1. Suppose the management decided to construct a hide, from which you could view wildlife at close quarters. If the only way to fund this project was an entrance fee, what would be the most you would be prepared to pay, per person, per visit?

   £0   £0.5   £1.0   £1.5   £2.0   £2.5   £3.0

   or other (please specify)

2. Currently a charge of £1 is made for use of the forest drive which is 7 miles long. If the only way to keep this drive open was to increase the fee, what is the most you would be willing to pay in total, per visit?

£0   £1.0   £1.5   £2.0   £2.5   £3.0   £3.5

or other (please specify)

3. The trees around one of the main centres are of high timber value. Suppose the management was faced with the choice of either felling the trees now or increasing the fee to the area. How much would you be willing to pay in total, per person, per visit, to save the trees?

£0   £0.5   £1.0   £1.5   £2.0   £2.5   £3.0

or other (please specify)

4. Suppose the management was considering selling the entire recreation area to a private company which would mean you would no longer be able to visit. If the only way to stop this was to raise money by selling day tickets to visitors, how much would you be willing to pay, per person, per trip?

£0   £0.5   £1.0   £1.5   £2.0   £2.5   £3.0   £3.5   £4.0

or other (please specify)

The contingent valuation results are shown in Table 8.2.

Table 8.2   The contingent valuation results

Item	WTP for hide (Q.1)	WTP for drive (Q.2)	WTP for landscape (Q.3)	WTP for forest (Q.4)
Mean (£)	0.84	1.58	0.8	1.25
Standard deviation	0.57	0.72	0.59	0.87
Range (£)	0–5.0	0–5.0	0–5.0	0–10.0

By taking the estimated 145 000 visitor days per annum and 6978 cars for the forest drive, we get figures for the aggregate willingness to pay for the hide, drive, landscape amenity and whole forest.

Resource	Aggregate WTP per annum (£)
Hide	121 800
Forest drive	11 026
Landscape value	116 000
Whole forest	181 250

Respondents were further questioned about their ages, income levels, travel distance, taste with regard to amenities and number of trips per year. Various multiple regression functions were specified in the light of equation (8.2) to estimate various parameters of the willingness to pay function. A semi-log function was found to fit best but explanatory power of all forms turned out to be low and some variables turned out to be insignificant.

Survey results obtained by way of the contingent valuation method appear to be consistent with demand theory (Schulze *et al.*, 1981b), but how do they compare with the figures obtained by hedonistic studies? The answer to this question was sought by Brookshire *et al.* (1982) who used both hedonistic property values and survey methods to value different air qualities in Los Angeles metropolitan area. The main reasons for the choice of venue were the existence of a well-defined air pollution problem and the existence of detailed property value data.

In the study 290 household interviews were conducted in 1978. Respondents were asked to state their willingness to pay for an improvement in the quality of air in their region. The quality of air was defined as good, fair or poor in the study area. Respondents in poor air regions were asked to value an improvement to fair quality. Households in fair air regions were asked to value an improvement to good quality. Those in good air areas were asked to state their willingness to pay for a region-wide improvement in air quality. As for the hedonistic price approach, 634 single family home sales in 1971 and 1988 were used in exactly the same area in which the contingent valuation was done. This revealed that people paid more for homes sold in better quality air regions. Hedonistic valuation results turned out to be higher than the contingent valuation survey figures, as had already been predicted in theoretical models (Brookshire *et al.*, 1982).

8.5 TRAVEL COST ANALYSIS

The origins used of travel cost analysis to value environmental attributes can be traced to Hotelling (1947). If a person travels to a recreation site which is provided at, say, a zero price, then he/she values it at a rate at least equal to the cost of getting there. If there is an admission charge this must be added on to the cost of travel to obtain the willingness to pay for the attribute. Since travel cost varies from person to person it is then possible to think in terms of a demand function.

Figure 8.2 shows the basic idea. For a recreation forest that attracts motorists from various locations, the cost of travel must be different for each motorist travelling from a different point. Travellers from a distant location pay higher costs than travellers from a closer area. Let us say that the cost of travel from a distant location is P_A and cost from a closer area is P_B on average. If we have a number of observations for travellers in two districts, *ceteris paribus*, we would obtain a downward sloping demand curve. People from both regions capture a consumer surplus. The surplus captured by travellers from the distant region is A whereas it is A+B+C for travellers from the region close to the forest. The willingness to pay by each group of travellers equals travel cost incurred, plus

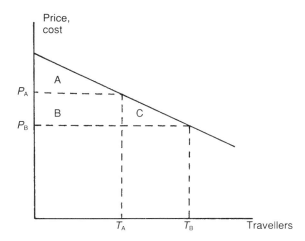

Figure 8.2   Consumer surplus and willingness to pay by travellers.

the relevant consumer surplus. The travel cost method gives best results when applied to well-defined recreational sites (Bateman, 1993).

The travel cost approach is different from the contingent valuation method in that the behaviour of subjects is observed in real markets rather than in hypothetical circumstances. The travel cost approach has been widely used in economic evaluation of transport projects. In Britain, the London–Birmingham motorway project, the M1, was the first major public sector project subjected to a rigorous cost–benefit analysis. This retrospective study was carried out by Beesley *et al.* (1960) under the auspices of the Road Research Laboratory. The M1 project, in the main, was a methodological study, but the principles established there have been widely used on many other transport studies, such as the Tay Bridge, the Severn Bridge, the London Victoria Underground Line and the Cumbrian Coast Railway Line.

Transport projects can be designed for a multitude of purposes such as to accommodate an increasing number of passengers and freight, to reduce journey time and to increase comfort and convenience. When a transport project is designed to save journey time the primary benefits are normally measured by the willingness to pay of the travellers in order to save that amount of time. A popular approach is to assume the value of an individual's time as being the money he/she could earn during the time saved. If a person's average hourly earning is £5, a transport project which saved 100 hours a year would be worth at least £500 per annum to that person. The rationale for this calculation derives from the individual's supply schedule of labour (Mishan, 1989). A similar logic applies to the valuation of environmental attributes when the travel cost method is used.

There are a number of problems associated with travel cost time, especially when putting a value on leisure time. In some cases leisure time values are obtained from studies of people's behaviour in situations where they could choose between slower and cheaper or faster but more expensive forms of transport. This generally relates to the revealed preference approach in consumer demand theory. In their economic evaluation of London's Victoria underground line, Foster and Beesley (1963, 1965) gave hourly work-time saved a price of 36.3 pence whereas leisure time was measured at a lower rate, 25 pence per hour, in terms of 1962 prices. The Department of the Environment in some studies used a figure for non-working time of 25% of average hourly wage rates (Harison and Taylor, 1969). Some recent studies tried to distinguish between those giving up possible earnings and those not (Bockstael *et al.*, 1987).

In addition to valuing leisure time there are other problems associated with the travel cost approach. Deciding on the functional form of the econometric model appears to be a serious issue. Hanley (1989), by using different specifications, obtains widely different results for evaluation of recreational benefits at the Queen Elizabeth Forest in Scotland. Furthermore, there is the question of how to apportion the travelling cost of a party when it contains individuals with different characteristics such as children, unemployed, pensioners, professionals, etc. In addition, when 'wanderers' – people who have no specific aim to visit a particular site – are encountered in the survey should we treat them the same as other respondents? It is further assumed that all users get the same total benefit from the use of the recreational facilities. In effect, their consumer surplus may not relate to the distance that they travel to an environmental attribute. Finally, the travel cost model does not readily capture changes in environmental quality.

Normally in the travel cost method the area surrounding the environmental attribute is divided into zones. Visitors are sampled to identify their zone of origin then for each zone visitation rates are calculated. A trip generation function is identified to fit the observations. A general function can be:

$$V_i = f(C_i, Y_i, \overline{C}_{ij}, Z_i),$$

where V_i is the number of visits from each zone; C_i is the travel cost; Y_i is the income of visitors in zones; $\overline{C}_{ij}$ is the cost of visiting substitute recreational sites from each zone; and Z_i are the socio-economic factors thought to be significant.

8.5.1 An example using travel cost analysis

As well as using the contingent valuation method to estimate the willingness to pay by visitors to the Queen Elizabeth Forest in Scotland, Hanley also used the travel cost approach for the same purpose. The study was carried out on the basis of respondents who intended to visit the area, and there were 319 in the sample response. 'Wanderers' were therefore excluded. A number of trip generating functions were specified; the most suitable one turned out to be a semi-log one which is:

$$\ln V_i/P_i = A + \alpha\, TC_i, \qquad\qquad (8.3)$$

where V_i/P_i is visits per capita from the ith zone; A is a constant; and TC_i is round-trip travelling cost plus entry fee from zone i.

Other explanatory variables such as income, substitute sites and socio-economic factors were not included. A travel time parameter was included which turned out to be insignificant. The results were:

$$\ln V_i/P_i = -2.6 - 0.6\, TC_i, \qquad \bar{R}^2 = 0.37.$$
$$\quad\ (6.06)\ \ (3.41)$$

Estimation of total consumer surplus turned out to be £160 744 per annum, which is reasonably close to the £181 250 estimate calculated using the contingent valuation method.

8.6   EXISTENCE VALUE

Existence value is not related to the actual or potential use of an environmental attribute but to its very existence. This was first suggested by Krutilla (1967) and was implicit in Weisbrod's article on option value which will be discussed at the end of the next chapter. The very fact that a natural attribute exists gives utility to people and therefore gives it a value. There are a number of interesting dimensions to this concept. First, even if people cannot actually visit an environmental phenomenon, say wildlife in Kenya, they nevertheless gain pleasure from pictures, films, broadcasts and writings. Some call this the vicarious use value in which people acquire benefits by indirect means. Randall and Stoll (1983) contend that although vicarious use value is a kind of use value, it is difficult to distinguish it from the existence value.

Second, existence value can be seen as an intrinsic value that is in the real nature of the attribute. At this point intrinsic value becomes a kind of sacred or divine concept. Even if a natural phenomenon is unrelated to humans it still has a value and humans are capable of appreciating this. Existence value is understood and well respected in many cultures. In the Western world the existence of protection funds for endangered species, threatened coasts, ancient forests, unique or historic landscape are an indication of this value. These funds often originate from the country or the region where the attributes are found, but sometimes they come from distant areas. For example, an Irish person who has never been outside Ireland and has no intention of doing so may willingly contribute to a wildlife protection fund in the Amazon. They may do this because they believe that the maintenance of biodiversity in the Amazon is essential for human survival and that one day an answer to one of our widespread diseases may be found there, or they may do it because they believe that all species are God's creatures and are thus sacred. All environmental attributes have intrinsic or existence value. Economists have not yet begun to do serious research in these domains in spite of the fact that existence value provides one of the building bridges between economics and the environmental scientists or indeed other groups such as moral philosophers, theologions, etc.

One way to identify existence value is to use contingent valuation surveys. However, it has to be confessed that a survey attempting to identify the global willingness to pay for environmental attributes is very difficult if not impossible. Existence value levels are inevitably influenced by religious and cultural positions and education. In some communities concepts such as biodiversity may be unheard of. People tend to assign high value to an attribute that exists locally, partly due to the fact that it relates to their lives, childhood sentiments, etc., but more importantly they tend to have good knowledge of it.

8.6.1 An example using existence value

In 1983 Brookshire and co-workers published the results of a survey aimed at identifying the existence value for big horn sheep and grizzly bears in the state of Wyoming. Both species, although highly valued by the community – especially by the hunters – were threatened in the region. The survey targeted hunters who could not be certain about the adequate supply of these species in future years.

A sample of potential users was asked to make bids corresponding to the cost of a hypothetical hunting licence in the area. Protection schemes of 5 and 15 years were assumed before hunting could take place. Potential hunters were asked to respond to changes in probability of supply. The results revealed that bids increased along with the probability of supply. The results of the survey were as follows:

Existence value	Bears		Sheep	
	5 years	15 years	5 years	15 years
Average bid $	24.0	15.2	7.4	6.9

8.7 BEQUEST VALUE

This concept produces a willingness to pay at present point in time in order to ensure that certain values are maintained and made available to future individuals. If these individuals are immediate descendants then the respondents would be fairly confident at guessing the nature of the beneficiaries' preferences. However, it would not be too difficult to make reasonably accurate guesses about the preferences of distant generations on basic issues such as clean air, clean water, maintenance of natural wonders, soil fertility, etc. In addition to environmental assets there can be little doubt that future generations would want to live in free, tolerant and just societies and thus institutions that support these values need to be maintained.

8.8   OPTION VALUE

There are a number of different interpretations of this concept, which was founded by Weisbrod (1964). In the main, the concept relates to the preservation of unique natural assets, i.e. the wonders of nature, which will be discussed in the next chapter. For a wider discussion on various interpretations of the option value see Johansson (1991).

8.9   RISK AND UNCERTAINTY IN VALUATION OF ENVIRONMENTAL ATTRIBUTES

A substantial degree of risk and uncertainty confronts economists as well as public and private sector policy-makers in assessing the value of projects involving environmental attributes. It is now standard practice in the public sector to evaluate investment projects that could create environmental externalities by using cost–benefit analysis techniques as part of a general fact-finding mission and aid to policy-making. As explained on pp. 237–8 environmental parameters have become an important part of cost–benefit analysis only recently. At project level, risk and uncertainty are relevant not only in determining overall financial viability but in putting value on wider environmental effects. For example, risk and uncertainty would be relevant in assessing whether or not a land drainage project would generate sufficient income for the farmers in the catchment area to meet its construction and maintenance costs. For various unforeseen reasons construction could be delayed, for example by labour or planning disputes, which would increase the cost; agricultural commodity prices could fall substantially halfway through the project's life, making earlier benefit estimates irrelevant; a large-scale drainage project could undermine recreational fishing in the affected areas. When a project's life is long, stretching up to 50–60 years (the River Maigue Drainage Scheme in Ireland has an estimated life of 58 years), forgone recreational benefits are also subject to risk and uncertainty. By looking at past trends it is possible to identify one set of figures but these could change drastically if the surrounding area becomes a major tourist attraction in later years.

In some part of the economic literature risk and uncertainty are used interchangeably but in fact they are different. The former relates to a situation in which the probability distribution of an event or the value of a variable is known. In other words, a risk situation would exist when the value of a variable, say benefit, is not known precisely but its probability distribution is known. For example, total benefits arising from an investment project may have 70% probability of being £1 million, 20% £1.5 million and 10% £0.5 million in a given year. Uncertainty relates to a situation in which the probability distribution is not known. In cost–benefit analysis a risky situation is more manageable than one with high uncertainty. Risk has been described as measurable uncertainty.

In the early days of cost–benefit analysis some practitioners handled risk and uncertainty by inflating the rate of discount. Let us say that for a relatively risk-free project a 5% discount rate has been used over its lifetime to calculate a net present worth. For a moderately risky project, a rate of 7%, and for a high risk one 10–12% or even higher figures may have been argued for. However, this practice assumes a specific kind of relationship between time and uncertainty. That is, the degree of risk increases exponentially with time. There may be no reason to believe that the risk profile of all projects takes this particular shape. It may be quite the opposite that the risky period for some projects could be the early stage, i.e. construction phase. Once a project is firmly established the rest could be relatively plain sailing.

Nowadays the practice of adding a risk premium on the rate of discount is not recommended (Pearce *et al.*, 1990). Instead, most practitioners believe that risk and uncertainty are better handled by way of adjustments to cost and benefit profiles leaving the underlying discount rate unadjusted for risk. Some even argue that communal projects should be taken as risk-free due to a large portfolio situation. As some projects fail to achieve their expected results in full, some others can over-materialize and in this way there will be an overall compensation.

A more acceptable practical approach to the treatment of risk and uncertainty is to test the sensitivity of the outcome of project evaluation to variation in the magnitude of key parameters (Brent, 1990). In order to be meaningful, it is best to use both pessimistic and optimistic values which differ from expected values by a factor proportional to the degree of risk and uncertainty. A shift in project feasibility as the analyst goes from optimistic and pessimistic projections implies existence of a break-even point between the two extremes. If the probability distribution of parameter values can be determined, it gives the probability of the parameters falling on the feasible side of its break-even value. Then the policy-maker must decide whether this probability is sufficiently large. If the probability distribution cannot be determined, a subjective evaluation of the parameter's value, lying in the acceptable range, may be made.

Decision-makers dealing with environmental projects are often faced with a wide range of problems such as predicting long-term environmental effects and assessing public attitude to those effects, as well as measures of environmental costs and benefits. Of course, some of these problems can be moderated by further research. This may or may not be feasible as research on environmental issues could be costly and time-consuming and decision-makers normally work under the pressure of time. On occasions policy-makers deal with development projects involving irreversible consequences, such as conversion of a tropical forest to agriculture involving destruction of plant and animal species which may have a great scientific value to future generations. Although the extent of these benefits may currently by shrouded in uncertainty, this may change as knowledge grows regarding the beneficial uses of tropical forest species. If conversion takes place this kind of genetic information would be gone forever.

Preservation of the forest, on the other hand, would create an option to under-stand the value of species. This is called the quasi option value. Fisher and Hanemann (1983) imply that quasi option value could be substantial in magni-tude to justify preservation. Taxation imposed on development projects could also be useful to moderate the pace of development to allow accumulation of information. Also it has been suggested that development projects can provide full information as to the potential value of species which may be understood in the distant future (Viscusi and Zeckhauser, 1976; Freeman, 1984; Miller and Lad, 1984).

Various other methods have been proposed to deal with risk and uncertainty involving environmental attributes. Risk–benefit analysis mainly focuses on the prevention of incidents carrying serious environmental risks. This may be viewed as the inverse of cost–benefit analysis as it starts by presuming no action (Winpenny, 1991). The cost of no action is the likelihood of a serious environ-mental accident occurring; the benefits would be saving the cost of preventative measures. Inaction would be justified if the costs were greater than the benefits.

Acceptable risk analysis is another criterion which tries to find out 'how safe is safe enough?' (Fischhoff et al., 1981). Some environmental projects involve a high degree of uncertainty due to ignorance. For example, we do not know pre-cisely how the ecosystem works. In sacrificing an environmental asset there could be a knock-on effect which may eventually lead to the breakdown of a system. In the case of a health project designed to prevent illness from, say, air pollution, epidemiological data may be imperfect or completely absent. Furthermore, the relationship between certain air pollutants and ill-health may be hotly disputed in the medical profession. Acceptable risk analysis seeks the opinion of various groups within a given profession, affected groups, and poli-cy-makers themselves. It has to be confessed that gathering this kind of infor-mation can be time-consuming.

Decision analysis criterion first identifies the objectives of the project, various ways of achieving these objectives, setting performance measures and identify-ing critical risk/uncertainty parameters. Decision analysis, like acceptable risk analysis, tries to incorporate the views of various groups such as specialists, non-specialists and decision-takers. Here the views are taken as subjective rather than objective factors. Winpenny (1991) contends that decision analysis can be incorporated into cost–benefit analysis. 'It would call for a more thorough and explicit analysis of possible outcomes and decision makers' preferences than is usual in traditional cost–benefit analysis, but this would greatly add to the credibility of cost–benefit analysis where there are significant environmental risks.'

9

Economics of natural wonders

In the case of the Hell's Canyon project, and quite probably in other similar proposals, both theoretical and empirical considerations suggest that it will not be optimal to undertake even the most profitable development projects there. Rather, the area is likely to yield greater benefits if left in its natural state.

A. Fisher, J. Krutilla, C. Cicchetti

9.1 SOME NATURAL WONDERS

There are certain features of nature which are generally accepted to be breathtaking. A number of terms are used to describe these resources: unique natural areas, outstanding natural phenomena, national parks, exquisite natural resources and natural wonders and a few examples are the Grand Canyon, Hell's Canyon, Yosemite, the Sequoia forests, the Yellowstone area, the Everglades, all in the United States; the Great Barrier Reef in Australia; the Fairy Chimneys of Cappadocia in eastern Turkey and the Giant's Causeway in Northern Ireland. This list is by no means exhaustive. Most nations consider parts of their territory as breathtaking. For example, in England and Wales there are 36 stretches of countryside designated as areas of outstanding natural beauty (Anderson, 1987). Strictly speaking they cannot be put into the natural wonders category as their form is largely the product of human management. A brief description of some natural wonders is given below.

9.1.1 The Grand Canyon

The Grand Canyon is part of the Colorado Plateau located at the north-west corner of Arizona in the United States. The canyon is cut by the Colorado river and is noted for its fantastic shapes and coloration. It is over 320 km long, 90 km of which lie within the Grand Canyon National Park, which was established in 1932. The canyon is over 1.5 km deep in parts and its width varies from 6 to 29 km from rim to rim. The Grand Canyon National Park, which is visited by millions of people every year, covers 800 km². From Toroweap Point, where the canyon is 900 m deep and 6 km across, there are excellent views of the inner gorge and the snake-like Colorado river below.

The Grand Canyon is the world's best exhibit of erosion, the result of cutting and grinding by fast-flowing mud and rock laden water, aided by frost, wind and rain. It has the world's most exposed geological timetable. The canyon wall from bottom to rim represents a period estimated at 700 million years. The whole panorama is a riot of colours from the mineral stains and mineral salts originally in the sediments. Figure 9.1 shows the structure of the canyon. The variable resistance of strong and weak rocks in the sedimentary formations has created an angular pattern of scarps and benches that are especially characteristic of the area.

The region is arid because it is shielded from oceanic influences by the Pacific mountains, and only occasionally does maritime air penetrate beyond these mountains. Generally speaking, the mean annual rainfall does not exceed 25 cm. This means that the canyon has been wonderfully preserved by the aridity of the climate.

Near desert conditions have also contributed to the restricted width of these spectacular land formations since normal weathering agents have had little opportunity to wear back the valley sides. Apart from being a major tourist

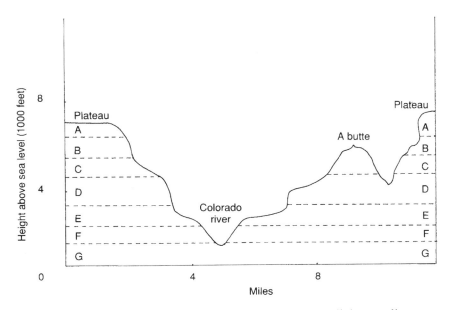

Figure 9.1 Geological section across the Grand Canyon. A, light grey limestone; B, grey sandstone; C, red shales and red sands; D, blue–grey limestone; E, green shales; F, coarse sandstone; G, granite.

Figure 9.2   A view of the Grand Canyon.

attraction the Grand Canyon is of great scientific interest to geologists and biologists as it gives a time scale of the earth's history and is also a wildlife laboratory containing numerous species of plants and animals. Figure 9.2 shows a section of the Grand Canyon.

9.1.2 Hell's Canyon

Hell's Canyon, the world's deepest gorge, is cut by the Snake river along the Oregon–Idaho border on the Columbia Plateau. The plateau extends between the northern Rockies and the Cascade Mountains and occupies parts of Washington, Oregon and Idaho states. It was formed by flows of dark volcanic lava and is trenched by the Columbia and Snake rivers. The Snake river is the largest tributory of the Columbia river and one of the most important streams in the Pacific north-west section of the United States. It rises from the mountains of Wyoming and flows through Utah, Nevada, Idaho, Oregon and Washington. From an elevation of 3000 m the Snake river descends to an elevation of 90 m and has upper, middle and lower sections. The lower Snake river, from Weiser to the mouth, flows through the 1600 m deep gorge of Hell's Canyon.

Hell's Canyon has a total length of 200 km. For 64 km it is more than 1.5 km deep, which makes this canyon a spectacular gorge. Parts of the canyon are coloured in shades of yellow, red and orange. In parts, the canyon walls rise vertically for several thousands of feet.

After much controversy between private developers and conservationists over damming the Snake river, mainly to control floods in the Columbia river, the Idaho Power Company built three dams in the area between 1959 and 1968. The rivers in this area are youthful and their terraces line the valleys. They often make good farm land, or benches to carry railways and roads. The area has a relatively long growing season, over 150 days are free from frost. Because of its power-generating and agricultural potential the canyon has been the centre of controversy between developers and environmental groups for many decades. Apart from the three power projects which have already been established, there were many new proposals in the 1960s and 1970s. However, in 1975 the United States Congress named much of Hell's Canyon as a wild and scenic area, stating that no dams would ever be allowed there. The theoretical rationale for this decision, which is the main purpose of this chapter, will be explained below.

9.1.3 Yosemite

The Yosemite National Park is one of America's premier natural wonders. Located in the Central Sierras in the state of California, it is a spectacular gorge, eroded mainly by a series of alpine glaciers. It was first discovered by Europeans in 1833, but it was not until 1851, when a battalion of miners pursued a band of marauding Yosemite Indians into the chasm, that its existence was finally confirmed and widely publicized. Within a few years the soaring cliffs and impressive waterfalls of the Yosemite had become the most famous natural wonder in the west of America.

There was immediate concern about the preservation of the area. In 1864 a small group of Californians, anxious to preserve the phenomenon from growing private abuse, persuaded their senator, John Canness, to propose legislation to set aside the Yosemite for public recreation. Eventually in that year a Yosemite park bill passed Congress and on 1 July 1864 President Abraham Lincoln signed the measure into law. Under this law the state of California received the area for public use, resort and recreation, provided that the land remained inalienable for all time. Indeed the Yosemite campaign marked the beginning of an idea of nationwide importance that governments should protect unique natural areas for the general public.

The Yosemite has a number of spectacular attractions such as steep rock walls, high waterfalls, huge domes and peaks. The greatest of these is El Captain, a granite buttress that rises over 1000 m from the valley floor. The plant life in the park changes with altitude. Lower levels are characterized by a mixture of scattered deciduous and coniferous trees. At the level of Yosemite Valley, there are large stands of coniferous trees. Higher up towards the tree line there are hemlock and lodgepole pine. The park's animal life includes black bears, chipmunks, deer and various squirrels. There is also a village near the falls.

9.1.4 Sequoia forests

Shortly after the discovery of the Yosemite, the Sierra Nevada Mountains yielded up another of its great natural secrets, the giant Sequoia (redwood trees). These are among the world's largest and oldest living things. Some trees are 3000–4000 years old, with a diameter of more than 10 m and a height of about 85 m. In 1852 some miles north of the Yosemite region, a hunter stumbled on a grove of these giant trees. After that the news spread rapidly across the country and even across the Atlantic to Europe, although in Britain many people discounted this news as a Yankee invention.

In 1890, to protect these trees, the region was set aside as a public national asset. The national park covers 4776 km², extending from Kings river in the north to the Mojave Desert in the south. The area provides timber, water, forage, wildlife and recreation. There are also canyons, caves, domes, mountain lakes and streams.

There are, in fact, two belts of Sequoia trees, one along the coast and the other midway up the western slopes of the Sierra Nevada, the Sequoia National Park. There is very little difference in appearance between the two belts of trees. In the early days of California's settlement these huge trees were targets for the lumbermen and many fell victim to their axes.

9.1.5 The Yellowstone area

The Yellowstone National Park, located in the north-western corner of Wyoming, was established in 1872. It is lacking in spectacular mountains but contains high-altitude lakes, waterfalls, canyons and geysers (hydrothermal hot springs). It is the nation's largest national park covering about 900 000 ha, mostly broad volcanic plateaux with an average elevation of 2400 m. Many of the geysers erupt to heights of 30 m or more. Old Faithful, the most famous, erupts fairly regularly every 30–90 minutes.

Most of the park is forested, the most common trees being lodgepole pine. The area is full of wild flowers, most of which blossom in the late spring. The dominant animal species are buffalo, deer, moose, bears, coyotes and rodents. Many hundreds of species of birds live in the park. The lakes and streams are stocked with fish. The Yellowstone river which rises on the slopes of Yount Peak enters the national park and feeds into Yellowstone lake. Below the lake it plunges into two spectacular waterfalls and enters the Grand Canyon of the Yellowstone. The Yellowstone lake has a shoreline of 180 km, a maximum depth of 90 m and a surface area of 360 km², lying 2300 m above sea level. It is the largest high altitude lake in North America.

9.1.6 The Everglades

The Everglades is a subtropical marsh region covering about 10 300 km² of southern Florida. It extends from Lake Okeechobee south-west to the Big

Figure 9.3   The Everglades.

Cypress Swamp and the Gulf of Mexico, south through the Everglades National Park on Florida Bay, and east to the vicinity of the Greater Miami metropolitan area. The area is only a few feet above sea level and is almost uninhabited. These low-lying swamplands consist mainly of black, salt-water muds in which many hygrophytic trees grow (Figure 9.3).

The state of Florida is an important citrus-growing region of the United States and the contribution of this activity to the regional economy is very substantial indeed. Over the years the areas of citrus production have gradually shifted southward from the interior towards the coast, mainly due to the frost-free conditions further south. The swampy soils and standing water of the Everglades have prevented most citrus production from moving further south into the effectively frost-free region of the state. In the past there have been many proposals by citrus and sugarcane growers to drain the swamps of Florida which have been opposed by conservation groups. The opposition was based on the widespread disruption to the complete ecological system of the vast marshlands that would result.

9.1.7 The Great Barrier Reef

This wonder is the largest coral reef in the world, extending more than 2000 km off the north-east coast of Australia. It is made up of many thousands of individual reefs, shoals and islets. This great system of coral reefs and atolls owes its origin in part to Pleistocene changes in sea level, but in the most part to long-continued subsidence, related to the faulting of the offshore region. This slow subsidence has enabled a great thickness of coral to develop, and it is on this basement that the present reefs and coral atolls have grown in the clear warm waters of the Coral Sea.

9.1.8 The Fairy Chimneys of Cappadocia

The Fairy Chimneys resulted from volcanic eruptions of Mount Erciyes and Mount Hasan in Cappadocia, eastern Turkey, about three million years ago. These eruptions covered the surrounding plateau with tuff from which the wind and rain have eroded Cappadocia's spectacular surrealist landscape. Soft but durable rocks were turned into tall pillars each wearing a large flat stone hat, often at a rakish angle (Figure 9.4). Colours ranging from warm reds and golds to cool greens and greys add to the surreal quality of the place. The area is spectacularly beautiful and undeniably strange – a Walt Disney landscape given an additional dash of science-fiction style just for good measure.

The area is fertile and has been settled since 400 BC. It is known that some of St Paul's earliest converts hollowed out rock chapels in the area, and others were added during the persecutions of Diocletian and Julian under Roman occupation. Despite human settlements the area is not spoiled. It is one of those rare regions in the world where works of people blend unobtrusively into the landscape.

9.1.9 The Giant's Causeway

The Giant's Causeway is a promontory of columnar basalt on the northern coast of County Antrim, Northern Ireland. Its prismatic, mostly irregular hexagonal forms were caused by the rapid cooling of the lava flows at their entry to the sea. The columns vary from 38 to 50 cm in diameter and some are 6 m in height. In places, the Causeway is 12 m wide and is highest at its narrowest part. The most remarkable of the cliffs is the Pleaskin, the upper pillars of which are 120 m high. Local folklore ascribes its formation to a race of giants who built it as a roadway to Scotland where a similar structure occurs.

9.2   A THEORY OF NATURAL WONDERS

What can we do with unique natural wonders like the Grand Canyon, Hell's Canyon, Yosemite, the Sequoia forest, the Yellowstone area, etc. Basically two things: we can preserve them in their original state or turn them over to developers. For example, we can drain the Everglades for agricultural and residential development or preserve it for its scientific, amenity and recreational value for present and future generations. Likewise, Hell's Canyon can be turned into a gigantic multipurpose river development project providing energy, irrigation, flood control and water sports facilities. Indeed, as mentioned above, there were many proposals of this nature in the 1960s and 1970s for both wonders. Similarly we

Figure 9.4   The Fairy Chimneys of Cappadocia.

can clear-cut the redwood forest for timber and farm the area. It may also be tempting to argue that a partial development would strike a compromise between the two extremes although conservationists would assert that a wonder is an indivisible unit.

There is a vast literature dating back to the 1930s on cost–benefit criteria for water resource projects, some of which involve natural wonders. Unfortunately, many economists who contributed to that literature said almost nothing about environmental matters. Only from the 1960s onwards did economists start to puzzle over matters regarding preservation. The theory which is presented here stems mainly from the works of Davidson *et al.* (1966), Krutilla (1967), Gannon (1969), Smith and Krutilla (1972), Fisher *et al.* (1972) and Arrow and Fisher (1974). This literature developed mainly in response to development proposals involving natural wonders and the balance of it seems to be on the side of conservation.

The supply of natural wonders, which provide unique recreation, amenity and scientific benefits, is virtually fixed. Once they are developed into industrial, agricultural or residential assets it may be impossible to restore them to their original state. In other words, development is likely to kill off preservation hopes for ever. Furthermore, artificial stocks of capital, technology, human knowledge and skill are incapable of reproducing natural wonders, nor can they enhance their beauty. Krutilla (1967) argues that wonders of nature do enter into individuals' utility function in the form of amenity, scientific and recreation benefits. Artificial capital and technology are only capable of increasing the volume of fabricated goods. He also points out that present tastes may change in the future in favour of natural wonders, quite substantially. One reason for this is an intertemporal externality which can be phrased as learning-by-doing. If facilities of natural wonders were to be diminished in the future then this would reduce opportunities for future generations to acquire skills to enjoy them. If, on the other hand, the wonders of nature were preserved by the present generations, the opportunity of acquiring and developing skills would also increase and consequently demand would rise over time as people learn to enjoy these facilities.

It should be obvious that whatever society does with natural wonders there is an opportunity cost involved. If these assets are preserved, the cost to society will be the loss of net benefits from development projects. If development takes place then a part of the cost will be the scientific and amenity benefits which would have resulted from conservation. The decision-maker must make up his mind in the face of these costs.

9.2.1 The model

The model is constructed by making a number of simplifying assumptions, similar to those used by Gannon (1969).

1. In a hypothetical environment there is a certain stock of natural wonders, *W*, and their supply is permanently fixed.
2. The scientific and amenity services provided by these wonders enter into individuals' utility function in the same way as other commodities. Society values natural wonders as perpetual consumption goods. Other items such as

fabricated goods, agricultural output and conventional services are put together into a different category (G).
3. Likewise, inputs are aggregated. Capital and labour are put together into one group and natural wonders into another, both of which are homogeneous and divisible. A natural wonder is an input but also has a potential of being an output. It is possible to envisage the natural phenomena input as land in traditional economic models.
4. Once natural wonders are used as inputs to produce G, the process is irreversible. That is, once committed to the production of G, society would not be able to recover the natural wonders and use them for their natural and scientific services. In other words, utilization of wonders as inputs would make restoration impossible.
5. If and when technological progress occurs it can only enhance the production capacity of G. Natural phenomena output, W, cannot be increased by advancing technology.

On the demand side we have a set of conventional indifference curves which imply that marginal utilities for both G and W are positive. When income levels are low the indifference curves are oriented towards G (a poor man would derive higher utility from the consumption of G compared with W). In other words, at low income levels the marginal utility of food, fabricated goods and conventional services are greater than the marginal utility of natural wonders. As society becomes better off, it would put a greater value on services provided by natural wonders. The situation is shown in Figure 9.5.

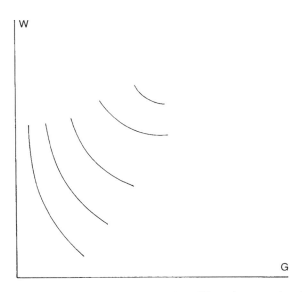

Figure 9.5   Social indifference curves for wonders (W) and conventional goods and services (G).

On the supply side there is the conventional production possibility curve, which is infinitely divisible and concave, indicating diminishing returns to each input aggregate. Any point on the production possibility curve, PP, indicates that inputs are fully used to produce a mixture of the two outputs, G and W. At a point like A the output mixture is $O\overline{W}$ of natural wonders and $O\overline{G}$ of goods and conventional services. It is important to note that in view of assumptions (1) and (4) above, once A is obtained, society cannot move north-west along the production possibility frontier as it would imply recovering natural wonders locked up in the production of G. However, society can always move south-east along the frontier right to the point P on the horizontal axis if it so wishes. Furthermore, when technology advances, the production possibility can only shift along the horizontal axis, as shown in Figure 9.6. That is, modern technology cannot recover nor can it reproduce the wonders of nature.

The next step is to put the indifference map together with the production possibility frontier (Figure 9.7). Given all the assumptions of the model, the socially optimal allocation policy is identified at A, a point where one indifference curve is tangent to the production possibility frontier. At point A, the slope of the indifference curve, $I_e I_e$, equals the slope of the production possibility frontier, PP. Society allocates resources in a manner that will yield socially optimal output levels of $O\overline{G}$ and $O\overline{W}$.

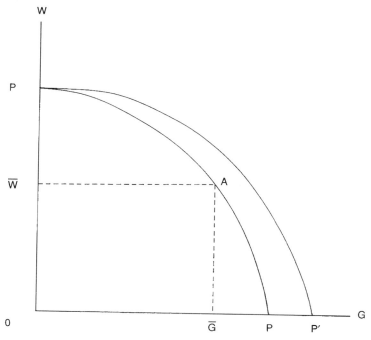

Figure 9.6   Production possibility frontiers for G and W.

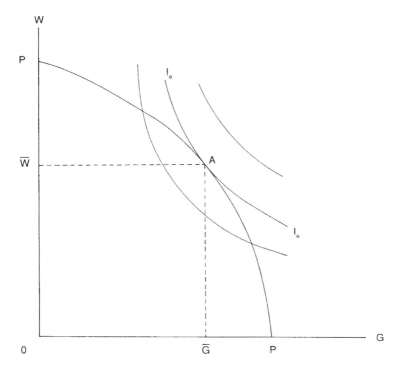

Figure 9.7   Socially optimal allocation of natural wonders.

Let us make A our starting point to explore two different cases: (1) a situation in which technological progress is taking place without any change in tastes: (2) technology is changing along with tastes, favouring natural wonders. Figure 9.8 shows the first case. The production possibility frontier expands along the horizontal axis. In the new situation on PP' is tangent to the I'I' indifference curve and this calls for an $O\overline{\overline{W}}$ level of natural phenomena, which is not available (assumption 4). Recovery of the $\overline{W}\overline{\overline{W}}$ level of natural wonders from the production process is ruled out by this assumption. The best that society can do is settle on A″ which contains $O\overline{W}$ of wonders. Note that the indifference curve I″I″ is lower than I'I'. Clearly, there are two expansion paths: optimal and available. The difference between the two can be termed the welfare gap, measuring the loss of potential well-being when the advancing technology pushes out the production possibility curve over time.

In the second situation technology is changing with tastes favouring the natural phenomena. If we imagine that, as the production possibility frontier expands along the horizontal axis, the indifference curves become less and less flat, then this is a case which will widen the welfare gap.

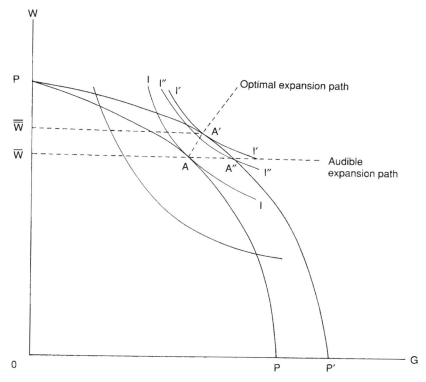

Figure 9.8   Socially optimal allocation of natural wonders over time with technological progress.

Gannon (1969) points out a parallel between a Malthusian conclusion and the one obtained in this analysis. A stagnant agricultural sector has been replaced by a stagnant natural phenomena sector. However, in Malthusian economics, history in many instances proved that the pace of technological progress increased the food availability over demand. A similar situation is unlikely to happen in the natural wonder sector.

There is a body of economists (Fisher *et al.*, 1972; Arrow and Fisher, 1974; Smith, 1977; Smith and Krutilla, 1972) who imply that as our ability to increase the volume of fabricated goods expands, our ability to maintain natural phenomena may decrease. The main reason for this is environmental pollution and congestion. In other words, with reference to the above figures, the production possibility frontier may not be fixed at P along the vertical axis, but may actually move southwards as the technology advances (Figure 9.9). This would, no doubt, increase the welfare gap.

In order to draw attention to negative aspects of technology, Page (1977, 1983) argues that in economics the tree of knowledge is generally considered to bear only good fruit, and the stock of industrial capital is assumed to be an

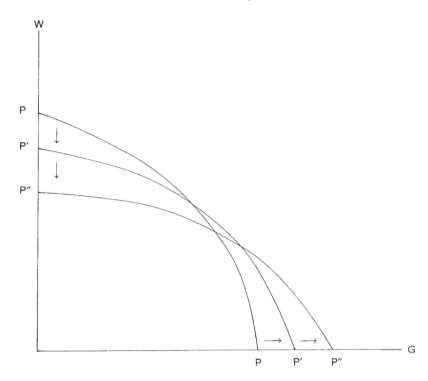

Figure 9.9   Shrinking natural phenomena sector due to technological progress and economic growth.

asset. The developing facts and events over the last two to three decades have proved that these common assumptions about technology and capital are rather debatable. Our present commitments to enhance the stock of knowledge and capital carry incalculable risks. Every year hundreds of new chemicals are created and every year a few of these are found to be potentially harmful and we try to remove them from the market. For example, DDT, thalidomide and CFCs were created, praised and used and now we know how harmful they can be.

Raucher (1975) estimates that more than 60% of human cancers are caused by environmental chemicals – many of them synthetic. The effects of many supposedly good substances on humans and the environment are often not realized until long after their introduction. Suspicion is growing with regard to our technological knowledge and the structure of the capital stock. For example, the full long-term effects of nuclear technology, which it is assumed will provide a sizeable proportion of our future energy demand, are beyond our comprehension at the moment. Recent developments in the field of genetic engineering seem to impose many unknown hazards. Destruction of the ozone layer may be one of many, yet unknown, serious environmental problems linked with the growing industrial structure and knowledge. Because of the expected benefits of modern technology we have already made many decisions which may involve unforeseen costs of great magnitude.

9.3   DEMAND FOR NATURAL PHENOMENA

It is well documented that there is increasing worldwide demand for natural phenomena both for scientific research and for leisure activities. Many researchers (e.g. Clawson and Held, 1957; Butler, 1959; Landsberg et al., 1963; Clawson and Knetsch, 1969; Howe, 1979) report a strong and growing demand for outdoor recreation. Especially since the Second World War, it is estimated that the demand for outdoor recreation has been increasing at about 10% per year. This is largely due to growing income and education levels which are not only confined to the Western world. Indeed the natural wonders, particularly those in the United States, have been visited by tourists coming from all corners of the world. Forecasting until the year 2000 and beyond, many economists argue that the increase in demand will continue strongly. In line with this prediction many popular areas have become crowded in recent years, thus pushing the recreation frontier into more remote areas. New areas have been set aside for conservation and attempts have been made to manage recreation not only in the United States but also in Europe.

Recreation projections can take many forms such as site-specific studies, e.g. for a particular natural wonder; activity specific studies; the number of recreation hours that will be created for a specific activity, e.g. boating, fishing, etc.; region-specific studies; expenditure on leisure activity, and so on. Most forecasting works are for planning purposes. When decision-makers are faced with a dilemma, such as conservation versus development, they need to find an answer to a wider question: what is the annual net benefit resulting from conservation for recreation purposes? Sometimes market prices provide handy measures for recreation services. In some cases, however, market prices do not exist because the services are publicly provided and entrance fees have been kept deliberately low for equity reasons.

The most popular way of estimating demand for recreation is through survey techniques as explained in the previous chapter. These may take the form of population surveys, site-specific demand estimates or willingness-to-pay approaches. Mostly willingness to pay is determined using on-site interviews with leisure takers, which could be straight questioning. What is the maximum amount you will pay to participate in a leisure activity? How many times would you visit per year? A more sophisticated interview would include a bidding game in which the leisure seekers could react to hypothetical increases in admission costs in the area. Bids can be raised or lowered, systematically, until the leisure seeker switches his decision from using one area to using another.

Using survey methods, some researchers have been able to construct willingness-to-pay functions for leisure seekers. For example, Davies (1964) estimated a willingness-to-pay function for game hunters in a large private forest in Maine. His function turned out to be log-linear in shape, similar to that of the Cobb–Douglas estimations. That is:

$$W = 0.74(L)^{0.76}(E)^{0.20}(Y)^{0.60}, \tag{9.1}$$

where W is willingness to pay of participants in game hunting; L is length of visit in days, i.e. a day measure for hunting; E is years of familiarity with the area; and Y is income of participants. All these variables are positively correlated with the willingness to pay and they are all inelastic. For example, a 1% increase in leisure seekers' income would call for a 0.6% increase in willingness to pay.

After the determination of the willingness-to-pay equation the leisure seekers were further sampled to determine the distribution of all three independent variable characteristics. For various intervals the average willingness to pay was computed and plotted against the estimated number of visits falling in that interval to yield the demand schedule shown in Figure 9.10 (ordinary curve), which

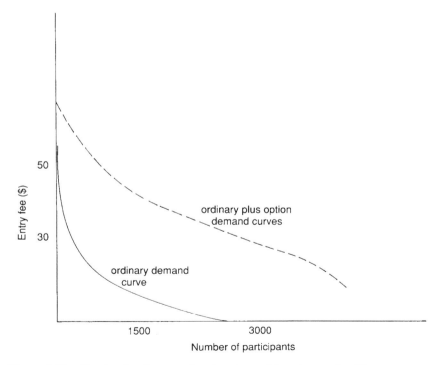

Figure 9.10 Simulated ordinary and option demand functions and willingness to pay. (Source: Davies, 1964.)

shows that the demand is highly sensitive to a change in price. The area under the demand cure estimates the total willingness to pay which is a true measure for benefits enjoyed by leisure seekers. In this case the figure is approximately $30 000 per year.

9.3.1 Option demand

There is an interesting theory which is highly relevant to the understanding of the amount of social benefits that natural wonders can yield. This theory, now widely known as 'option demand', was founded by Weisbrod (1964). Following the publication of this theory a heated discussion took place between Lindsay (1967), Kahn (1966), Long (1967), Byerlee (1971) and Cicchetti and Freeman (1971) as to whether option demand was a new concept or part of the classical and well-known concept of consumer surplus. Weisbrod advances his argument by considering an extreme case of a commodity, the purchase of which is infrequent and uncertain, and production of which cannot be re-initiated at any cost once it has been halted and inputs devoted to other uses. However, in recent years the theory has been used widely to study cases other than the extreme ones.

The option demand theory is most forceful in cases such as unique natural wonders where the market may fail to allocate resources optimally between competing uses for the simple reason that certain kinds of economically significant votes never get taken in the ballot box. In the marketplace resources are normally allocated on the basis of decisions by customers in small individual market transactions. These, unfortunately, do not include the appraisal of consumers' desire to keep the goods and services available for future use.

For example, on the question of conservation versus development of Hell's Canyon, the former may not look feasible because the revenue collected from visitors may not be large enough to cover the opportunity cost of preserving it (Weisbrod uses the example of visits to the Sequoia National Park). If Hell's Canyon is redeveloped or the Sequoia trees cut down it will be extremely difficult, or even impossible, to restore them to their original state. However, some potential visitors may be willing to pay a sum for the option of consuming the services of these wonders in the future. This sum may be large enough to preserve these resources.

The implication of the option demand theory in the economic evaluation of natural wonders is enormous. Referring back to Figure 9.10, let us say that total willingness to pay, which is about $30 000 per annum, by leisure seekers is not enough to make a case for preservation and the developers are pressing hard in their discussions with the government for planning permission. Assume that some economists have constructed a willingness-to-pay function on the basis of the option demand theory. Their study covers not only those who are interested in using the phenomenon now, but those who are interested in using it at an unspecified future date and are willing to pay something towards that aim.

Needless to say, a demand curve based on the new willingness-to-pay function will be further to the right, as depicted by the dashed curve in Figure 9.10, implying a greater total willingness to pay.

Further studies on option demand exposed a number of fine points. One case is the risk that preservation as well as development can bring hazards such as floods, forest fires, construction failures, etc. The net option value of preserving a wonder could then be negative (Schmalensee, 1972; Henry, 1974). Others point out that option value is essentially a gain from being able to learn about future benefits that would be made impossible by development (Conrad, 1980). This makes the concept of option value a kind of value of information.

A development project that passes a conventional cost–benefit test might not pass a more sophisticated analysis that takes into account option value. There can, of course, be a number of counter-arguments in favour of development. For example, development of a natural wonder does not have to be an irreversible process; dams on Hell's Canyon can always be knocked down. True, there will be a difference between the restored canyon and the original one, but as far as the average sightseer is concerned this difference may not be very great. Another argument may be that tastes can change in favour of development projects to yield a higher volume of agricultural output, fabricated goods, etc. In effect governments throughout the world are keen on material advancement; most elections in the free world are contested on economic issues such as growth.

10
Economics and policies in nuclear waste disposal

The containment and storage of radioactive wastes is the greatest single
responsibility ever consciously undertaken by man.

Senator Howard H. Baker

The controversy over the comparative merits of nuclear energy and fossil fuels
has been raging since the end of the Second World War. This issue has important economic, environmental and ethical dimensions puzzling policy-makers as
well as the general public throughout the world. Problems created by the use of
fossil fuel are widespread, causing concern in developed as well as developing
countries. As mentioned previously, apart from the scarcity issue, some serious
environmental complications such as the continuing build-up of carbon dioxide
in the earth's atmosphere and acid rain have now become a matter for public
debate. The effects of global warming could be catastrophic to the world system
as we know it. Agriculture may be greatly disrupted, maritime life may be
affected by the changed climate and ocean levels might rise, flooding many
coastal regions and cities. It is well documented that acid rain is damaging
forests, freshwater fish stocks, crops, historic monuments, human health and
even plumbing systems in buildings.

Nuclear power as a substitute source of energy has equally puzzling problems.
A breeder reactor both produces and consumes plutonium and thus this highly
toxic material, on occasions, needs to be transported over long distances, even
across national frontiers. In the past, many nuclear scientists argued that the
probability of a significant accident such as a core meltdown at a power station
was negligible, but the view on this has changed following the accidents at
Three Mile Island in the United States and Chernobyl in the Ukraine.
Furthermore, the prospect of sabotage in power stations and possible theft of
plutonium for terrorism remind us that present generations too are exposed to
substantial risks.

The most serious problem with nuclear energy is the safe disposal of the
radioactive waste. It has to be transported over long distances, stored and monitored for a very long period of time. Some highly active wastes can be captured

in glass or ceramic containers, but their long-term stability cannot be guaranteed as substances tend to migrate towards the surface of these containers. If they come into contact with groundwater they could be carried away from their intended locations.

Currently there seem to be two conflicting views on the disposal policy. One contends that in view of the rapidly increasing inventory of wastes in temporary storage there is no alternative but to figure out, quickly, some means of isolating them geologically. The other view is that a far safer policy would be to keep wastes above ground at their present sites so that they can be easily retrieved as the scientific knowledge on disposal makes progress. The nature of the problem here is that it is virtually impossible to estimate when enough will be known to proceed to the final stage. Some environmental pressure groups argue that just as governments brought top scientific minds together to start the nuclear age, they must do the same again to help to devise a safe way out (Ulrich, 1990).

10.1 A BRIEF HISTORY OF NUCLEAR POWER

In 1895 Roentgen's discovery of the X-ray and one year later Becquerel's iden-tification of natural radiation, marked the beginning of the 'nuclear age'. In 1939 Hahn and Strassmen achieved the splitting of the uranium atom in Berlin which started fission technology. In 1942 at the University of Chicago, Fermi proved that the fission chain reaction in uranium nuclei could be sustained and con-trolled, making it feasible to harness this energy. From this, nuclear technology developed along two paths: the production of nuclear bombs and the develop-ment of reactors for commercial purposes. During the Second World War attempts to create nuclear bombs intensified in the United States at Oak Ridge, Tennessee, and Hanford, Washington. In July 1945 the first nuclear bomb was dropped on Hiroshima to end the war.

The creation of atomic bombs was the original driving force behind for the development of fission technology but at the same time the United States gov-ernment decided to use the nuclear technology for peaceful purposes. After the war, efforts to create nuclear energy reactors intensified and in 1957 the coun-try's first nuclear power plant at Shippingport, Pennsylvania, began operation. The UK, Canada, West Germany and France followed suit. By 1990 there were 423 nuclear reactors in operation providing about 20% of the world's electricity supply and a further 105 units were under construction or consideration (Traiforos et al., 1990). In some countries nuclear power provides more than half of the energy requirement, for example, in France the figure is about 70%.

However, the tide was beginning to turn against nuclear power in the late 1970s and early 1980s as accidents at the Three Mile Island plant in the USA and Chernobyl in the Ukraine increased public anxiety. Furthermore, governments' inability to find a satisfactory solution to the waste disposal problem was putting the brakes on the global expansion of nuclear energy. After the accident at Three Mile Island many nuclear power generating units, including the already completed units, were cancelled in the United States. Recently the Swedish government decided to phase out nuclear power by the year 2010, even though nuclear power currently provides almost half of the country's energy requirement.

10.2 NUCLEAR WASTES

Nuclear wastes are the by-product that results from using radioactive materials and are associated with three distinct activities: research, energy generation for peaceful purposes, and creation of military devices. There are four categories of nuclear waste: high-level, mid-level, low-level and mill tailings.

1. *High-level wastes.* These are spent fuel from nuclear power plants and the waste from defence activities. High-level wastes are the most radioactive category; they usually decay or lose radioactivity rapidly. However, some high-level material may also contain elements that decay very slowly and may remain radioactive for thousands of years.

   Spent fuel rods are removed from nuclear reactors when they can no longer efficiently contribute to the chain reaction. They can be reprocessed to separate the high-level wastes from the remaining fissile and fissionable material in the rods. When spent fuel is reprocessed, a much smaller volume of high-level wastes will need to be transported and stored. In addition, reprocessing is an efficient way of using uranium which is a scarce commodity.

   The question of whether or not to reprocess fuel rods is a difficult policy issue as bomb-grade plutonium is created during reprocessing, which increases the security requirement of the operation. In 1977 the United States government banned commercial reprocessing, much to the dislike of the nuclear power industry. High-level wastes radiate intense energy and emit alpha, beta and gamma radiation.

2. *Mid-level wastes.* These consist of synthetic elements formed as a by-product of reactor operation. They emit less penetrating radiation than high-level wastes but decay slowly. Some species of mid-level wastes remain active for millions of years. Most mid-level material results from reprocessing nuclear fuel as part of defence-related activities.

3. *Low-level wastes.* Low-level wastes typically contain small amounts of radioactivity dispersed in a large amount of material and pose little potential hazard. They result from commercial, industrial and medical processes and consist of discarded protective clothing, papers, filters, resins, rags, etc. Low-level wastes do not require extensive shielding, but some protective shielding may be needed for handling certain low-level waste. In the United States commercial low-level wastes are currently disposed of by shallow land burials in containers at three federally licensed sites in South Carolina, Washington and Nevada.

4. *Mill tailings.* These are naturally radioactive rock and soil that are the by-products of mining and milling uranium. They contain small amounts of

radium that decays to radon, a radioactive gas. Tailings are not a major radiological hazard and thus a common method of disposal is to cover them with enough soil in isolated locations to protect the public from radon.

10.3 TYPES OF RADIATION

Radiation is a natural part of life. It occurs naturally in the form of radon gas, which has already been mentioned, and in rays from the sun. Many households are also familiar with artificial radiation, for example, microwave ovens containing radiation to cook our food. Most atoms in nature, which are the building blocks of all matter, are stable, i.e. they retain their form and substance for ever. But some atoms are unstable and change to different forms; these are called radioactive. During the changing process, changing atoms release energy in the form of electromagnetic waves or fast-moving particles. Radiation can disturb the biological function of living tissues.

There are three kinds of radiation: alpha, beta and gamma.

1. Alpha radiation consists of positively charged particles. Although the most energetic, they are the least penetrating, and can be stopped by a sheet of paper or even the outer layer of skin.
2. Beta radiation consists of high-speed electrons. They can pass through about 2 cm of water or human tissue. Most fission products in spent fuel emit beta radiation but they can be stopped by a thin sheet of aluminium.
3. Gamma radiation consists of high-energy, electromagnetic waves that are extremely penetrating and thus can pass through the human body. Due to intense penetrating powers this type of radiation can damage living creatures, including humans. Gamma-emitting substances are handled by remote control mechanisms. Dense material such as concrete, steel, lead and deep pools of water can provide shielding.

Figure 10.1 shows the penetration power of all three forms of radiation. Nuclear wastes contain higher than natural concentrations of radioactive atoms. For example, when spent fuel is removed from a reactor it contains fission products, radioactive forms of uranium and radioactive atoms. Fission products are the remains of split atoms. Spent fuel becomes less dangerous over time as the radioactive elements decay, during which it emits alpha, beta and gamma radiation. Eventually, the level of radioactivity declines to levels found in ore.

10.3.1 Inventory

Substantial quantities of nuclear wastes exist in nuclear countries. The United States has the highest quantity. In 1990 over 100 nuclear power plants were being operated by utilities in 33 states. The fuel for a nuclear electricity generating unit consists of pellets of enriched uranium which are sealed inside metal rods; after about three years of use, the fuel no longer contributes efficiently to

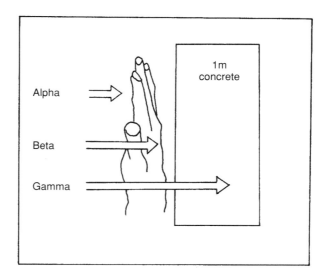

Figure 10.1   Penetrative power of radiation.

the nuclear chain reaction and is removed as spent fuel. Nearly all spent fuel is currently stored in deep steel-lined concrete pools filled with water at the reactor sites. In 1988 18 000 tonnes of spent fuel were in storage; by the year 2000 the accumulation is expected to reach over 40 000 tonnes (US Department of Energy, 1988). Table 10.1 shows where nuclear fuel is stored.

Defence high-level wastes result from the reprocessing of specially designed and irradiated fuel to obtain materials for nuclear weaponry. Over 10 000 tonnes of defence high-level wastes are currently stored temporarily at the Department of Energy sites in South Carolina, Washington and Idaho. In 1987 there were 379 300 m³ of high-level defence wastes. Figure 10.2 shows the proportions. The volume of defence wastes has been rapidly increasing since the end of the Cold War due to the dismantling of nuclear weapons.

10.4   DISPOSAL METHODS

Various solutions have been proposed for waste disposal, some of which are undergoing extensive study by various governments and independent research bodies in various parts of the world. Some of the most prominent disposal methods are outlined below.

1. *Space disposal.*   Space has long been considered by some scientists as the ultimate disposal area, especially for high-level nuclear wastes. Since the early 1970s NASA (National Aeronautics and Space Administration) has been studying several space disposal options. The main idea involves trans-porting waste canisters to a low-orbit transfer point by using a special space

Table 10.1 States where spent fuel is stored and projected amounts by the year 2000

State	Spent nuclear fuel (tonnes of uranium) 1987	2000 (estimated)
Alabama	992	2 010
Arizona	34	925
Arkansas	328	709
California	435	1 643
Connecticut	844	1 651
Florida	882	1 766
Georgia	386	1 369
Illinois	2 406	5 718
Indiana	49	49
Iowa	153	294
Kansas	48	305
Louisiana	69	605
Maine	324	538
Maryland	436	859
Massachusetts	338	554
Michigan	712	1 743
Minnesota	385	805
Mississippi	101	398
Missouri	83	358
Nebraska	258	600
New Hampshire	0	224
New Jersey	537	1 493
New York	1 136	2 290
North Carolina	722	1 823
Ohio	94	645
Oregon	200	432
Pennsylvania	1 175	3 298
South Carolina	1 014	2 459
Tennessee	160	751
Texas	0	846
Vermont	271	433
Virginia	701	1 490
Washington	53	312
Wisconsin	557	979
Total	15 903	40 293

Source: US Department of Energy (1988).

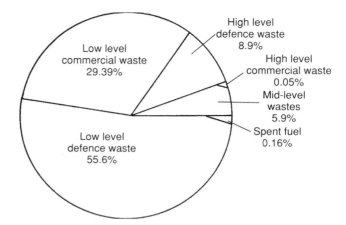

Figure 10.2   Percentages of radioactive wastes produced in the United States by volume. (Source: US Department of Energy, 1988.)

shuttle and then repulsing them into outer space. Once waste canisters are sent into outer space their potential for environmental impact and human health effects would be nil.

Space disposal looks attractive at first glance but it has very serious technical and economic problems (Lipschutz, 1981). The US Department of Energy suggests that the cost of space disposal would be prohibitively high, increasing the cost of electricity to consumers by as much as 50% due to space shuttle costs alone (US Department of Energy, 1978). In addition there would be serious risks involved in a space disposal programme; over the years space missions have had spectacular failures. The failure of the Apollo mission a few years ago, captured by television cameras, is still vivid in our minds. Indeed an accident of this kind in a space disposal mission could contaminate large areas of the earth, affecting the populations of many countries. It is doubtful that any democratically elected government would now take space disposal seriously.

2. *Seabed disposal.*   It has been suggested that nuclear waste could be isolated from the biosphere by placing it in the seabed. There are a number of attractive features to this proposal which is under study, especially by scientists in Japan and the United States. The seabed in the deep ocean is remote from human activities; large areas are available, and it provides fewer intractible political problems than land, space and ice disposal. The waste canisters could be dropped in needle-shaped penetrometers to enter about 50–100 m deep into the sediments at the bottom of the ocean. Although the water column is not regarded as a barrier, it would inhibit human intrusion and would help to dilute any waters that might seep from the canisters.

A number of areas have been proposed as suitable: deep ocean sediments, ocean trenches and sub-sediment bedrock. Red clay sediments in the centre of sub-tectonic plates which are about 3–5 km below the ocean surface present suitable areas (Kerr, 1979). Particles have been settling on the ocean floor evenly and continuously for millions of years. Furthermore, such areas appear to be seismically and tectonically stable.

Ocean trenches plunge deep into the earth where vulcanism and seismic instability exist. Canisters dropped there would be drawn into the earth's crust and safety isolated. There has been less enthusiasm for the sub-sediment bedrock disposal. A major obstacle to ocean disposal is international law. The London Convention of 1972 forbids the dumping of high-level nuclear wastes in international waters. However, the Convention did not ban disposal of low-level wastes in territorial waters, a common practice by the UK and Japan.

3. *Ice disposal.*   This method has been proposed mainly for the disposal of high-level radioactive material in continental ice sheets. In 1958 Bernard Philberth, who pioneered the idea, received a German patent. Ice disposal provides a number of attractive features. It is away from human settlements, it could provide long-term isolation, it does not require the development of any 'exotic' technology as a relatively simple operation is required.

Three concepts have been proposed: meltdown, anchored emplacement and storage at surface facilities. In the meltdown concept hot waste canisters placed in holes would be allowed to sink right to the bedrock at a daily rate of 1–2 m. In this way canisters would reach bedrock in 5–10 years. Anchor emplacement is similar to the first concept except that it would allow retrievability for a certain period of time, approximately 200–400 years. This method limits the sinking of canisters to a desired level by a cable. Eventually snow and ice would cover all facilities and it is estimated that the system would reach bedrock in about 30 000 years (Lipschutz, 1981). Surface disposal would also allow retrievability for a few hundred years.

Ice disposal has a number of shortcomings. First, transportation to remote areas could be expensive and hazardous. Second, ice sheets are subject to periodic surges on a scale of thousands of years which may cause dislodgement of canisters into the sea. Third, disposal of too many hot containers in an area could cause the temperature there to rise, creating widespread melting. Fourth, the Antarctic Treaty prevents waste disposal in the Antarctic, but does not extend to Greenland. However, Denmark, who has some jurisdiction over Greenland and the people who live there, is unlikely to agree to the use of their territory as a nuclear rubbish dump.

4. *Deep hole placement.*   This involves placement of nuclear waste canisters in holes as deep as 10 km. Sedimentary or crystalline rocks in areas of seismic and tectonic stability are thought to be suitable. It may not be economically feasible to dispose of high volumes of waste in this manner. Furthermore, our geological knowledge about the earth's crust at such depths is incomplete at present.

5. *Shale disposal.*   Shales – sedimentary rocks that can produce oil – have very low permeability and good sorption properties. Shales selected for

injection are likely to be the those that would retain the material within the shale bed. Wastes mixed with cement could be injected into fractured shale deposits and be allowed to solidify there at depths of 300–500 m.

6. *Deep injection disposal.* This is thought to be suitable mainly for watery wastes injected into porous or fractured strata at depths of 1000–5000 m. However, at this point in time not enough is known about the mobility of wastes through the host rocks.

7. *Melted rock disposal.* In this, high-level wastes in slurry form with the possible addition of a small volume of other wastes, could be placed underground. After the water had evaporated the heat would melt the rock and in about 1000 years the waste–rock mixture would resolidify, trapping the radioactive material.

8. *Tuff disposal.* Tuff is a volcanic rock noted for its durability and sorptive qualities. The Yucca Mountain site in the state of Nevada is an area that contains tuff layers created in a volcanic eruption about 11 million years ago. During the eruption large quantities of gases were released into the air and eventually fell to the surface as ash. Still hot, the ash welded together and became compressed under its own weight, forming layers as thick as 1800 m. Some scientists believe that this area is suitable to become a repository for high-level wastes.

9. *Salt disposal.* Underground salt deposits are thought to be highly suitable for the disposal of wastes, particularly mid-level wastes. The main problem with geological disposal is the contact with circulating groundwater which could carry radioactivity into the human environment. The existence of aged salt deposits implies that the area has been dry for long periods of time as salt is highly soluble in water. Furthermore, salt has self-healing properties; fractures and entry holes would be closed up soon after emplacement.

10. *Do nothing yet.* The no-action alternative would leave nuclear wastes at surface facilities in or around nuclear reactor sites for an indefinite period of time until a lasting and less controversial solution of disposal is discovered. The most attractive aspect of this option is easy retrievability. On the other hand, it is impossible to predict when enough will be known about nuclear waste storage to allow steps to be taken to deal with it.

10.5 THE UNITED STATES' FIRST NUCLEAR WASTE REPOSITORY

The Waste Isolation Pilot Plant (WIPP) is the United States' first nuclear waste repository, located in Los Medenos in south-eastern New Mexico about 40 km east of the town of Carlsbad. It is an underground geological repository used for the disposal of mid-level wastes resulting from the nation's defence activities and can hold approximately 185 000 m^3 of material. The WIPP is a facility of the United States Department of Energy, which was authorized by the Public Law of 96–164, 1979. The law made the WIPP exempt from licensing by the Nuclear Regulatory Commission (NRC) and is self-regulated by the Department of Energy. However, the Department of Energy must demonstrate compliance with the 'Standards for the Management and Disposal of Spent Nuclear Fuel, High Level and Transuranic Radioactive Wastes' promulgated by the United States Environmental Protection Agency (EPA). It has been pointed out by Neill and Chaturved (1991) that the EPA does not have the authority to assess the WIPP's compliance with the above-mentioned standards, nor is the EPA seeking such authority. As a consequence, the Department of Energy will determine whether or not the WIPP meets compliance with the EPA standards.

The Department of Energy's interest in the Carlsbad area, which contains large underground salt deposits, began in the early 1970s when the Department's earlier efforts to develop a repository in the salt mines of Lyons, Kansas, failed. A great deal was already known about the geology and other properties of the Carlsbad salts as a result of potash mining activities there. In the early 1970s the collapse of the mining industry in the area created unemployment and economic hardship for the community and ideas were sought to revive the region. With support from the local business community and politicians the Department of Energy initiated a study in the region. After extensive tests Los Medenos was chosen as a potential repository site.

One of the most attractive aspects of the Los Medenos site is its dryness. In a geological repository the main problem is contact with groundwater which can provide potential pathways for waste transport away from the repository and eventually bring toxic substances to the human environment. Because salt is highly soluble in water, aged salt deposits imply that the area has not been in contact with circulating groundwater for long periods of time. The main body of the WIPP is about 600 m underground in the lower part of a large Permian age salt deposit estimated to be 225 million years old. Locations at a depth of greater than 450 m below the surface are likely to remain unaffected by erosion provided that there is no human intrusion.

The initial construction of the WIPP began in 1983 and the main structure was completed within the first five-year period. The Department of Energy had plans to start shipping waste to the WIPP in October 1988. As the deadline approached it became clear that the Department had not completed all the necessary preparations to start the activity. At the time of writing in 1992 the operation is still postponed. When completed, the WIPP will consist of eight panels with seven rooms in each panel designed to be 91.5 m long, 10 m wide and 4 m high. The bulk of the wastes would be emplaced in 55-gallon drums stacked three high in the rooms. In addition, there will be odd-sized and shaped waste containers. About one-third of the waste has already been generated and temporarily stored at the Department of Energy Weapon Laboratories awaiting transfer to the site. The end of the Cold War has increased the urgency of this project because of rapidly increasing wastes resulting from the reduction in nuclear weapons.

The Department of Energy has plans to experiment with the wastes in a study which is likely to take at least five years to demonstrate safe disposal. Only then will the Department of Energy determine if the WIPP may be used as a permanent repository. The main facilities both above and below ground have been completed including seven out of the planned 56 waste chambers.

The sealing of the repository is likely to take place around the year 2025 provided that operational experiments turn out to be successful. There are a number of technical problems such as fast closure of the salt on chambers and determining the structure of the brine reservoir which exists in and around the salt deposits. However, many believe that tests and other operations are likely to be successful, although there may be a few hitches from time to time.

10.6   REPOSITORY COSTS

Disposal of nuclear wastes in a permanent repository is an extremely costly operation to initiate and maintain. The main items involved are: research and administration including characterization of proposed locations; construction; operation; transportation of wastes to the repository; emplacement of wastes and the sealing of the repository; and monitoring. Since many countries already have a large inventory of highly toxic nuclear wastes, construction and management of repositories will soon become a global business requiring taxpayers to meet the costs without cutting corners. In addition, there are distant costs such as health detriments if the long-term integrity of the repository is somehow undermined; these would fall upon future generations. Some of the costs are outlined below.

10.6.1   Research and administration to identify a site

In the United States studies for isolating radioactive wastes have been in progress since 1957 when the National Academy of Sciences first recommended deep geological disposal. Investigation of salt deposits as repository sites started in the 1960s. In the 1970s studies began into tuff on federally owned reservations and later these studies were extended to cover granite. After the passage of the 1982 Nuclear Waste Policy Act (NWPA) (US Congress, 1982) the Department of Energy has been conducting extensive research on many locations with particular reference to high-level wastes. The NWPA provides financial aid to offset the impact from siting, developing and decommissioning a repository. Various financial packages are available to states or Indian tribes willing to host a repository.

The Nuclear Waste Policy Act also established a Nuclear Waste Fund to enable the federal government to recover all of the costs of developing a disposal system. The Act requires that utilities pay a fee into the Nuclear Waste Fund, currently set at 0.1% kWh, for electricity generated by commercial nuclear power plants. The Department of Energy is required by law to evaluate the adequacy of the fee. The cost of defence wastes will be met by the federal government.

Efforts to identify suitable sites in which to isolate the United States' nuclear wastes have been going on since 1957 involving many government departments, universities, other scientific institutes and the private sector. Especially during the last two decades numerous reports, some of which are voluminous site characterization reports, have been published. A large number of the individuals involved in research, administration and publishing are highly trained professionals. Therefore, the opportunity cost of labour and material involved in research and administration regarding identification of repository sites must be quite high.

10.6.2 Construction, operation and closure

In this category costs are siting, preliminary design development, testing, regulation, administrative work associated with repositories, engineering work, construction, operation and, finally, closure and decommissioning. Construction is the largest cost item, and includes site preparation, the construction of surface facilities including the functional testing of the waste-handling building, the installation of utility networks, the construction and outfitting of shafts and ramps, and the excavation and construction of underground support areas. Operating costs include all staffing, maintenance, supplies, waste packages and utilities. The final part of the cost is closure and decommissioning which contains all costs associated with backfilling and permanently sealing the underground repository and decommissioning the surface facilities.

Work on nuclear waste repositories is well advanced in the United States. Each year a comprehensive cost analysis of nuclear waste management systems is performed as a reference document that aids in the financial planning by the Department of Energy. The cost analysis follows the most current programme strategy, plans and policies. The Department of Energy's estimate of a single spent fuel and high-level waste repository cost is $8.7 billion in terms of 1988 prices. As for the WIPP it has been estimated that since 1975 the United States government has spent about $1 billion on construction and operating, in terms of 1991 prices.

10.6.3 Transport costs

As required by the Nuclear Waste Policy Act of 1982 the US Department of Energy is developing transportation capability necessary to support the waste management system. In accordance with the 1982 Act the Department of Energy is using the private sector to the maximum extent possible in the development of the transportation system.

Transportation includes purchasing waste containers, operating costs in accepting the waste and providing all the transport services needed to support the Department of Energy's waste management system, including the construction, operation and decommissioning of each maintenance facility. The development of other transport support facilities are continuing as needs and functional requirements become clear.

The US Department of Energy (1990) estimates that for a single spent fuel and high-level waste repository case the total transportation cost is likely to be £2.8 billion in terms of 1988 prices. A 20% contingency factor has been included in this estimate to cover uncertainties, interest charges and to accommodate further refinements.

For the WIPP project in 1981 the state of New Mexico entered into an agreement with the Department of Energy for road upgrading and repair. During the 1984 financial year eight projects for preliminary engineering were authorized and funds were created for use by the state of New Mexico Highway Department. Between 1984 and 1991 expenditure on roads amounted to just under $100 million in 1991 prices. In addition the Department of Energy earmarked a further $300 million for upgrading the transport infrastructure nationwide in relation to WIPP. However, some proposed legislation aims to double this figure. The main collection points are expected to be Argonne National Laboratory, Illinois; Oak Ridge National Laboratory, Tennessee; Mound, Ohio; Savannah River Site, South Carolina; Rocky Flats, Colorado; Los Alamos Laboratory, New Mexico; Lawrence Livermore Laboratory, California; Hanford Site, Washington; and Idaho National Engineering Laboratory, Idaho.

10.6.4 Monitored retrievable storage facility

The United States' Nuclear Waste Policy Amendments Act of 1987 (US Congress, 1987) authorized the Department of Energy to locate, construct and operate a facility for monitored retrievable storage. The costs involved in this exercise would be design, construction, operation and decommissioning. Following this Act the Department of Energy initiated a three-part action plan to site, construct and operate an integrated monitored retrievable storage facility with a target for receiving spent fuel in 1998. The first phase will allow limited amounts of waste beginning in 1998. The second phase will consist of the full capacity ready to receive and store spent fuel starting in the year 2000.

At present the Department of Energy is investigating a number of monitored retrievable storage design concepts which would allow waste acceptance in 1998. No final decision has yet been made on a particular system. A conservative estimate by the US Department of Energy (1990) puts the cost of monitored retrievable storage in the neighbourhood of $1.9 billion.

10.6.5 Other short- and medium-term costs

Today, the US government is increasingly focusing on the Yucca Mountain site with a view to making it a permanent repository for spent fuel and high-level nuclear wastes. The WIPP project is designed primarily for mid-level defence wastes. The Nuclear Waste Policy Amendments Act of 1987 allows the

Secretary of Energy to enter into benefits agreements with the state of Nevada regarding a repository and a state or Indian tribe regarding a monitored retrievable storage site. In return for money, the state or Indian tribe waives its rights to disapprove of the recommendation of a site for a repository or a monitored retrievable storage facility.

Under the provisions of the Amendments Act the annual payment to the state government of Nevada would be $10 million on execution of a benefits agreement. It would stay at that level for each year until the beginning of waste acceptance at the repository, then annual payments would go up to $20 million. Annual payments would continue until the closure and decommissioning of the repository. For the monitored retrievable storage, the annual payment to the host Indian tribe or state government would be $5 million upon execution of the agreement. This payment would double once waste acceptance begins at the facility. It would remain at that level until the completion of decommissioning.

The US Department of Energy calculates that in the single repository scenario a total for benefits payments would be in the neighbourhood of $0.7 billion, including compensation for the retrievable storage facility.

Each repository region has unique features in terms of human as well as natural geography. A repository could disrupt economic and other activities in the region. In the case of the WIPP project, for instance, the area is rich in potash and hydrocarbon deposits. In effect, it is one of the United States' most extensive potash mining districts. In the absence of the WIPP potash mining would continue. Based upon estimates by Cummings *et al.* (1981), commercial exploitation of potash deposits could generate about $2.5 billion for the region in terms of 1991 prices. In addition, hydrocarbon deposits in the form of natural gas, oil and distillate exist 8 km north-east of the WIPP site. A reserve estimate by the same researchers reveals a forgone income of about $0.5 billion (1991 prices).

The US Department of Energy publishes figures for the total cost of a radioactive waste management system over its complete life cycle (US Department of Energy, 1990). One estimate for the Yucca Mountain site, including a facility for retrievable storage and transportation system amounts to $26 billion, expressed in constant 1988 dollars. In the event of a second repository the total system cost is estimated to be $34–35 billion, depending upon the quantity of spent fuel and high-level waste requiring disposal. These figures do not include the WIPP project. It should be emphasized that cost estimates are constantly revised by the Department of Energy in the face of changing circumstances. Estimates are for the purpose of financial planning and also to ensure that the revenue-producing fee on nuclear power utilities required by the Nuclear Waste Policy Act of 1982 is adequate to meet the waste disposal cost. Delays in further legislation, safety guidelines, site identification, construction, accidents in research laboratories, power generating units, transportations and emplacement would increase the costs substantially. However, what is clear is that nuclear waste disposal is a costly business and thus one can't help wondering how the impoverished countries of the former communist world can afford to dispose of their waste without cutting corners.

10.7 LONG-TERM COSTS

The most controversial and sensitive aspect of the nuclear waste storage is the future health costs that could fall upon generations yet to be born. Some species of wastes in repositories will remain active for millions of years. For example, iodine-129, one of the longest living species, has a half-life of 17 million years; neptunium-237, 2 million years; plutonium-239, one of the most dangerous types, 24 000 years. Performance of a repository could be undermined by changes in precipitation, erosion, infiltration of water, tectonic folding, deformation of the earth's crust, volcanism, meteorite impact and human intrusion. If wastes were exposed to the human environment then they would become a serious threat to the health and safety of future generations.

10.7.1 A long-term cost study for the WIPP

Health

The economics of the future health costs of the WIPP project was considered by Logan *et al.* (1978) who considered two broad classes of health risks from exposure to nuclear wastes: cancer and genetic deformities. Their work was based upon the BEIR Report (1972), the most accepted source of information on health and the effects from radiation exposure. The BEIR Report gives estimates of the number of deaths per year by cancer from a hypothetical increase in continuous exposure of 0.1 rem to the United States population. A rem (roentgen equivalent man) is the measure of a dose of radiation and its effects on human tissues. The Department of Energy estimates that in the country the average person receives an effective dose of about 1 mrem per day, mostly from natural sources.

According to the BEIR Report a continuous exposure of 0.1 rem per year would create an average excess death of about 5100 for all cancers for the United States population. The Report also contains estimates for genetic effects, including chromosomal and recessive diseases, congenital abnormalities and regenerative problems. Calculation of the excess deaths for genetic disorders encounters greater uncertainty than for cancer as it depends on the mean age of reproduction as well as exposure rate. Logan *et al.* (1978) estimate an annual excess death of about 5050 for the US population from genetic disorders. Such figures, however, are debated by the scientific community. For example, it has been reported by Shrader-Frechette (1991) that some studies suggest that the cancer risk from radiation is 3–4 times higher than previously thought (also see Raloff, 1989).

The region within a radius of about 150 km from the repository in New Mexico and Texas is considered for health effects with the assumption that most of any release of material would be dispersed within this area. In order to provide a realistic dispersal, which requires non-uniform spread, the area is divided into seven zones. The study region is projected to have a population of 1.88 million by the time the repository is sealed. This population, engaged mainly in agricultural pursuits, is assumed to remain constant after the decommissioning of the repository. The major population centres surrounding the repository are Midland, Odessa, Hobbs, Rosewell and Carlsbad.

The methodology for calculating health risks using the BEIR Report is in the following form:

$$N_k \begin{bmatrix} M_{1,1,k} & \cdots & M_{1,7,k} \\ & & \\ M_{25,1,k} & \cdots & M_{25,7,k} \end{bmatrix} \begin{bmatrix} D_1 \\ \\ D_7 \end{bmatrix} = \begin{bmatrix} TD_{1,k} \\ \\ TD_{25,k} \end{bmatrix}$$

where N_k is the population in the kth zone at time t; M_k is a matrix of doses (i,j,k) for the kth zone in mrems per person at time t, for the ith nuclide (there are 25 radionuclides) impacting the jth body organ (there are seven); D is a vector of increased deaths per 0.1 rem per million population; and TD_k is a vector representing deaths associated with the dose from each nuclide in the kth zone at time t.

In calculating health costs a cut-off period of 1 million years is assumed, although some wastes will remain active well beyond that period. Logan *et al.* (1978), Cummings *et al.* (1978) and Schultze *et al.* (1981a) envisage some repository disruptions that could result in a release from the disposal site either directly to surface or to surrounding groundwater. These disruptions are: a severe earthquake, volcanic action and meteorite impact. A severe earthquake could lead to reactivation of old faults around the repository or even create new ones in the region, fracturing the repository and bringing wastes into contact with circulating groundwater. In order to assign a probability to earthquake damage Logan *et al.* (1978) studied tremors within 10 000 km^2 around the repository between 1964 and 1976 and identify 12 serious cases. Given the geological structure in the region, the probability of a repository fracture is estimated to be $1.4/10^7$ per year.

The likelihood of volcanic action which may affect the repository is estimated to be $8.1/10^{13}$ per year, a figure based upon observations that an average of one new volcano every 20 years has occurred during the last 225 years. By using a random probability from the WIPP area relative to the total area of the earth, the figure was computed. As for meteorite impact, a figure of $1/10^{13}$ per year is obtained based upon the same logic as volcanism. Note that the likelihood of volcanism and meteorite impact is extremely small and thus the calculation by Logan *et al.* (1978) is dominated by earthquake risk. Risks such as the occurrence of ice ages are not considered.

On the basis of risks such as earthquake, volcanism and meteorite impact the total health effects calculated are shown in Figure 10.3 which represents average

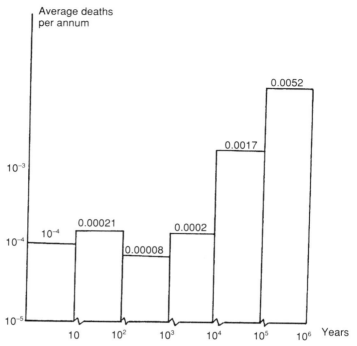

Figure 10.3   Average death rates from all cancer and genetic effects for the study region around the WIPP. (Source: Logan *et al*, 1978.)

rates of death per year over segments of repository lifetime considered. The increasing rate of mortality in later years is due to breakthrough of neptunium. The rate subsequently levels out as neptunium contribution peaks around year 1 000 000. The average annual excess death rate from cancer and genetic illness over the cut-off period is 0.0048.

The next step is to assign monetary values to the increased health risk, for which compensation variation is used. This considers the amount of compensation required to induce individuals to accept a situation where the probability of death is increased. The main problem with this method is that as the probability of death approaches unity, the compensation would approach infinity. However, this method is extremely useful in cases where the probability of death is increased only slightly, which is appropriate in this case (Cummings *et al.*, 1981).

Thaler and Rosen (1976) studied the United States labour market to identify cases in which individuals may accept a small increase in risk to life in return for higher earnings. In this a wage–risk analysis is conducted to ascertain how much individuals would require in additional income if they were to voluntarily accept jobs involving a marginal increase in risk of death. In order to estimate wage levels as a function of risk, Thaler and Rosen consider the existence of a job market for safety with the implication that job-associated risks and wage rates are positively correlated. Their research reveals that in the United States labour market, jobs with an extra risk of 0.001 paid between $176 and $260 extra per annum in 1967. Logan *et al.* (1978) note that these figures underestimate the value of risk to life as Thaler and Rosen's database includes an exceptionally high-risk set of job classification which may attract fewer risk-averse individuals. In the event, they express a preference for the high figure that is that each person would be willing to work for $260, in 1967 prices, per year less if the extra death probability were reduced from 0.001 to 0.0. By using OECD's labour statistics we can update the annual value to risk to life to $1264 in terms of 1991 prices.

Since the community is prepared to pay $1264 in terms of 1991 prices, to reduce the risk of death by 0.001, then the cost of reduction by 0.0048 would be $6067. The aggregate undiscounted health cost would then be $6 067 000 000 over one million years. Assuming that the repository will be sealed in the year 2025, the sum total of discounted costs to 1991 would be $9940, using a 5% social rate of discount.

Monitoring

In early stages of nuclear technology some scientists believed that once nuclear wastes are safely isolated in geological repositories there may not be a need for monitoring. Furthermore, time scales that are involved here are so long that they dwarf the span of recorded history. No human community has survived for periods anything like the time scales involved here. Nevertheless, today most people believe that it would be highly irresponsible to bury wastes in the ground and forget about the whole thing. As long as our societies remain in existence we must monitor nuclear waste repositories with the utmost diligence. In this section a 10 000 year cut-off period is assumed for monitoring costs, which is in line with recommendations by the US Environmental Federation Agency (see p. 295).

One proposed solution is the construction of a mausoleum to attract the attention of communities to the disposal site. However, experience with such structures indicates that they are unlikely to remain intact for long periods of time (Lipschutz, 1981). It was reported in Channell *et al.* (1990) that the Gnome site, located 13 km from the WIPP, was an underground test site in 1962. Since then the area has been cleaned up and a monument was erected, which has degraded over the years. Although measurable radioactivity still exists at the surface, the access to the site is open and there is little awareness by the locals of the significance of the test which took place only about 30 years ago.

Monitoring is also essential in order to prevent human intrusion in the form of search for natural resources. As mentioned earlier, the region is rich in potash and also contains sizeable oil and gas deposits. Channell *et al.* (1990) estimate an intrusion rate amounting to three boreholes per square kilometre per 1000 years on the basis of historic records for the WIPP area. Drilling in the region could create serious problems. First, it could alter the hydrology of the district. Second, boreholes going through the repository could bring wastes into contact with the brine reservoirs under the repository. Some of these reservoirs contain several million cubic metres of brine and sufficient pressure to flow to the surface.

A recent study by the Bureau of Land Management in the area found widespread non-compliance of private industry with the government regulations to plug abandoned holes securely. In October 1990 it was rediscovered that a gas well located just outside the WIPP site had slant drilled under the site in 1983 and produced gas from it until 1988 and the lease is still active. In 1987 the Department of Energy signed an agreement with the state of New Mexico prohibiting slant drilling under the area. Neither the Department of Energy nor the Bureau of Land Management knew of the existence of the well until a newspaper reporter received information from an anonymous source, then the event became public knowledge. It is possible that in the absence of monitoring, drilling in a remote area may occur in the future.

At the WIPP site only the operational experiments are imminent; sealing of the repository and monitoring are some distance away, beyond the year 2025. Many believe that as the time approaches towards the sealing operation there is likely to be a stringent regulation introduced for monitoring, which may even require more than one authority. Since the procedure to watch over the site is unknown at this point in time it is difficult to identify the cost of this operation. The state of New Mexico Environmental Evaluation Group, an independent technical appraisal body for the WIPP, guesstimates that the annual cost of monitoring should be $1–2 million (1991 prices) provided that no extraordinary event such as a determined threat of sabotage takes place. The figure of $1.5 million is used as an estimate for the annual monitoring cost over the study-period, which is 10 000 years. The total undiscounted monitoring cost is-$15 000 000 000. Its net present value, using a 5% social rate of discount, becomes about $5 700 000.

10.8 NUCLEAR WASTE DISPOSAL POLICIES

10.8.1 United States of America

Given the fact that the United States was the first nuclear country, militarily as well as commercially, it is no surprise to find that it has accumulated the largest quantity of wastes. Serious efforts to dispose of nuclear wastes by way of technical studies and legislation also started first in the United States. In 1980 the government announced that resolving waste management problems shall not be deferred to future generations.

In 1982 the Nuclear Waste Policy Act was passed which was a major milestone in the management of the nation's nuclear wastes. Signed into law by the President on 7 January 1983, and amended by the Nuclear Waste Policy

Amendments Act of 1987, this legislation established a national policy for storing, transporting and permanently isolating spent nuclear fuel and high-level radioactive waste. The Act also established the Office of Civilian Radioactive Waste Management within the Department of Energy to implement the policy and to develop, manage and operate a safe management system to protect public health and the environment. The main responsibilities of the Department of Energy are:

- to site, construct and operate geological repositories;
- to site, construct and operate one monitored retrievable storage facility; and
- to develop a system for transporting the waste to geological repositories and to monitor retrievable storage facilities.

The Nuclear Waste Policy Act specifies the process of selecting a site, constructing, operating, closing and decommissioning the repository. The United States Congress approved geological disposal as a reasonable method of disposal to protect the public and the environment from the dangers of nuclear waste and spent fuel. After a repository is closed, waste isolation will be achieved by a system of multiple barriers – natural and engineered. The former include geologic, hydrologic and geochemical. The latter consists of a waste package and underground facility. The waste package includes waste, container and material placed over and around containers. The underground facility consists of underground openings and backfill material that are used to further limit groundwater circulation around the waste containers.

In February 1983, the Department of Energy carried out the first requirement of the Nuclear Waste Policy Act by formally identifying nine sites as potentially acceptable for the first high-level waste repository. They were:

1. Vacherie, Louisiana (domal salt)
2. Cypress Creek, Mississippi (domal salt)
3. Richton, Mississippi (domal salt)
4. Yucca Mountain, Nevada (tuff)
5. Deaf Smith County, Texas (bedded salt)
6. Swister County, Texas (bedded salt)
7. Davis Canyon, Utah (bedded salt)
8. Lavender Canyon, Utah (bedded salt)
9. Hanfort Site, Washington (basalt rock).

The locations of these sites are shown in Figure 10.4.

After identification of these sites the Department of Energy published draft General Guidelines for the Recommendation of Sites. The final guidelines were published in December 1984. The Nuclear Waste Policy Act requires the Department of Energy to nominate at least five sites for site characterization (a formal information-gathering process including experiments on site). The Department must then recommend not fewer than three of those sites as

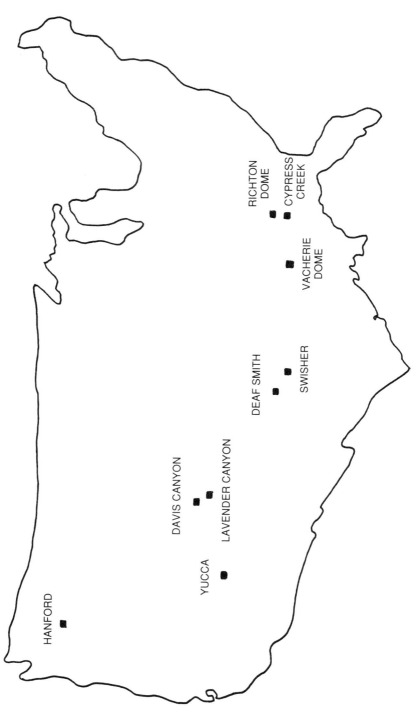

Figure 10.4 Proposed high-level nuclear waste repository sites, USA.

candidates for the first repository. After the site characterization is completed, one site will be recommended for repository development.

Under the provision of the Nuclear Waste Policy Amendments Act a Nuclear Waste Technical Review Board was established to evaluate the technical and scientific validity of activities of the Department of Energy regarding nuclear waste disposal. The Board must report to the Congress and the Department of Energy at least twice a year until disposal begins in a permanent repository.

In May 1986 the Department of Energy recommended three sites to the President for site characterization. They were the Hanfort site, Washington; Deaf Smith site, Texas; and the Yucca Mountain site, Nevada. In the Amendments Act, Congress directed the Department of Energy to conduct site characterization at the Yucca Mountain site as the candidate for a repository and to terminate site-specific studies at all other locations. At the Nevada site the Department of Energy is carrying out environmental and socio-economic studies in addition to technical characterization which includes two shafts more than 300 m deep. If the Yucca Mountain site turns out to be unsuitable all activities will be terminated and the Congress will be notified. If the site proves to be suitable for a repository it will be recommended to the President and to Congress for approval. The Nuclear Waste Policy Act permits the state of Nevada to submit a notice of disapproval to the Congress within 60 days. This will prevent the use of the site unless Congress passes a joint resolution to override the state's disapproval within the next 90 days, which would make the site a repository. If the state of Nevada enters into a compensation agreement (p. 287) then it waives its right to disapprove of the site recommendation.

If the Yucca site is chosen, about 2300 ha will be taken up to house the repository. The operation will look like a large mining complex, having a surface area of 60–160 ha surrounded by a 3-mile (5 km) controlled area. The repository would be constructed about 300 m below the surface. Ramps would be built to transport waste from the surface to the underground facility and vertical shafts would be built for ventilation, equipment and personnel. Provisions are being made to provide for the retrievability of the waste canisters for up to 50 years after emplacement. After that the repository would be sealed.

The United States Environmental Protection Agency (1985) recommend that a nuclear waste repository must be demonstrated to be capable of isolating radionuclides from the environment for at least 10 000 years. In this there must be reasonable assurances not only that future geological and hydrological changes at the site will not result in a leakage, but also that the construction and operation of the repository meets the highest standards of safety. However, Malone (1990) argues that a number of factors could affect the repository's performance during this period. Furthermore, the new time scale of 10 000 years does not consider the changing nature of the waste over a sufficient period of time (Kirchner, 1990). Based upon analysis, including isotopic composition of the waste, and cycling in the biosphere, Kirchner shows that wastes involve risks over a period at least double the 10 000 year period.

As for the Waste Isolation Pilot Plant in New Mexico, which is likely to become the United States' first nuclear waste repository aimed at storing transuranic, or mid-level, wastes generated by defence activities, it is a facility of the Department of Energy. The Public Law of 1979, 96–164, exempted the WIPP from licensing by the Nuclear Regulatory Commission. This does not apply to the high-level defence wastes as the Nuclear Regulatory Commission must regulate them. The WIPP is self-regulated by the Department of Energy. However, for the long-term stability of the WIPP and compliance with the EPA standards must be documented. The Department of Energy rather than the EPA is expected to do this.

The Department of Energy initially plans to perform experiments with a small quantity of waste before determining whether the WIPP is to be a permanent site. The Department had planned to start shipping waste to the WIPP in October 1988 but as the deadline approached it became clear that the site was not ready to receive waste. After that various target dates were established and none turned out to be viable. At the time of writing, shipping for experimental purposes is still postponed.

The Nuclear Waste Policy Act also requires the development and implementation of a safe transport system for nuclear wastes. Prototype spent fuel containers have passed various tests in which they were subjected to accident conditions. New, larger shipping containers are being designed to increase the amount of fuel carried to minimize journeys to repositories. These must undergo extensive testing and will require certification by the Nuclear Regulatory Commission before they can be used.

Existing regulations and laws on shipments enforced by federal, state and local agencies will be followed during shipments. In addition, the Department of Energy is developing procedures for inspection guidelines, route selection and other transport issues in collaboration with the affected states, local governments and Indian tribes and the local public. The Department of Energy also provides technical assistance and funding to states for training public safety officials in procedures for safe transport and emergency response situations for nuclear waste transportation.

10.8.2 United Kingdom

In September 1976 the Royal Commission on Environmental Pollution published its sixth report, *Nuclear Power and the Environment,* cmnd 6388. The Commission recommended that the responsibility for developing the strategy to deal with nuclear waste disposal should not lie with the body responsible for developing and promoting nuclear power. In its White Paper, cmnd 6820, the government accepted that an overall long-term strategy is needed and stated that the Secretaries of State for the Environment would in future be responsible, together with the Secretaries of State for Scotland and Wales, for radioactive waste management policy. Some of the stated objectives are: to secure the programmed disposal of waste accumulated at nuclear sites, to ensure that there is adequate research and development on waste disposal, to secure the disposal of waste in appropriate ways, at appropriate times and in appropriate places.

Up until 1989 nuclear power was a priority for successive post-war governments in Britain, both in terms of national energy policy and government research and development expenditure. For example, expenditure on research and development of the first breeder reactor programmes totalled £4 billion at 1988 prices (UK Energy Committee, 1990). There have been three reactor development programmes:

1. Magnox technology developed in the early 1950s formed the basis of the first UK civil reactor programme.
2. Advanced gas-cooled technology was announced in 1964. This involved a 20-fold scaling-up from a small prototype. By the early 1970s problems surrounding the construction of advanced gas-cooled reactors were compounding. Nevertheless, in 1978 the government decided to order two more advanced gas-cooled reactors.
3. The proposed use of pressurized water reactors was announced in 1979. This was intended to produce 15 gigawatts of capacity and cost £15 billion in terms of 1979 prices. However, the UK Energy Committee (1980/81) expressed strong reservations about the justification for this programme. After a long public enquiry, in 1987 the government authorized the construction of Britain's first pressurized water reactor, Sizewell B, which is officially scheduled for completion in 1994. At the time when Sizewell B was authorized, the pressurized water reactor scheme was scaled down to a small family of four identical reactors. On 9 November 1989 plans for the three other reactors were abandoned, at least until 1994 when the government intends to review the nuclear situation.

The government statement of 9 November 1989 was indeed an abrupt reversal of post-war nuclear policy. One reason for the reversal was failure to privatize the nuclear power stations. In July 1988 the Energy Committee expressed concern about the cost of nuclear power by noting that the government had glossed over the industry's economics and ignored the industry's external costs. Serious doubts were expressed about the wisdom of privatizing nuclear power. Likewise, in September 1987 Kleinwort Benson, one of the government's financial advisors, expressed doubts about the feasibility of privatizing the nuclear industry. Main areas of concern were decommissioning power plants, management of spent fuel and waste, and insurance risks such as plant damage and injury to third parties.

As for the disposal of Britain's nuclear wastes, UK Nirex, the industry's waste disposal company, is studying the area near Sellafield in Cumbria, in the north-west of England. One major reason for the choice of this location is that most of the waste is generated there. If Sellafield proves to be unsuitable, Nirex could focus on Dounray in the north of Scotland, in which case costs will more than double, largely because of transportation. The main object of the Sellafield study is to construct a deep underground repository for the nation's low and mid-level defence and civilian wastes.

Early tests at the site revealed that the geology and hydrology in the region are more complex and less favourable than Nirex had originally hoped. However, the company's manager, Michael Folger, in a letter to the local councils in 1992, emphasized that geological studies had confirmed the overall suitability of the area for an underground repository. It is expected that after further tests Nirex will apply for planning permission in 1994. If everything goes well, the date of completion for the repository would be the year 2005. Sealing of the repository could take place around 2050. The original cost estimate was £2.5 billion in terms of 1992 prices but delays in acquiring planning permission, construction, emplacement and decommissioning would increase the cost.

10.8.3 European Community

The desire to create the safest possible conditions associated with nuclear activities is expressed in the preamble to the 1957 Treaty of Rome, when the European Atomic Energy Community, Euratom, was created. In it the onus is on the member nations to ensure safety, in view of their direct involvement in the field and their national legal and administrative regulations. However, it is one of the tasks of the European Commission to establish and apply a common safety policy in order to protect the people and the environment from nuclear wastes. The Community's policy on nuclear safety mainly comprises research and implementation of regulatory measures. The results of completed and ongoing research are made available to all interested parties.

On the transportation of wastes across national frontiers, the International Atomic Energy Agency (IAEA) in Vienna, drew up regulations in 1959. These largely cover the packages and containers on which the safe transportation of radioactive substances depends. Containers must be specifically designed to keep the level of ionizing radiation as low as possible during normal transport and during accident scenarios. Also, the packages must be designed to prevent any possibility of triggering a chain reaction. The European Community co-operates with the IAEA and other bodies concerned with transportation in an effort to approximate the national laws of the member states. Emphasis is placed upon prevention of theft.

As the traffic of radioactive wastes is increasing, regulation is becoming more and more urgent. The European Community is intensifying efforts to harmonize licensing procedures and transport formalities and the establishment of emergency services capable of coping with accidents by road, rail, water and air.

Co-operation between member states to manage nuclear wastes was called for in 1973 by the Council of Ministers. Then the Community embarked on research that takes two forms. First, the Community runs five-year action programmes which it partially finances and which it co-ordinates in the framework of a shared-cost project with various national laboratories. Second, it undertakes its own research at the Joint Research Centre, mostly in Italy.

In 1980 the Council of Ministers adopted a proposal laying down a plan of action for the management and storage of radioactive wastes. This plan extended from 1980 to 1992. There are numerous research projects underway, most of them following on from earlier research work including the need to diminish the volume of waste generated. The Commission is also involved in efforts to store wastes in underground repositories in many member countries. The Commission is also promoting large-scale exchange of information on the decommissioning projects carried out in member and non-member states. Nuclear safety policy is in a continuous state of development as technology progresses and political conditions change.

11

Ordinary and modified discounting in natural resource and environmental policies

Justice is the first virtue of social institutions, as truth is of systems of thought. A theory, however elegant and economical, must be rejected or revised if it is untrue; likewise laws and institutions, no matter how efficient and well arranged, must be reformed or abolished if they are unjust.

J. Rawls

Let us make no mistake about it, discounting is one of the most fundamental factors affecting policy decisions about destructible resources. In Chapter 3 it was explained, at some length, that the magnitude of the discount rate is one of the most crucial factors in determining the economic viability of afforestation projects. Furthermore, it is one of a number of crucial variables in identifying the optimum cutting age (p. 106–17). In the economics of fisheries, when the discount rate is not zero the owner of the fishery would be faced with an intertemporal trade-off, that is, *ceteris paribus*, a positive rate of interest implies a larger harvest this year and a smaller one the next year (p. 55–8). As for policies regarding the depletion of mining, petroleum and natural gas deposits, it is the discount rate which determines forcefully the price/output path for the extractive industry (equation (5.9)).

There are a number of important aspects of discounting such as the distinction between private and social rates of discount; distinction between social time preference and social opportunity cost rates; determining the correct magnitude of discount rate; discounting on behalf of present and future generations and the gap between ordinary modified discounting methods. This chapter will touch on all these issues.

When destructible resources are owned by private individuals the market rate of interest will be a powerful guiding force in their business decisions. But when resources are owned by society, policy-makers will want to use the social discount rate as opposed to the market rate. It is at this point that most of the controversies arise. What is the theoretical foundation of a social rate of discount? What is the correct magnitude for the communal rate? How can the claims of future generations be incorporated into the social rate of discount? How different should the social rate be from the market determined one? What reasons can be given for the refutation of ordinary discounting in favour of modified discounting? These are some of the questions that economists ask themselves. These are the questions that generate controversy.

11.1 PRIVATE VERSUS SOCIAL RATE OF INTEREST

In the economic literature there is a long-winded discussion about whether the market rate of interest is equal to the social rate of interest. A substantial body of economists argue that in a world of perfect competition, which requires a set of heroic assumptions, a single interest rate equates the marginal time preference of savers with the marginal rate of return on capital. Preference patterns for consumers include desired distribution of expenditures over time. By borrowing and lending at the market rate of interest they arrange their expenditure in such a way that total satisfaction during the entire period for which their plan extends is at a maximum, as judged by their present preferences. Firms invest up to a point where the rate of return on marginal investments is equal to the interest rate. Consumers' plans to save are brought to equality with producers' plans to invest and the ruling interest rate reflects both the time preference of consumers and returns which can be earned on capital projects. In this situation the optimal rate of investment and saving is achieved and there should be no divergence between the social and private rates of discount.

Figure 11.1 illustrates the case. Curve II shows the community's consumption preference pattern between two periods, t_0 and t_1. This social indifference curve is obtained by aggregating individual indifference curves. PP' shows the community's production possibility frontier which is in normal shape. A movement from P to P' indicates that, *ceteris paribus*, the marginal productivity diminishes as we employ more capital. The slope of the transformation schedule at any one point is related to the net marginal social productivity of capital at that point. Optimality requires a state of tangency between one indifference curve and the transformation schedule. This is established at D where angle α_2 shows that both curves have equal slopes. The state of optimality is obtained by a saving level which is CP. On the other hand, if the amount of saving was, say, EP, this would create two interest rates and the community would be placed on a lower level of welfare, shown as I'I'. At this point, where saving is suboptimal and the social indifference curve cuts the transformation schedule at H, the slope of I'I' is less than the slope of PP'. This is seen by the difference between angles α_2 and α_3 ($\alpha_2 > \alpha_3$). Here society is effectively expressing a preference for a higher level of saving and investment. In the absence of restrictions, the economy should be able to move from H to D as the saving/investment level

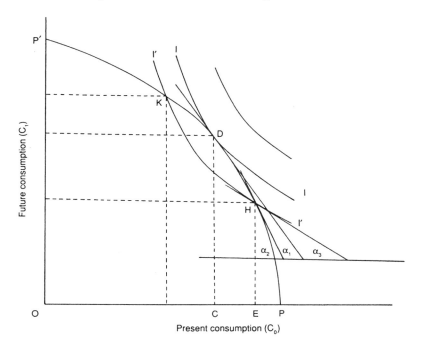

Figure 11.1   Optimum and suboptimum levels of saving and resulting social rates of discount.

increases towards C where the discrepancy between the two rates eventually disappears.

Point K is the reverse case where the community is investing too much and as a result the social time preference rate becomes greater than the social opportunity cost rate, which also puts the community on a lower level of welfare. Again in a world of perfect competition, society should be able to move from K towards D by increasing the level of consumption at the expense of saving and ultimately the gap between the two interest rates will disappear at D.

11.1.1 Barriers to equilibrium

A number of economists (e.g. De Graaff, 1957; Feldstein, 1964; Mishan, 1971) argue that the point of optimality in which a single interest rate equates community's desire to save with investors' plans to invest is unlikely to be achieved because of a number of reasons.

Inequality in distribution of wealth

When individuals lend and borrow money in the market they are likely to make errors of judgement or become victims of unfortunate events. The wealthier the borrower the better placed he/she is in any problem arising from misjudgement or misfortune. Therefore, given the differences in wealth it would be irrational for lenders to lend as much to the poor as to the better off members of society, or at least to lend the same amount on the same terms to each. Therefore, in order to have a single interest rate, the ideal market also requires that 'people be equally wealthy and wise'. In the real world, the situation is very different indeed, as wisdom and wealth are unequally distributed. Every loan is a gamble and thus the interest charge must reflect not only the pure rate of discount, but compensation for risk taking. This gives rise to varying rates of interest depending on the position of the borrower.

Ignorance and interdependence

In the perfect market it is assumed that a private individual in his/her intertemporal allocation of consumption must foresee their future income as well as the future prices of all goods and services. Individual saving rates depend on expected future prices, which are influenced by the saving rates of others. No one individual acting separately has any means of knowing what other individuals' intentions are. In other words, individuals cannot have the necessary information to determine optimally their saving levels. On the investment side the revenue from one project depends on the investment decisions on other projects. The individual investors, just as the individual savers, cannot have the information necessary for rational intertemporal decision-making.

Institutional barriers

Baumol (1968, 1969) argues that there are institutional barriers such as taxation which prevent society from attaining the optimal level of equilibrium corresponding to point D in Figure 11.1. In order to see this let us establish a simple model by making a number of assumptions. First, in a society all goods and services are provided by corporate firms. Second, these corporations finance their projects by equity issues. Third, corporate income is subject to a uniform tax rate of 50%. Fourth, there is a unique interest rate r at which the government borrows money to finance public sector projects, which require input resources. Since corporations are the only owners of inputs they can only be transferred to public use by taking them out of the hands of corporate firms. The opportunity cost of input resources can then be calculated simply by determining the returns which would have been obtained if they had been left for corporate use.

In a riskless world investors would expect the same rate of return on money invested in either the private or the public sector. This means that corporations must return $r\%$ to their shareholders. But a 50% tax on corporate earnings means that corporations must provide a gross yield of $2r$ for their shareholders. In other words, resources if left in the hands of the private sector would have yielded $2r$ in real terms. This proposed calculation has significant consequences for public policy. With a 5% rate of interest on government bonds, the rate of discount on government projects is not 5%, but is in the region of 10%. Baumol further argues that whether resources are transferred from private to public sector by borrowing or by taxation does not matter. Equally, it makes no difference whether the resources are drawn to the public sector from private investment or from private consumption; all that matters is that the transfer must take place through the agency of the corporation. The transferred inputs would have brought corporations $2r$ as a result of consumers' marginal valuation of those commodities.

Natural barriers

In order to open the road to the point of optimality one may think that the government should abolish corporation and other distortive taxes. Unfortunately, this would not be sufficient to solve the problem because there is another barrier, risk. Since the national investment portfolio consists of a large number of projects, it is reasonable to expect that some will underachieve whereas others will exceed their estimates and in this way there will be an overall compensation. Society benefits from the entire set of investment projects, whether they are public or private. The transfer of an investment from private hands to the government does not affect its flow of benefits to society, nor does it mean that risks can be offset to a greater or lesser degree against the other projects.

Unlike Samuelson (1964) and Arrow (1966) who argue that the risk premium should be excluded from the discount rate, Baumol (1968, 1969) concludes otherwise. Since risk does exist from the individual investor's point of view, it plays exactly the same role as corporation tax by driving a wedge between the rate of time preference and rate of return. Firms will invest only up to a point where expected returns are higher than they would be in the absence of risk. Consequently the economy will be stuck at the point of low investment and low state of welfare.

11.2   FOUNDATION FOR THE CHOICE OF A SOCIAL RATE OF DISCOUNT

There is an equally long-winded argument in economic literature about this issue. There are, basically, three ideas: the government borrowing rate, the social opportunity cost rate and the social time preference rate.

11.2.1   The government borrowing rate

The use of the long-term government borrowing rate as the appropriate social discount rate in public sector economics was the earliest practice. During the 1930s in the United States, when cost–benefit analysis was first used to evaluate the worth of various land-based projects, the discount rates employed by the appraisal agencies were very much in line with the long-term government borrowing rate. There were two justifications for using such a rate. First, some economists related the government borrowing rate to the risk-free rate of return in the economy. Their argument was that since public sector projects can be considered as risk-free, because of a large portfolio situation, only a risk-free interest rate was suitable in cost–benefit analysis. Second, the government borrowing rate represents the cost of capital used in the construction of public sector projects. However, under close scrutiny these arguments proved to be flawed and thus this school has lost its appeal.

It is not correct to claim that government bonds are risk-free. In fact, they are subject to two types of risk. First, variations in the purchasing power of money affect all real interest rates, including those promised by bonds. Second, the market value of a bond is also affected by fluctuations in the interest rate. For example, take a perpetual bond of £1000 issued at a fixed interest rate of 5% at the time when the ruling market rate was also 5%. If the market rate goes up to 10%, then nobody will pay £1000 for this bond as one could earn twice as much by putting £1000 into a bank account. Therefore, a doubling of the market interest rate would halve the market value of the bond. As for the second alleged merit of the bond rate this too is a dubious argument. In reality, governments raise only a small proportion of their revenue by borrowing, the bulk of their revenue comes from taxation. Therefore, nowadays most economists do not advocate the use of government borrowing rate in public sector projects.

11.2.2 The social opportunity cost rate

A large number of economists contend that, since capital funds are limited, a public sector investment project will displace other projects in the economy. Therefore, in cost–benefit analysis the appropriate rate of interest must be the one that reflects the social opportunity cost of capital. This rate measures the value to society of the next best alternative investment in which funds might otherwise have been employed. Generally, the next best alternative investment is thought to be in the private sector (Krutilla and Eckstein, 1958; Hirshleifer et al., 1960; Kuhn, 1962; Joint Economic Committee of the US Congress, 1968).

There are some serious problems associated with this approach. First, the private sector's view of rate of return on capital can be very different from society's view of profitability. In private profitability accounts, external costs such as noise, pollution and congestion which a project may generate are not normally considered in calculations. Another point is that private profits may be quite high, not as a result of an efficient operation, but as a result of market imperfections such as monopoly, cartel, oligopoly, etc., all of which work against the public interest. Most economists believe that the private profits and resulting rates of return on capital require a substantial social adjustment before they can be used in public sector project evaluation.

As explained in the forestry chapter, the British social discount rate is based upon the opportunity cost rate concept. In determining the 'correct' magnitude of this rate the British government has largely looked upon the private sector profitability in the country. For example, in the spring of 1989 the government revealed its new policy on discounting in a parliamentary exchange when Mr Boswell rose to ask John Major, then the Chancellor of the Exchequer, whether the government would make a statement on new investment in nationalized industries and the discount rate used for appraising investment in other parts of the public sector. In response, Mr Major stated that

> The government have reviewed the level and use of discount rates in the public sector. These were last reviewed in 1978. Since then the rate of return in the private sector has risen to around 11 percent. In the light of this the government have decided to raise the required rate of return for nationalised industries and public sector trading organisations from 5 percent to 8 percent in real terms before tax. The new required rate of return of 8 percent will be an important factor in setting new financial targets, but there will be no impact on pricing during the life of the existing financial targets.
>
> (*Hansard*, 1989)

At present, the choice of discount rate is a matter for individual nationalized industries or trading bodies to decide in consultation with sponsor departments and HM Treasury. The government's main concern is that industries' approach should be compatible with achieving the required rate of return on the programme as a whole. In appraisal of new capital projects risk should be given proper attention. The effect of full allowance for risk will often be implicitly equivalent to requiring a higher internal rate of return on riskier projects. The British government also decided that the discount rate to be used in the non-trading sector should be based on the cost of capital for low risk projects. In the conditions that prevailed in 1989 this indicated a rate of not less than 6% in real terms. Risks were to be analysed separately and projects that were more risky were required to demonstrate correspondingly lower costs or higher benefits. This would ensure that projects in the non-trading sector would be as demanding as those in the trading sector. In particular, they would provide a comparable basis for the consideration of private participation in public sector activities by taking account of the full economic cost of the public sector option. Table 11.1 gives the details of the rates of return in various sectors of the British economy which formed the basis of the government's conclusion.

11.2.3 The social time preference rate

Another school of economists believes that the correct rate of discount for public sector investment projects is the social time preference rate (STPR), also called the consumption rate of interest (CRI). The rationale for their argument is quite simple. The purpose behind investment decisions is to increase future consumption, which involves a sacrifice on present consumption.

What constitutes a social time preference rate? There seem to be two rational factors: diminishing marginal utility of increasing consumption, and risk of death. As regards the former, it is argued that in most societies the standard of

Table 11.1   Rates of return in various sectors of the British economy, 1960–86

| | Rates of return | | |
| | | | |
Year	All sectors	Manufacturing non-North Sea	Manufacturing companies
1960	13.1	13.1	14.5
1961	11.3	11.3	12.0
1962	10.3	10.3	10.8
1963	11.4	11.4	11.5
1964	12.0	12.0	11.8
1965	11.4	11.4	11.5
1966	10.0	10.1	9.5
1967	10.1	10.2	9.6
1968	10.2	10.3	9.3
1969	10.1	10.2	9.6
1970	8.9	8.9	8.0
1971	9.1	9.1	6.8
1972	9.5	9.5	8.1
1973	8.9	9.0	7.9
1974	5.2	5.4	4.0
1975	4.0	4.3	2.6
1976	4.4	4.4	2.9
1977	7.5	6.8	5.7
1978	7.9	7.1	5.9
1979	7.4	5.7	4.2
1980	6.4	3.9	3.1
1981	6.2	3.0	2.0
1982	7.5	3.8	3.6
1983	9.1	4.8	4.1
1984	10.7	5.6	4.8
1985	11.4	7.2	6.1
1986	10.0	8.9	7.2

Source: *British Business*, 9 October 1987.

living enjoyed by individuals is generally rising. As the income of a person increases steadily, the satisfaction gained also increases but at a slower rate. That is, each addition to his/her income yields a successively smaller increase to his/her economic welfare. Therefore £1 now should mean more to an individual than £1 later on. As regards the latter, it is argued that the risk of death is a powerful factor affecting an individual when making intertemporal consumption decisions. His/her preference for present consumption over future consumption of the same magnitude is perfectly sensible since he/she may not live to consume it in the future.

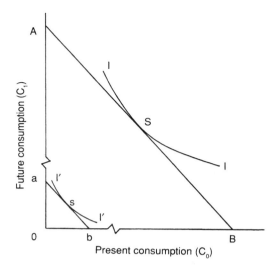

Figure 11.2   Communal indifference curve (*II*) and indifference curve for Mr/Ms Average (*I'I'*).

The author has, over the years, developed models to estimate social time preference rates for a number of countries (Kula 1984a, 1985, 1986a, 1987b). The simplest is as follows. Technically speaking, in a two-period analysis the STPR corresponds to the marginal rate of substitution of consumption along the social indifference curve. In Figure 11.2 the trade-off between present and future consumption is shown along the social indifference curve II. If the community is located at a point such as S, then the STPR will be equal to the marginal rate of substitution of consumption (MRSC) at this point, minus one. That is:

$$S = |MRSC_{0,1} - 1|, \tag{11.1}$$

where S is social time preference rate, MRSC is marginal rate of substitution of consumption, and subscripts 0 and 1 refer to present and future, respectively.

In order to arrive at an operative formula, the following argument is suggested. Let us assume a typical individual, Mr/Ms Average, who represents the community at a given point in time. He/she has an indifference curve, shown as I'I' in Figure 11.2, which is a miniaturized version of the social indifference curve. Along this we locate him/her at point s, where the marginal rate of substitution is the same as point S on the social indifference curve, i.e. OA/OB = Oa/Ob. Then this time preference equals the STPR. He/she has, say, a two-period consumption utility function, u, which has a constant elasticity, and its net present value (NPV) is:

$$\text{NPV}(u) = \frac{(A)(C_0)^{1-e}}{(1-e)} + \frac{(A)(C_1)^{1-e}}{(1-e)(1+m)}, \tag{11.2}$$

where C is real consumption with subscripts from year 0 to year 1; A is a constant; e is elasticity of marginal utility of consumption; and m is risk of death from one period to the next.

The marginal rate of substitution is a ratio which is arrived at by differentiating equation (11.2) with respect to C_0 and C_1 and dividing $d\text{NPV}(u)/dC_0$ by $d\text{NPV}(u)/dC_1$, that is:

$$\frac{d\text{NPV}(u)}{dC_0} = (A)(C_0)^{-e}$$

$$\frac{d\text{NPV}(u)}{dC_1} = \frac{(A)(C_1)^{-e}}{(1+m)}$$

$$\text{MRSC}_{0,1} = \frac{d\text{NPV}(u)}{dC_0} \Big/ \frac{d\text{NPV}(u)}{dC_1} = (1+m)\left(\frac{C_1}{C_0}\right)^e.$$

Then by definition the STPR (S) is:

$$S = (1+m)\left(\frac{C_1}{C_0}\right)^e - 1. \tag{11.3}$$

The growth rate of the individual's real consumption between these two time points is:

$$g = \frac{(C_1 - C_0)}{C_0},$$

where g is growth rate of real consumption.

This yields:

$$(1+g) = \frac{C_1}{C_0}.$$

Substituting this in equation (11.3) we get:

$$S = (1+m)(1+g)^e - 1. \tag{11.4}$$

By linear approximation equation (11.4) can be written as:

$$S = e(g) + m. \tag{11.5}$$

11.2.4 Estimates of social time preference rates for the UK, United States and Canada

What are the likely estimates of social time preferences rates for the UK, the United States and Canada on the basis of equation (11.5)? Let us take the

Table 11.2   Annual crude death rates (%) in the UK, US and Canada

Country	Year						
	1946	1950	1955	1960	1965	1970	1975
UK	1.16	1.17	1.17	1.15	1.16	1.20	1.18
US	0.99	0.99	0.99	0.99	0.99	0.99	0.99
Canada	0.99	0.99	0.99	0.99	0.99	0.99	0.99

Source: Kula (1987b).

mortality-based pure time discount rate, m, first. In 1985 the UK's population was 55.9 million and the number of deaths was 655 000, giving, on average, a death rate of 1.17%. The crude death rates calculated in this fashion for these countries are shown for some selective years in Table 11.2. The figures for the United States and Canada have remained virtually unchanged at 0.99%, whereas the figure for the UK is around 1.17%. These numbers are taken to represent the risk of death faced by Mr/Ms Average in these countries. That is:

Country	Mortality based pure time discount rate, m (%)
UK	1.17
United States	0.99
Canada	0.99.

The growth rate of real consumption, g, is obtained by fitting the equation

$$\ln(C) = A + gt \tag{11.6}$$

to the series in Table 11.3 where C is per capita real consumption, A is a constant and t is the number of years between 1954 and 1976 inclusive. The results are:

Country	Growth rate of real consumption, g (%)
UK	2.0
Unites States	2.3
Canada	2.8

The third component parameter, the elasticity of marginal utility of consumption, e, a relatively simple model, which was initiated by Fisher (1927) and Frisch (1932) but more recently developed by Fellner (1967), can be used. In this model the elasticity of marginal utility of consumption is measured by the ratio of income elasticity to pure price elasticity of the food demand function, that is,

$$e = \frac{y}{\hat{p}}, \tag{11.7}$$

Table 11.3 Series for equation (11.6)

Year (*t*)	Per capita real consumption (*C*)		
	UK, £ (1970 prices)	US, $ (1967 prices)	Canada, $ (1967 prices)
1954	415	1810	1258
1955	431	1913	1315
1956	431	1972	1371
1957	440	1976	1371
1958	449	1945	1375
1959	466	2031	1412
1960	480	2043	1415
1961	487	2036	1402
1962	493	2100	1446
1963	512	2152	1488
1964	526	2235	1539
1965	529	2340	1594
1966	537	2436	1638
1967	545	2472	1701
1968	556	2560	1735
1969	558	2592	1771
1970	571	2597	1797
1971	585	2657	1894
1972	617	2801	2010
1973	642	2892	2117
1974	636	2842	2201
1975	632	2947	2265
1976	630	2983	2376

Sources: UK: *Economic Trends*, Central Statistical Office 1955–1979, HMSO, London.
US: *Statistical Abstract of the US*, 1957–1976, Government Printing Office.
Canada: *Canada Yearbook* 1955–1976, Ministry of Supply and Services.

where y is the income elasticity of food demand function and $\hat{p}$ is the pure price elasticity which is obtained by eliminating the income effect from the total price elasticity, p.

In order to find the required elasticity parameters in equation (11.7) the food demand functions are specified as:

$$\ln D = \ln \beta + y \ln Y + p \ln P + tT \qquad (11.8)$$

for the UK and

$$\ln D = \ln \beta + y \ln Y + p \ln P \qquad (11.9)$$

for the United States and Canada. D is the adult equivalent of per capita food demand, Y is the real income measured in the same adult equivalent unit and P is the relative price of food in relation to non-food. The series for regression equations (11.8) and (11.9) are shown in Table 11.4. The results are:

Country	y	p
UK	0.39	−0.54
United States	0.51	−0.37
Canada	0.50	−0.41

Then, in order to estimate the pure elasticity, $\hat{p}$, the following procedure, supported by Stone (1954) and Fellner (1967), is employed:

$$\hat{p} = y - (b)(p), \tag{11.10}$$

where b is the share of food in consumer's budget (a percentage). Equation (11.10) is also the standard Slutsky equation expressed in terms of elasticities. The budget share of food, b, is 20% in the UK and United States and 17% in Canada. Substituting these together with other estimates in equation (11.10) gives:

$$\text{UK } \hat{p} = -0.47$$
$$\text{United States } \hat{p} = -0.27$$
$$\text{Canada } \hat{p} = -0.32$$

Then by using equation (11.7) we get:

Country	Elasticity of marginal utility of consumption, e
UK	−0.70
United States	−1.89
Canada	−1.56

Substituting these estimates in equation (11.5) we get:

Country	Social time preference rate (%)
UK	2.6
United States	5.3
Canada	5.4

11.3   ISOLATION PARADOX

This analysis is about the interdependence of welfare between generations and the resulting social discount rate, which first appeared in the writings of Landauer (1947) and Baumol (1952). They argue that an individual may be perfectly willing to make a sacrifice on his/her present consumption to benefit future generations if others do so. A single person would not make the sacrifice alone since he/she knows that his/her own loss would not be compensated for by a future gain.

	Food demand, D, per capita adult			Income, Y, per capita adult			Relative food price, P		
Year T	UK (1970 £)	US (1967 $)	Canada (1961 $)	UK (1970 £)	US (1967 $)	Canada (1961 $)	UK (1970 = 100)	US (1967 = 100)	Canada (1961 = 100)
1954	121	504	285	481	2144	1512	101	104	101
1955	123	511	301	498	2271	1581	110	102	101
1956	130	483	315	501	2343	1650	110	101	101
1957	131	480	315	509	2351	1652	108	101	102
1958	131	469	315	521	2316	1659	107	103	104
1959	134	471	323	541	2420	1704	108	100	99
1960	136	533	327	559	2436	1708	105	99	99
1961	137	525	319	568	2428	1699	102	99	100
1962	137	529	320	575	2501	1745	97	99	101
1963	137	527	318	597	2561	1893	101	99	104
1964	138	535	325	613	2656	1850	101	99	105
1965	136	554	328	617	2774	1911	99	100	106
1966	138	557	321	628	2880	1957	99	102	110
1967	138	555	328	638	2917	2022	99	100	108
1968	137	560	324	650	3007	2054	98	99	106
1969	135	551	328	652	3041	2086	99	99	107
1970	134	560	336	667	3049	2108	100	98	106
1971	132	558	353	682	3095	2214	102	97	104
1972	128	588	359	718	3255	2343	104	98	109
1973	126	572	360	742	3352	2460	112	108	111
1974	125	552	355	738	3282	2552	115	113	125
1975	121	571	363	731	3283	2619	116	112	124
1976	118	590	385	731	3435	2743	121	110	118

Source: Kula (1984a, 1987b).

Sen (1961), building on the ideas of Landauer and Baumol, argued that the present rate of saving, and hence the resulting discount rate, not only influences the division of consumption between now and later for the same group of people, but also that between the consumption of different generations, some of whom are yet to be born. If democracy means that all the people affected by a decision must themselves take part in the process, directly or through representatives, then clearly there can be no democratic solution to the optimal level of saving. He then questions whether the rate of discount revealed by individuals in their personal choices is an indication of views of present generations of the weights to be attached to their own consumption or that of the consumption of future individuals. The isolation paradox proves a negative answer to this question. Assume a situation in which a man, in isolation, is facing a dilemma of choosing between one unit of consumption now and three units in 20 years' time. He knows, for some reason, that he will be dead in 20 years' time, and may well decide to consume the unit. But now imagine that a group of men come along and tell him that if he saves the unit, they will follow suit. Then it may not be irrational for the first man to change his mind and save the unit. This is because the gain to future generations would be much greater than his loss and he could bring this about by sacrificing only one unit of consumption. Therefore, without any inconsistency, he may act differently in two cases. This indicates that although individuals are not ready to make the sacrifice alone, they may well be prepared to do so if others are ready to join in. Sen concludes that the saving and investment decision, and hence the resulting social discount rate, is essentially a political decision and cannot be resolved by aggregating isolated decisions of individuals.

Later Sen (1967) pointed out that the earlier view by Baumol (1952) was not the same as his concept of the isolation paradox. In isolation individuals would like to see others invest but abstain personally unless forced to participate in a joint action. Otherwise they would choose to be free riders. In order to secure a collective investment a compulsory enforcement is needed. In Baumol's case the problem is one of assurance; that is, individuals are willing to invest if they feel others will do the same. In other words, for each individual to invest as a separate unit, faith is enough and compulsory enforcement is not necessary.

Marglin (1962, 1963b), dealing with similar issues, argues that private investment decisions in which benefits appear only after the investor's death are undertaken because of the existence of a market which makes it possible to change the returns which occur after the investor's death into consumption benefits before the investor's death. But why do governments require citizens to sacrifice current consumption in order to undertake investments which will not yield benefits until those called upon to make the sacrifice are all dead, despite the fact that there is no market by which one generation can enforce compensation on the next? Is it because there is a difference between the way individuals view their saving versus consumption decisions collectively and individually?

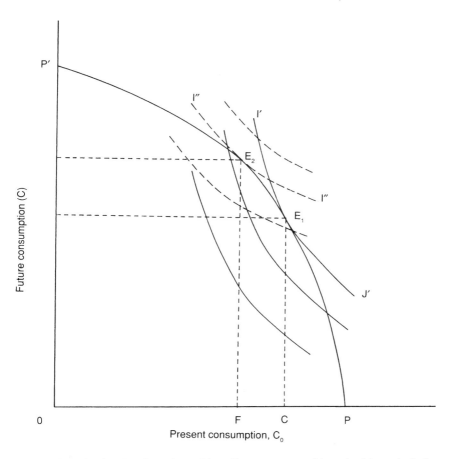

Figure 11.3   Saving levels and resulting discount rates with and without isolation paradox.

One possible explanation is that individuals have dual and inconsistent time preference maps: one map representing the selfish side of our character and the other the responsible citizen's. For example, an individual acting as a citizen may favour a strictly enforced system of traffic laws, but when caught as an offender may attempt to bribe the arresting officer. Figure 11.3 illustrates the situation. Indifference curves depicted as solid lines are derived by aggregating the unilateral decision of individuals acting in isolation in the marketplace. The optimal point of equilibrium is established at point E_1, where I'I' and PP' schedules are tangent, requiring a saving level of CP. However, if individuals' preferences were aggregated with the assurance that others would participate in the collective decision the resulting indifference curves would be less present-orientated *vis-à-vis* those of the curves derived by aggregating the isolated decisions. In Figure 11.3 these curves are shown as dashed lines, and the optimum equilibrium point is E_2, requiring a saving level of FP. Clearly, the equilibrium reached by one indifference map is not optimal by another one. Of course, the most important question is, which preference map is relevant? One can argue that since actions speak louder than words the preference patterns revealed in the marketplace are more genuine and should be considered by the decision-making body.

Marglin's analysis triggered off further discussion and it will be revealing to formalize it by setting up a simple model with a number of assumptions. Time is divided into two periods, present and future. All members of the present community die at the end of present, and their places are taken by individuals who come into existence fully grown at the beginning of future. The same investment opportunities are open to all individuals. Members of society are alike in the sense that they have similar time preference maps. The utility function of one individual is:

$$U_i = f(C_i, C_f, C_p - C_i), \qquad (11.11)$$

where U_i is the utility function of the ith individual; C_i is the consumption of the ith individual; C_f is the consumption of future individuals; and $C_p - C_i$ is the consumption of ith individual's contemporaries.

Differentiating equation (11.11) we get:

$$dU_i = \frac{\partial U_i}{\partial C_i} dC_i + \frac{\partial U_i}{\partial C_f} dC_f + \frac{\partial U_i}{\partial (C_p - C_i)} (dC_p - dC_i) \qquad (11.12)$$

where

$$\frac{\partial U_i}{\partial C_i} = 1; \quad \frac{\partial U_i}{\partial C_f} = \alpha; \quad \frac{\partial U_i}{\partial (C_p - C_i)} = \beta > 0. \qquad (11.13)$$

That is, the marginal utility of each item in the utility function is positive. The marginal utility of the ith individual's consumption is unity, that of future generations is α and his contemporaries is β. Assume now that the marginal rate of transformation between present and future is $\bar{r}$. That is, at the margin £1 of present sacrifice adds £$\bar{r}$ to the consumption level of the next generation.

The individual is willing to invest £1 as long as

$$dU_i \geq \alpha \bar{r} - 1. \qquad (11.14)$$

But instead of him/her investing, if somebody else invests he/she will still get satisfaction because the utility of future generations will increase to the same extent, but the loss in his/her eyes would be only β instead of unity. Therefore, the ith person will be pleased if somebody else invests so long as

$$dU_i \geq \alpha \bar{r} - \beta. \qquad (11.15)$$

A third person is also guided by equation (11.15). This means that nobody wants to invest personally, though each would like to see others invest. Each individual would be willing to invest provided others did so, for in this case the gain from the investment of others would outweigh the loss on one's own investment. If there are n individuals in the community, investment will take place if the following holds:

$$dU_i \geq n\alpha\bar{r} - 1 - \beta(n-1), \tag{11.16}$$

where $n\alpha\bar{r}$ is the utility gain to the ith person via future generations' benefit; 1 is utility loss to the individual for his/her sacrifice; and $\beta(n-1)$ is utility loss to individuals due to contemporaries' sacrifice.

Each is made better off so long as

$$n\alpha\bar{r} \geq 1 + \beta(n-1). \tag{11.17}$$

If the numerical values for the parameters were

$$\alpha = 0.10; \beta = 0.15; \bar{r} = 2$$

we would need at least 17 people ($n = 17$) to participate in a joint action to satisfy equation (11.7).

Tullock (1964), Usher (1964) and Lind (1964), commenting on Marglin's analysis, argue that his model is extremely sensitive to changes in the assumed values of the parameters. In Marglin's model a less altruistic value of α, say, 0.074, with other parameters being the same leads to a completely different result. No matter how large n there would be no collective investment because equation (11.7) will not hold. Instead, the sacrifice should be made towards contemporaries rather than posterity. In effect, in making charitable contributions we normally require to know by how much the recipients are worse off than ourselves. An increase in the consumption level of a person who is already consuming more than we are does not improve our state of satisfaction. It is quite reasonable to assume that even at the market determined rate of saving, and discount rate, the next generation is going to be better off than the present one. A collective saving of the kind which is recommended by Landauer, Baumol, Sen and Marglin suggests taxing the poor in order to help the rich. Individually, some wealthy members of our generation might make charitable gifts to future generations on the grounds that their consumption levels, on average, are likely to be lower than the consumption levels of those who make the contribution. Alas, even this argument does not seem to be convincing. Since the poor have always been with us it is more than likely that the next generation will also have some poor and needy. Therefore, some wealthy members of the present generation may choose to make a charitable contribution on the condition that the recipients should be the less fortunate members of future society. The existence of many foundations set up by the rich is the obvious indication for this choice.

Marglin's evidence for future-orientated altruism is dubious as he asks: why do governments want citizens to undertake sacrifices, the returns from which will not come about until a long time interval has elapsed? One can also argue about the necessity of a present-orientated altruism by putting forward equally hearsay arguments, such as governments are too forceful, income tax is excessively burdensome, etc.

11.4 ORDINARY DISCOUNTING

The market method of discounting rests on the following argument. If an individual has a marginal time preference rate of, say, 5% per period (year), then he/she is indifferent between the alternatives of one extra unit of current consumption and $(1 + 0.05)$ units of consumption in year 1, $(1 + 0.05)^2$ in year 2, and so on. Reversing this process, an extra unit consumed in year 1 has a present value of $1/(1 + 0.05)$ in year 0. Although each individual's time preference rate is subjective, it is through the operation of capital markets that they are revealed. Assuming perfect competition and the absence of risk, all intertemporal traders will be able to borrow and lend at the market rate of interest. If an individual's time preference rate is greater than the market rate he/she will choose to borrow; if it is lower, he/she will choose to lend.

The ordinary discounting formula is:

$$\frac{1}{(1 + r)^t} \text{ or } e^{-rt},$$

where r is the market rate of interest and t is time (0, 1, 2, ...), normally expressed in terms of years. If we assume a long-term market project which throws off £1 net benefit every year, the discounted values for each year can be obtained by using those formulas. Appendix 1 gives the ordinary discount factors between years 1 and 50 for interest rates 4–12% and 15%. These numbers give the net present value of £1 net benefit expected at any one year in the future. For example, the current market value of £1 which is expected in year 30 is £0.0573 at 10% interest rate.

Ordinary compounding is the opposite of ordinary discounting, and is given by:

$$(1 + r)^t \text{ or } e^{rt}$$

That is, £1 invested at 10% interest rate will become £17.45 in 30 years' time.

The use of ordinary discounting in economic analysis is a very widespread practice indeed. Of course, it is perfectly legitimate and proper to use this method in private transactions through the market mechanism. Samuelson (1976) was very lucid on this point:

> When what is at issue is a tree whose full fruits may not acquire until a century from now, the brute fact that our years are numbered as three score and ten prevents people planting the trees that will not bear shade until after they are dead – altruism, of course, aside. To argue in this way is to fail to under-

stand the logic of competitive pricing. Even if my doctor assures me that I will die the year after next, I can confidently plant a long-lived tree, knowing that I can sell at a competitive profit the one year old sapling.

The next section explains that to argue in the same way for collective projects is to fail to understand the nature of public sector economics. In the meantime some project evaluation methods which are widely used in economic analysis are outlined.

11.4.1 Net present value

This is a simple method in which the costs and benefits generated by a project over time are discounted and then summed to obtain an overall figure. If this figure is greater than zero, then the project in question becomes feasible, otherwise it fails, indicating that it does not generate discounted benefits large enough to offset discounted costs. The formulas are:

$$\text{NPV} = \frac{B_0 - C_0}{(1+r)^0} + \frac{B_1 - C_1}{(1+r)^1} + \dots + \frac{B_n - C_n}{(1+r)^n} \qquad (11.18)$$

or

$$\text{NPV} = \sum_{t=0}^{n} \frac{1}{(1+r)^t}(B_t - C_t) \qquad (11.19)$$

or

$$\text{NPV} = \int_0^n (B_t - C_t)\, \bar{e}^{rt}\, dt, \qquad (11.20)$$

where t is time, 0, 1, 2, ..., n; n is the project's life; B_t is the social benefit at time t; C_t is the social cost at time t; r is the discount rate; and $[1/(1+r)^t]$, e^{-rt} are ordinary discount factors.

11.4.2 Internal rate of return

This is the twin sister of the net present value method. The internal rate of return IRR yields a zero net present value figure for the project under consideration, that is:

$$\frac{B_0 - C_0}{(1+x)^0} + \frac{B_1 - C_1}{(1+x)^1} + \dots + \frac{B_n - C_n}{(1+x)^n} = 0 \qquad (11.21)$$

or

$$\sum_{t=0}^{n} \frac{1}{(1+x)^t}(B_t - C_t) = 0 \qquad (11.22)$$

or

$$\int_0^n (B_t - C_t)\, e^{-xt}\, dt = 0, \tag{11.23}$$

where x is the internal rate of return (unknown) which satisfies the equations.

Having calculated the internal rate, the policy-maker compares it with a pre-determined rate. If the internal rate proves to be greater than the predetermined rate then the investment project becomes acceptable.

11.4.3 The benefit–cost ratio

The benefit–cost ratio (BCR), the ratio of the project's discounted benefits to discounted costs, must be greater than one. That is:or

$$\text{BCR} = \frac{\sum_{t=0}^{n} \dfrac{1}{(1+r)^t} B_t}{\sum_{t=0}^{n} \dfrac{1}{(1+r)^t} C_t} > 1 \tag{11.24}$$

or

$$\text{BCR} = \frac{\int_0^n B_t e^{-rt}\, dt}{\int_0^n C_t e^{-rt}\, dt} > 1. \tag{11.25}$$

All these three methods, i.e. NPV, IRR and BCR, have been used universally to evaluate public as well as private sector projects. Note that they are all based on ordinary discounting, which is $1/(1+r)^t$ or e^{-rt} for NPV and BCR and $1/(1+x)^t$ or e^{-xt} for IRR. Furthermore, some of these methods have been redeveloped for a 'better' use in public sector economics. Let us look at two examples: the normalized internal rate of return and the shadow price algorithm.

Normalized internal rate of return

This was devised by Mishan (1967, 1971, 1975) who advocates that the benefits generated by a public sector project at time t should be compounded forward to a terminal date, usually the end of the project's life, to obtain a terminal value, that is:

$$\text{TV}(B_t) = B_t(1 + \bar{x})^{n-t}, \tag{11.26}$$

where $\text{TV}(B_t)$ is the terminal value of benefit arising at time t; $n-t$ is the number of years between time t and the terminal date, n; and $\bar{x}$ is the appropriate rate of discount.

The encashable benefits should be compounded by r, the social rate of return on capital, because the money used in public projects has an opportunity cost.

The non-encashable benefits conferred on the community, on the other hand, should be compounded at the social time preference rate, s. If we assume that b is the encashable and $1 - b$ the non-encashable part, the terminal value for benefit at time t can be rewritten as:

$$TV(B_t)^1 = bB_t(1 + r)^{n-t} + (1 - b)(1 + s)^{n-t}B_t \qquad (11.27)$$

Then we try to find which rate equates the total terminal benefits of the project, $TTV(B)^1$, to the initial capital cost, K. That is:

$$K = \frac{TTV(B)^1}{(1 + x^*)^n} . \qquad (11.28)$$

By rearranging we get:

$$x^* = \frac{\sqrt[n]{TTV(B)^1}}{K} - 1. \qquad (11.29)$$

Since $TTV(B)^1$ and K are positive, x^* is a single valued parameter which may be called the normalized internal rate of return.

One factor that motivated Mishan to develop the normalized internal rate of return method was the problem of multiple roots which exists in the internal rate of return criterion.

The shadow price algorithm

This was devised by Feldstein (1972) who argued that in a world of second best where the social opportunity cost and the social time preference rates are different, we need two types of price in public sector investment appraisal: the relative price of future consumption in terms of current consumption and the relative price of private investment in terms of current consumption. The shadow price algorithm contains these two prices. In this method the costs and benefits of public sector projects are transferred into the equivalent consumption and then discounted back to the present at the social time preference rate. The logic of doing so is that every investment decision, private or public, is taken in order to increase the future stream of consumption. One pound investment today results in a future stream of consumption. The opportunity cost of £1 displaced private investment is its expected future consumption stream discounted to the present at the social time preference rate.

In a situation where the social rate of return on private investment exceeds the social time preference rate, the opportunity cost of £1 displaced private investment exceeds one. Let S denote the shadow price of £1 forgone private investment, reducing private investment by £1 is therefore reducing the present value of private consumption by S pounds. On the other hand, the shadow price of displaced present private consumption is its face value. Let us assume that the proportion of funds that would have been invested is p and that $1 - p$ would have been consumed in the private sector, then the shadow price algorithm would be:

$$\text{NPV} = \sum_{t=0}^{n} \frac{B_t - [(p)S + (1 - p)]C_t}{(1 + s)^t} \tag{11.30}$$

which transfers the costs and benefits into equivalent consumption then deflates by the social time preference or the consumption discount rate.

11.5 MODIFIED DISCOUNTING VERSUS ORDINARY DISCOUNTING

The modified discounting method (MDM) is for investment projects and natural resources which are owned publicly. It should not be confused with private sector or market-based projects. There are two main characteristics of this method; the first relates to the nature of public sector projects and the second to ethics. The nature of public sector projects is that they provide a fixed and unalterable stream of net returns to society, the constituent members of which continuously change by birth and death over time. Since there is no market for the beneficiaries to exchange their share of returns from a publicly owned concern, they can only wait to acquire what is due to them. For discussion concerning the characteristics of public projects that provide an unalterable stream of consumption benefits to society which cannot be exchanged or assigned away, see Broussalian (1966, 1971). As for the ethical dimension, the public sector policy-maker acting on behalf of future as well as present generations treats all individuals, whether presently alive or yet to be born, equally.

Ordinary discounting, on the other hand, carries out a systematic discrimination against future members of society. At this stage in order to have an intuitive idea of how this happens, imagine a public sector project which yields a fixed sum of annual net benefits to society at large over, say, a 200-year period. When the decision-maker uses ordinary discounting, $1/(1 + r)^t$, where r is the social rate of discount and t is time, he/she gives considerably less weight to the benefits which belong to future generations *vis-à-vis* those of present ones. For example, let us assume, for the sake of simplicity, that all benefits as they become available are to be equally distributed between the existing members of society at the time. Under conventional discounting, the first year benefits will be discounted by $1/(1 + r)$. The benefits that will arise in 200 years time will be discounted by $1/(1 + r)^{200}$.

Now consider two individuals, one born now and the other born at the beginning of year 200. The former will receive £1 net benefit at the end of year 1 and the latter £1 at the end of year 200. Let us say that r is 5%, the discounted values to these individuals become:

$$\text{Present individual, } 1/(1.05) = 0.952381$$

$$\text{Future individual, } 1/(1.05)^{200} = 0.000058$$

The discrimination against the future person is clear when one remembers that these two individuals will wait an identical length of time to obtain their £1, that is one year.

The development of the modified discounting method (Kula, 1981, 1984b, 1986b, 1988a) led to a number of debates in journals such as *Environment and Planning A*, *Journal of Agricultural Economics* and *Project Appraisal* (Kula, 1984c, 1987a, 1989a; Price, 1984b, 1987, 1989; Thomson, 1988; I. Bateman, 1989; Rigby, 1989; Hutchinson, 1989; Bellinger, 1991). Below, after scrutinizing ordinary discounting, the principles of modified discounting are established and the contents of these debates summarized.

11.5.1 Conceptual and moral flaws in ordinary discounting

When ordinary discounting is used in public sector economics it is assumed, although implicitly, that society resembles a single individual who has an eternal, or very long, life. This is an amazingly flawed assumption which lies at the heart of the discounting problem in public economics. The fact of the matter is that society consists of mortal individuals with overlapping life spans and it is these mortal souls who are the recipients of net benefits which stem from public assets. This conceptual error leads the public sector policy-maker to a morally indefensible discrimination against future generations. Below I shall demonstrate precisely, with reference to the net present value criterion, how this happens.

The net present value is the most popular project evaluation method, widely used in cost–benefit analysis as well as economic appraisal of natural resources. A few writers have noticed that despite its general popularity, this method has serious limitations when it comes to evaluating long-term projects, particularly those in which the costs and benefits are separated from each other by a long time interval, a feature of many public sector projects. The discount factor used to deflate future benefits or costs is a decreasing function of time and approaches zero over time. Multiplying distant costs and benefits by the relevant factor reduces them to insignificant figures. With ordinary discounting the distant consequences of public assets become immaterial and thus do not play a decisive role in the decision-making process. Some economists argue that by following the path of ordinary discounting, present members of society could pass great costs on to future generations by undertaking projects (nuclear waste storage) or failing to undertake projects which may have a low establishment cost now but yield large benefits in the years to come (afforestation) (Nash, 1973; Page, 1977).

Now let me outline, by making a number of simplifying assumptions, a simple model which will enable the reader to understand how the net present value criterion is flawed in communal decision-making.

1. It is assumed that a society consists of only three individuals, namely, person A, person B, and person C.

2. The life expectancy of the individuals in this hypothetical society is three years and the individuals are of different ages. Person A is the oldest member of the society being two years old and having one more year to live; person B is one year old and has two more full years to life; person C is the new arrival and has three full years ahead of him/her.

3. The size of the population and the life expectancy of its members are stable, that is, there are always three members and the life expectancy of each member is three years at birth. Individuals are mortal, but the society lives on. In the first period there exists three individuals, A, B, and C. At the beginning of the second period person A dies and is replaced by a newcomer, person D. At the beginning of the third period person B dies and person E joins, and so on.

4. In the first period, society decides to undertake a project, the costs and benefits of which extend over many generations. The project has an economic life of, say, five years and, as a consequence, in addition to the decision-making population, future members of society who are yet to be born will be affected.

5. The initial construction cost of the project, say £300, is shared equally by the existing members at the time, that is, A, B, and C. These costs, $\overline{C}$, are regarded as negative benefits accruing to them, that is, $\overline{C}_A = -£100$, $\overline{C}_B = -£100$, $\overline{C}_c = -£100$. It is also assumed that the money comes from a source displacing each individual's private consumption by that amount.

6. From the second year onwards, the project yields net consumption benefits over its lifetime. The certainty equivalent of these benefits is £300 every year. When these net benefits become available at the end of each period they are distributed equally between the existing members of society who are alive at the time. Each beneficiary consumes his/her share of benefit immediately it becomes available. At the end of the fifth year the project is written off with no cost.

7. At the beginning of the first period, the project is evaluated under the rules of the net present value criterion. Since it is assumed that the project is financed through a reduction in individuals' private consumption, and the resulting benefits are also consumed by the individuals when they become available, the analyst needs to know the social time preference rate to deflate the consumption stream and obtain a net present value for this project. It is also convenient to assume that all members of society, at any point in time, have the same time preference rate of, say, 10%. Then the social time preference rate would be 10% at all times, a figure reflecting all individual rates.

By using the net present value criterion, which is:

$$\text{NPV} = \sum_{t=1}^{5} \frac{1}{(1 + s)^t} NB_t, \tag{11.31}$$

where NB_t is the net benefit arising at time t; s is the social time preference rate, or the consumption rate of interest, which is 10% throughout; and t is the time which goes from year 1 to the end of year 5.

Substituting numerical values gives:

$$\text{NPV} = \frac{300}{(1.10)^1} + \frac{300}{(1.10)^2} + \dots + \frac{300}{(1.10)^5}$$

$$\text{NPV} = £592.6.$$

Now in view of the above assumptions if equation (11.31) is rewritten in detail by taking into consideration who receives the benefits and pays for the cost over time we get:

$$
\begin{aligned}
\text{NPV} = & \left[-\frac{100}{(1+s_A)} - \frac{100}{(1+s_B)} - \frac{100}{(1+s_C)} \right] + \left[\frac{100}{(1+s_B)^2} + \frac{100}{(1+s_C)^2} + \frac{100}{(1+s_D)^2} \right] \\
& + \left[\frac{100}{(1+s_C)^3} + \frac{100}{(1+s_D)^3} + \frac{100}{(1+s_E)^3} \right] + \left[\frac{100}{(1+s_D)^4} + \frac{100}{(1+s_E)^4} + \frac{100}{(1+s_F)^4} \right] \\
& + \left[\frac{100}{(1+s_E)^5} + \frac{100}{(1+s_F)^5} + \frac{100}{(1+s_G)^5} \right].
\end{aligned}
\tag{11.32}
$$

This takes into account births and deaths and who receives how much and when; s_i is the time preference rate of the ith person alive at that time. Since by assumption 7 all individuals' time preference rates are 10%, these rates are substituted into equation (11.32) to give the same numerical result, £592.6.

Equation (11.32) which gives an identical result to equation (11.31), brings out a very important argument, upon which is based the net present value criterion. In the first year, the members of society who actually pay for the project (persons A, B and C), wait one year to incur their share of cost. Their subjective valuation of the cost, seen from the date at which they became aware of it, is expressed in the first square bracket of equation (11.32). The worth of this cost to society, seen from the beginning of year one, is $(-1/1.10)300$.

In the second year, however, person A is no longer in the society. He/she died at the end of the first year, immediately after incurring his/her share of the cost, and is replaced by a newcomer, person D. The worth of the second round of benefits to the society seen from the beginning of the first year is expressed in the second square bracket. Persons B and C, indeed, wait two years for these benefits which they acquire at the end of year two. For person D, however, equation (11.32) assumes that he/she also waits two years to get his/her share. In fact he/she does not: he/she waits one year for this.

In the third year the situation is the same. The decision-maker looks at the third round of benefits from the vantage point of the first year and assumes that each recipient waits three years for his/her share. Person C's share of £100 is deflated by the discount factor $1/(1+s_C)^3$ assuming that he/she waits three years, which indeed he/she does. On the other hand, persons D and E wait two years and one year, respectively, to acquire their shares which arise at the end of the third year.

The same treatment is given to the individuals who receive their shares of the benefits at the end of the fourth and fifth years, that is they all wait four and five years, respectively, which is not true.

Table 11.5 summarizes the relationship between the project and the members of society over time. Column (6) illustrates the number of years each beneficiary waits in order to acquire his/her share of benefits. Column (7) on the other hand shows the power to which the beneficiaries discount factor rises in the net

Table 11.5 Time association between the project and population

Period (1)	Total net benefits arising (£) (2)	Persons born at the beginning of each period (3)	Persons alive at each period (4)	Individuals net share of benefits (£) (5)	Number of years each person waits to acquire his/her share (6)	Power of $(1+s)$ in NPV (7)	Weights given to the share of each beneficiary (8)
1	300	C	A	−100	1	1	$1/(1.10)$
			B	−100	1	1	$1/(1.10)$
			C	−100	1	1	$1/(1.10)$
2	300	D	B	100	2	2	$1/(1.10)^2$
			C	100	2	2	$1/(1.10)^2$
			D	100	1	2	$1/(1.10)^2$
3	300	E	C	100	3	3	$1/(1.10)^3$
			D	100	2	3	$1/(1.10)^3$
			E	100	1	3	$1/(1.10)^3$
4	300	F	D	100	3	4	$1/(1.10)^4$
			E	100	2	4	$1/(1.10)^4$
			F	100	1	4	$1/(1.10)^4$
5	300	G	E	100	3	5	$1/(1.10)^5$
			F	100	2	5	$1/(1.10)^5$
			G	100	1	5	$1/(1.10)^5$

present value criterion to deflate the benefits which are accruing to them so that an overall net present value can be obtained.

Once we dispense with the assumption that society is not like a single individual but consists of mortal souls, the discrimination against future generations becomes very clear as the model illustrates. The more distant future individuals are from current generations the less weight is attached to their shares. For the benefits arising in the fourth year, person F waits only one year for his/her share and is given a weight of $1/(1.1)^4$. For the benefit arising in the fifth year, person G, who is a more distant member than person F, also waits one year to acquire his/her share, but this is given a weight of $1/(1.1)^5$.

How can such an assumption be justified; that is, why should the welfare of future generations be less important than that of the present ones? This practice of undervaluing future generations' shares in comparison with present ones, is in conflict with the views of a number of writers who have made explicit statements on this subject. Pigou (1929) argues that 'it is the clear duty of government, which is the trustee for unborn generations as well as for its present citizens'. Marglin (1963b) argues likewise: 'since the generations yet to be born are every bit as important as the present generations'. These authors indicate that, in their view, the welfare of each generation, present and future, is equally important, and thus the government should give equal weight to each generation's share of benefits (or losses).

More recently, Rawls (1972) developed a theory of social compacts as a basis for ethics. Rawls devotes most of his efforts to the intratemporal case with only passing reference to the intertemporal one. In his argument regarding intertemporal fairness and determination of just saving principles over time, Rawls treats each generation equally. In the original position, representatives from all generations are assumed to know the same facts and to be subjected to the same veil of ignorance. Also they will probably be given one vote to cast their opinion during the possible discussions with regard to choosing one of many methods of carrying out the policies that are agreed on. In the original position it is hard to imagine that any method like the net present value criterion, which gives unequal weights to the receipts of different generations, will be accepted as fair. Apart from the question of fairness, it is unlikely that representatives will accept this evaluation method. This is because under the rules of this method it is possible for one generation to pass costs on to the succeeding ones. Since delegates would not know to which generation they belonged, they may be at the receiving end of the costs. In order to safeguard every interest, it is more likely that they would promote some other project evaluation methods giving equal weight and importance to the welfare of all generations. In the end, what is good for one would be good for everyone. It is not considered that in the original position the delegates would deny discounting. It was one of the conditions that in the original position the representatives should know the general facts of human society and laws of human psychology. Since individuals, in normal circumstances, discount their future utilities and disutilities and would not be indifferent between their preference for a present consumption benefit over a future one of the same magnitude, there would, therefore, be no grounds for delegates to disregard this preference of individuals.

In view of the above model, one can hardly claim that by using the ordinary discounting method public sector policy-makers give equal weights to the well-being of future generations compared with present ones.

11.6   THE MODIFIED DISCOUNTING METHOD (MDM)

In the above project, there is no quarrel with the net present value criterion as far as the treatment it gives to persons A, B, and C. The actual time that those individuals wait for their share of benefits and costs is properly taken into account in equation (11.32). If the same procedure is repeated for all individuals and the discounted net consumption benefits are summed, this would give a different overall figure. This method is called modified discounting, or the sum of discounted consumption flows, which measures the overall worth of a project to individuals who are actually related to it.

Table 11.5 shows the appropriate length of time and resulting discount factor for each member of society who is associated with the project. Table 11.6 shows the correlation between the hypothetical project and the individual members of society over time. Column (1) shows recipients of net benefits arising from the investment projects. Columns (2)–(6) show the division of net consumption benefits between individuals on a yearly basis. Column (7) gives the net under-counted share for each individual associated with the project. Columns (8)–(12) show the discounted net benefit accruing to each individual by taking into account the correct length of time for every person.

The figure in the bottom right-hand corner, £729.5, is the total discounted consumption value of the project to society. It can be obtained by summing either the numbers in the last column or the numbers in the second half of the last row. Note that by using the net present value method a figure of £592.6 was obtained whereas the modified discounting method yields £729.5. That is, the new method shows a much higher consumption value due to the fact that future individuals are given exactly the same treatment as those who are currently alive.

11.6.1 The formula

This simple illustrative example enables the reader to understand the nature of the modified discounting method. One way to arrive at £729.5 is to sum the annual discounted net benefits over the project's life. The figures in the second half of the last row are obtained by summing vertically the annual discounted net benefits of the project. Columns (8) and (9) are different from columns (10)–(12). This is because in the third year a point is reached where, from a discounting point of view, the relationship between the project and the individuals becomes repetitive. Therefore, summation can be separated into two parts: sum-

Table 11.6 The undiscounted and discounted net benefits on the basis of the MDM

| Person (1) | Undiscounted net benefits | | | | | | Deflated net benefits on the basis of modified discounting | | | | | |
	t_1 (2)	t_2 (3)	t_3 (4)	t_4 (5)	t_5 (6)	Total (personal) (7)	t_1 (8)	t_2 (9)	t_3 (10)	t_4 (11)	t_5 (12)	Total (personal)
A	−100	–	–	–	–	−100	$\dfrac{-100}{(1.1)}$	–	–	–	–	−90.9
B	−100	100	–	–	–	0	$\dfrac{-100}{(1.1)}$	$\dfrac{100}{(1.1)^2}$	–	–	–	−8.3
C	−100	100	100	–	–	200	$\dfrac{-100}{(1.1)}$	$\dfrac{100}{(1.1)^2}$	$\dfrac{100}{(1.1)^3}$	–	–	66.9
D	–	100	100	100	–	300		$\dfrac{100}{(1.1)}$	$\dfrac{100}{(1.1)^2}$	$\dfrac{100}{(1.1)^3}$	–	248.7
E	–	–	100	100	100	300			$\dfrac{100}{(1.1)}$	$\dfrac{100}{(1.1)^2}$	$\dfrac{100}{(1.1)^3}$	248.7
F	–	–	–	100	100	200				$\dfrac{100}{(1.1)}$	$\dfrac{100}{(1.1)^2}$	173.5
G	–	–	–	–	100	100					$\dfrac{100}{(1.1)}$	90.9
Total (project)	−300	300	300	300	300	1000	−272.7	256.1	248.7	248.7	248.7	729.5

ming until the beginning of the third year and summing from year three onwards. That is:

$$\text{NDV(MDM)} = \sum_{t=1}^{n-1}\left[\sum_{k=2}^{t}\frac{NB_t}{(1+s)^{k-1}} + (n+1-t)\frac{NB_t}{(1-s)^t}\right]$$

$$= \sum_{t=n}^{n}\left[\sum_{k=2}^{n}\frac{NB_t}{(1+s)^k}\right], \tag{11.33}$$

where NDV (MDM) is the net discounted value based on the modified discounting method; NB_t is the net benefit accruing to each individual at time t; n is the life expectancy of individuals (3 in this example); t is the time or age of investment in terms of years (1–5 in this example); k is the number representing each cohort recruited into the population; and s is the social rate of interest.

It must be obvious by now that the MDM will be influenced by a number of factors such as the magnitude of the discount rate, growth of population and the life expectancy in the community. These variables differ from one country to another. In the UK, over the last 20 years the population has remained around the 56 million mark and there is no reason to believe that it will change substantially one way or the other in the foreseeable future. Therefore, it will be reasonable to assume a zero growth population for the UK model. With regard to the expected lifetime of individuals, it should be noted that the life expectancy of UK citizens has moved slowly upwards over the years. Although it is conceivable that these figures may continue to increase over the years, the statistics indicate that this is likely to happen only very slowly. For the time being a constant life-expectancy figure of 73 years as an average for both sexes is assumed in the UK. One further assumption is that the structure of the population consists of equal cohorts, and that recruitment into that population (necessary with the assumption of constant population) equals deaths.

The next step is to imagine a situation in which a perpetual public sector project yields £1 net benefit every year (in the case of net cost, it can be expressed as –£1) which is divided equally between the members of society who are alive at the time. It would also be convenient to think that all net benefits as they become available will be consumed by the recipient. Then calculate a net discounted value for each single individual who receives net benefits from the project. Finally, by putting the UK population into 73 single age groups and dividing the annual £1 net benefit by that number we get the following formula for the modified discount factor:

$$\text{MDF} = \frac{1}{n}\left[\frac{1}{(1+s)^t}(n+1-t) + \sum_{1}^{t-1}\frac{1}{(1+s)^t}\right], \text{ when } t \le n \tag{11.34a}$$

$$= \frac{1}{n}\sum_{1}^{n}\frac{1}{(1+s)^t}, \text{ when } t > n, \tag{11.34b}$$

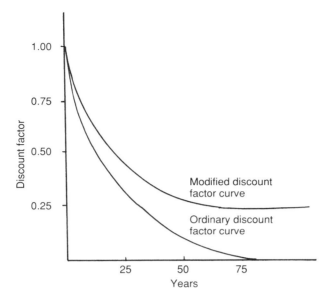

Figure 11.4   Ordinary and modified discount factors at a 5% discount rate.

where n is the life expectancy of the UK citizens (73 years) and also represents the single age groups into which the population is divided, i.e. generations; t is the age of public sector project, which may go from year one to infinity, yielding £1 every single year; and s is the social time preference rate.

Equation (11.34a) corresponds to the first three years of the hypothetical investment project and equation (11.34b) beyond year three. Appendix 2 gives the modified discount factors calculated for interest rates 1–15% over a 73-year period. Factors beyond that period will be equal to the figure for year 73.

Figure 11.4 illustrates the difference between the ordinary and modified discount factors at a 5% discount rate. It must be clear to the reader that the MDM contains numerous net present value calculations, one for every mortal individual who is associated with the project. The association continues as long as the individual and the project survive. Note that gaps between the ordinary and modified discount factor curves are narrow at the early years, but widen as time passes.

11.7   DEBATES ON MODIFIED DISCOUNTING

There has been an ongoing debate in the literature between the critics and the followers of the modified discounting method. This is clearly a healthy development which will bring greater insights into the new method. Below, the various arguments on this issue are explained briefly.

Price (1984b, 1987, 1989) debates a number of issues, mostly in relation to afforestation projects involving many generations. First, when forestry projects are viewed by different generations from their vantage points the decision-maker on the ground may be tempted to realize the investment at an earlier age than the one agreed originally, which could even diminish the value of the original decision of planting, felling and replanting. Second, even when the initial decision is honoured by all generations, this would put individuals who initiate afforestation projects at a disadvantage. According to Price, justice and efficiency will be achieved by using a low discount rate, based solely on the diminishing marginal utility of consumption, in the usual way.

Thomson (1988) contends that the well-being of all generations is not in fact of equal importance to decision-makers, and it is quite understandable that they favour the present and the near future. Thomson also queries why date of birth is chosen as a point of reference for each generation. Rigby (1989) asserts that traditional cost–benefit methods ignore the distributional consequences of projects over time by assuming that losers will somehow be compensated. It is indeed the case that ordinary discounting implicitly treats society as an individual and Kula is correct in making this the issue. However, his modified discounting makes a strong assumption that projects' costs and benefits are distributed equally within a given generation. In practice, since ordinary discounting wipes out distant consequences of communal projects, often a cut-off period is taken. The new method would make decision-makers consider all costs and benefits over much longer time periods.

Hutchinson (1989) argues that the modified discounting method would increase the size of the public sector investment portfolio. This has obvious implications for the allocation of an economy's limited resources between the private and public sectors. Bateman (1989) points out that the new method should not be seen as complete rejection of ordinary discounting. Both methods adopt a positive time preference to deflate future costs and benefits. Bellinger (1991), in an attempt to broaden the scope of modified discounting, points out that the Rawlsian covenant which underpins the modified discounting may be too restrictive to prevent future generations from exercising free choice at times when they are technologically able to do so.

In a theoretical debate, Price's argument that an ongoing investment decision such as afforestation should be viewed at different points by different decision-makers is misleading. There will be times when each decision-maker will have an unfair advantage in relation to future generations. If each generation misuses that time advantage then all generations will be deprived in the end. Unlike the conventional appraisal criteria, the ground rules for the new method are laid down in the original position by representatives from all generations, not by the present foresters who are here now and at a point of advantage in relation to future generations.

As for Price's recommendation to use a low discount rate for communal projects based upon diminishing marginal utility of consumption, there is a serious flaw. An interest rate applied over the whole time span of the project by using ordinary discounting would mean that those individuals who will be born at some future date are going to experience diminishing marginal utility on their consumption starting from now! Let us say that a 2% appropriately low discount rate containing only one parameter, the diminishing marginal utility of increasing consumption, is to be employed on a long-living project. Referring back to the early example on p. 322, i.e. a public sector project which yields a fixed sum of annual net benefits to society over a 200-year period, the ordinary discount factor over 200 years will become 0.019. This figure will be used to deflate the net benefits accruing to individuals at that date. Now consider those two individuals; one is born immediately and the other will be born at the beginning of year 200. The discount factor for the current individual is 0.98 compared to 0.019 for the future person. Remember, the correct interest rate advocated by Price is based solely on the individual's experience of diminishing marginal utility of increasing consumption. It must follow that the future individual who will be born in year 200 is going to experience diminishing marginal utility on his/her consumption starting from now!

Thomson's (1988) argument, which implies that governments do not really care much about future generations, is hard to accept. One can give many examples to illustrate that democratic governments which are accountable to the general public do care about future individuals. Take the Forestry Commission which was established in 1918 with a view to creating forests almost exclusively to benefit future generations. Indeed, the Commission is still planting trees every year which will be felled in 40–60 years' time, after most of us have passed on (this is despite the hindrance of ordinary discounting which was explained in Chapter 3). It is true that governments have so far used the ordinary discounting method, but perhaps this is just because the modified discounting method has not been available until now.

As for the issue of discounting future individuals' share of benefits to their date of birth, I believe that this is a proper procedure. Imagine a large government project with a long life. Present individuals will be affected as soon as a decision is taken to go ahead with it. Future individuals will be affected at the point of their birth as they will either receive a benefit from it or incur a cost. To give an extreme example, consider those babies born with leukaemia as a result of radioactive contamination created by a publicly owned nuclear power generating unit nearby. However, one may wish to modify my factors by changing the point of reference from birth to, say, the voting age, but this will not alter the nature of the new method.

Rigby (1989) correctly points out that although the modified discounting method does take into account the intergenerational distributional aspects of public sector projects, it does not deal rigorously with the intergenerational distribution. The assumption that £1 throw-off is equally distributed between the existing population is necessary for the computation of general modified discounting method factors. For projects that redistribute income unevenly between the existing individuals, this can easily be taken into account in numerators of evaluation methods.

I.Bateman (1989) sees modified discounting as an extension of ordinary discounting in view of the fact that both methods adopt a positive time preference to deflate the future net benefits of an investment project. It is certainly true that the ordinary discounting method is an integral part of the modified discounting method as it is used to deflate the net benefits which belong to mortal individuals. The fact of the matter is that the summation of these ordinarily discounted individual net benefits over time yields results that are very different from the conventional factors. As Samuelson (1976) puts it 'let us make no mistake about it, the positive interest rate is the enemy of long-lived investment projects'. This is not necessarily the case with modified discounting.

Hutchinson's (1989) analysis, which implies that the use of the modified discounting method would increase the size and scope of the public sector, would only be correct under the assumption of an unlimited budget. But when the government is tied by budgetary constraint instead of a larger expenditure there will be a quite different project mix in the public sector portfolio. This will include some previously rejected long-term projects at the expense of 'quick and dirty' ones.

I welcome Bellinger's (1991) attempt to expand the scope of the modified discounting method to encompass private sector projects that generate intergenerational externalities. What is the social value of private projects which will affect future generations? This indeed is a fair and topical question. For example, a private nuclear power station creates serious intertemporal externalities in the form of highly toxic and long-living radioactive wastes which could affect the health and safety of numerous generations to be born (Chapter 10). Bellinger, by taking my method one step further, makes an appreciable contribution.

At one point Bellinger argues, that treating 'rich' (future generations) less favourably than 'poor' (present generations) would be more in line with the Rawlsian theory. The Rawlsian difference principle allows inequalities, or unequal treatment, only if such practices are to the advantage of the least well-off, whose better prospects act as incentives, so that the economic process is more efficient and innovation proceeds at a faster pace. The modified discounting method is not at odds with the difference principle as it differentiates between income levels in the process of discounting. In order to see this clearly we must remember that the social time preference rate is the appropriate discount rate in the new method, equation (11.4); $S = (1 + m)(1 + g)^e - 1$, where e is the elasticity of marginal utility of consumption, g is growth rate of per capita consumption and m is a mortality-based time discount rate. According to calculations for the United States these figures turned out to be:

$$e = 1.89$$
$$g = 2.3\%$$
$$m = 1.0\%$$

making the social time preference rate (STPR) S:

$$S = (1 + 0.01)(1 + 0.023)^{1.89} - 1$$
$$= 5.4\%.$$

Suppose that the growth rate of real consumption is zero and thus future Americans are not going to be better off than present ones; with no growth in real incomes, $g = 0$, the social time preference rate would be:

$$S = 1\%.$$

Figure 11.5 illustrates modified discount factors with and without income growth. Clearly the new method differentiates between income levels. For example, the weight used for utility of the richer generations, say, 1000 years from now, is 0.244. The figure would be 0.707 if they were predicted to be at the same income levels with present ones. Note that the ordinary discounting method, by using either rate, gives a weight of 0.000 to utility level 1000 years ahead.

Now, let us turn our attention to the decision-making process carried out by different generations over time and the Rawlsian 'covenant' is no sacred cow. Bellinger argues that his general approach allows future generations to alter or

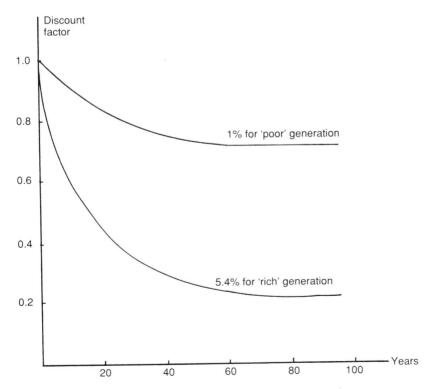

Figure 11.5   Modified discount factors for 'rich' and 'poor' generations.

continue a policy which has been in progress for many years. I would have no problem with this provided that major policy changes were made within the Rawlsian framework by using the original position and the veil of ignorance. Only this would result in a just decision criterion which would then be accepted by all generations. As long as human nature and the general conditions of life do not change, I do not think that a policy decision made by one generation would need to be altered by other generations. However, Bellinger's general approach may be useful in a situation in which, say, due to resource depletion and environmental contamination, conditions of life may deteriorate in the future, making a previously just policy unworkable, or even unjust. Rawlsian theory, in my opinion, is flexible enough to rectify this situation.

What if humankind is essentially selfish in its pursuit of ends such as domination, wealth and prestige, come what may? It would not be difficult to see that, with humanity's ability to change the natural environment, each successive generation would inherit fewer natural resources and more and more contaminated surroundings, eventually leading to self-destruction. Rawls acknowledges that such tendencies may exist but queries their primacy; in justice as fairness one does not take people's propensities and inclinations as given, whatever they are, and seek the best way to fulfil them. Rather, their desires and aspirations are restricted from the outset by the principles of justice, which specify the boundaries that humanity's systems of achieving ends must respect. In a well-ordered society, an effective sense of justice belongs to a person's good, and so tendencies to instability and excess are kept in check, if not eliminated. Ordinary discounting, which creates excess in the hands of those who are in a position of advantage in relation to future generations, is hard to defend in a just society.

11.8   SOME APPLICATIONS OF THE MODIFIED DISCOUNTING METHOD

By now it must be obvious to the reader that the use of the new method in policy-making for destructible resources will yield very different results from those obtained by means of the conventional criteria. The difference will be most marked in projects with long time horizons. Figure 11.4 illustrates the difference between the ordinary and modified discount factors at a 5% discount rate. Any economic policy which takes into account the time dimension on natural resources and the environment will be affected by the modified discounting method. A few examples are given to illustrate the potential impact of the modified discounting method on resource policies.

11.8.1   Cost-benefit analysis of afforestation projects

Table 11.7, which combines Table 3.1 and 3.2, shows the net benefit profile for the afforestation project explained in Chapter 3 (p. 93). Columns (3a) and (3b) show the discounted net benefits on the basis of the ordinary method. Columns

(4a) and (4b), on the other hand, give the results on the basis of the modified discounting method. A 5% discount rate, widely used by the Forest Service during 1986/87 in economic evaluation of afforestation projects in Northern Ireland, is employed to obtain the deflated net benefit figures in columns (3a), (3b), (4a) and (4b). The resulting net present values are:

Table 11.7 Undiscounted and discounted (ordinary as well as modified methods) net benefits of afforestation project, 1986/87 prices, Northern Ireland

Year (1a)	Net benefits (undiscounted) 1986/87 prices 1 ha (2a)	Discounted net benefit at 5% Ordinary (3a)	Modified (4a)	Year (1b)	Net benefits (undiscounted) (2b)	Discounted net benefit at 5% Ordinary (3b)	Modified (4b)
0	−355	−337	−337	26	−23	−6	−9
1	−23	−21	−21	27	−13	−6	−9
2	−23	−21	−21	28	−23	−6	−9
3	−23	−19	−20	29	−23	−5	−8
4	−23	−18	−19	30	256	59	90
5	−23	−18	−18				
6	−23	−17	−17	31	−23	−5	−8
7	−23	−16	−17	32	−23	−5	−8
8	−169	−115	−117	33	−23	−5	−7
9	−23	−15	−15	34	−23	−4	−7
10	−23	−14	−15	35	241	43	77
				36	−23	−4	−7
11	−23	−13	−14	37	−23	−3	−7
12	−23	−13	−14	38	−23	−3	−7
13	−23	−12	−13	39	−23	−3	−7
14	−23	−12	−12	40	265	37	80
15	−42	−20	−22				
16	−169	−78	−89	41	−23	−3	−7
17	−23	−10	−11	42	−23	−3	−7
18	−23	−10	−11	43	−23	−3	−7
19	−23	−9	−11	44	−23	−3	−7
20	−75	−29	−33	45	2582	284	697
21	−23	−8	−10				
22	−23	−8	−10				
23	−23	−8	−9				
24	−23	−7	−9				
25	191	57	74				

NPV (ordinary discounting) = –£435 per ha
NDV (modified discounting) = £89 per ha.

The afforestation project fails decisively at 5% interest rate when the ordinary discounting method is used in evaluation. But it succeeds with the modified discounting method.

11.8.2 The optimum felling age for publicly owned forests

The main objective of a private entrepreneur is to maximize the net cash benefits arising from his/her project. In Chapter 3 it was explained how profits can be maximized in forestry by choosing the optimum cutting age in a model which contains numerous chains of cycles of planting and replanting on a fixed acre of land. This yields a much shorter rotation than the maximum sustainable yield, which has traditionally been advocated by foresters. Samuelson (1976) strongly argues that a commitment to the maximum sustainable yield will severely undermine profits from afforestation projects.

Do private owners in practice use Samuelson's method to fell their forests? To answer this question we need empirical research. Lönnstedt (1989) has studied the cutting decisions of private small forest owners in Sweden and discovered that owners deliberately postpone the felling age well beyond the commercial criterion. In some cases this may even exceed the maximum sustainable yield. The need for this study arose because of low cutting intensity prevalent among small forest owners in Sweden, which is also observed in other Nordic countries, and in France, Germany and the United States. Lönnstedt concludes that most forest owners have a long time perspective and prefer to hold their estates in trust and hand on to the next generation. That is to say, most owners look upon the estate as an inheritance from their grandparents and a loan from their children. Instead of cash benefits a mature forest passed on to the next generation tends to give owners more satisfaction that they are fulfilling their duties to their offspring as well as forefathers.

As for the public sector forestry, in addition to maximization of the value and volume of domestically grown timber, forestry authorities have many other objectives such as to provide recreational, educational and conservation benefits and enhance the environment through visual amenity. Many foresters believe that the maximum sustainable yield is a much more suitable criterion to achieve these multiple objectives than the commercial felling age which aims to maximize cash benefits only.

How would the modified discounting method affect the optimum cutting age as compared with the ordinary discounting method? Let us assume, for illustrative purposes, that the decision-maker aims to maximize the net revenue from a public sector afforestation project by using the modified discounting method. He/she has a maximization problem similar to the one described in Chapter 3. It was also explained in Chapter 3 that, *ceteris paribus*, a low interest rate will

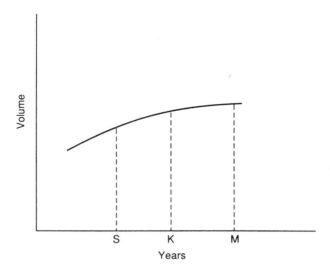

Figure 11.6 Various ages for felling.

extend the optimal rotation point. Since the modified discount factors are greater than the ordinary factors for a given interest rate, the new method will act in exactly the same way as a low interest rate and thus prolong the cutting age. Therefore, the modified discounting method will bring about a solution which is longer than the one obtained by way of ordinary discounting.[1]

Figure 11.6 shows the three criteria for the felling age. Point M is the foresters' maximum sustainable yield which they believe is the best solution. Point S, on the other hand, may be called Faustman–Ohlin–Samuelson solution which is obtained by using the ordinary discounting method. The new solution, K, which is longer than S, will be obtained by way of the modified discounting criterion.

How close would the new solution be to the maximum sustainable yield? The answer will largely depend on the magnitude of the social interest rate. If this rate is low then the new solution will be quite near to M.

11.8.3 Modified discounting in fishery policies

In Chapter 2 two conflicting criteria for fishery management were described: the maximum sustainable yield (advocated by some marine biologists) and the

---

1. Currently the author is working on a mathematical model to calculate precisely the optimum felling age by using the modified discounting method. Details of this model are not yet finalized but will become available in due course.

optimum sustainable yield (advocated by economists). The former relates to the greatest yield that can be removed each period of time without impairing the capacity of the resource to renew itself. The optimum sustainable yield, on the other hand, is the one where the difference between the costs of the fishing operation and the value of the catch are greatest. It is important to emphasize once again that economists recommend a greater conservation measure than marine biologists by advocating a level of activity which is well below the maximum sustainable yield. Turvey (1977) contends that economists and marine biologists should, in fact, be complementary professionals but that they quarrel as though they were substitutes, especially when they do not understand each other's language clearly.

This section will demonstrate, briefly, how the modified discounting method would affect the optimum sustainable yield if used in policy decisions. Figure 11.7 is similar to Figure 2.8 where the cost of a fishing operation is inversely related to the biomass level. The growth rate of biomass increases in the early stages when nutrients are abundant in the aquatic environment, then beyond a certain point it declines, becoming zero at K, the maximum carrying capacity.

In the absence of discounting, which implies that fishermen are indifferent between earning their money now or at a later date, the optimum sustainable yield is Q_1, where the difference between sustainable revenue and the cost of

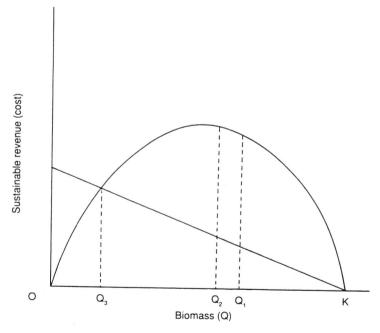

Figure 11.7   The optimum level of fishing would be between Q_1 and Q_2 with modified discounting.

catching fish is greatest. A positive interest rate implies a biomass level between Q_1 and Q_3. The higher the interest rate, the more impatient the fishermen are for harvest at present. Consequently, larger harvests now mean smaller biomass levels and less fish tomorrow.

Let us say that in a communally owned fishery the economically optimal level of activity is identified at Q_2, with a, say, 5% social discount rate by using the ordinary discounting method. At the same interest level the socially optimum activity will be to the right of Q_2, when the modified discounting method is used to identify what is best from the community's viewpoint. Intuitively this must be easy to grasp when we remember that the modified discount factors remain everywhere above the ordinary discount factors at a given interest rate (the lower the discount rate, the lower the fishing activity and thus the optimal biomass level shifts from Q_2 towards Q_1).

To summarize, compared with the biological criterion, the optimum sustainable yield recommends a reduction in fishing activity and hence maintenance of larger biomass. Compared with ordinary discounting the modified discounting method requires further reduction in the level of fishing. In this way a much larger biomass level will be maintained for present and future generations.

11.8.4 Modified discounting and economic evaluation of long-term costs of nuclear waste storage

The development of a safe and permanent method of isolating highly radioactive wastes is an unprecedented problem in human history. Never before has it been necessary to devise methods to isolate from the biosphere highly toxic nuclear wastes which will remain active for millions of years and thus could affect the health, safety and civil liberties of future generations. Chapter 10, after reviewing various proposed disposal methods, focused on the long-term health and monitoring costs of the Waste Isolation Pilot Plant (WIPP), which is likely to become the United States' first purpose-built nuclear waste repository.

The study was confined to estimates of limited health costs that could be created as a result of some repository disruptions in the form of earthquakes, volcanic action and meteorite impact. The time period was assumed to be 1 000 000 years, starting from the sealing of the repository in year 2025. The annual health cost in terms of 1991 prices was calculated to be $6067, amounting to $6 067 000 000 (6.067 billion dollars) which was then reduced to $9940 net present value in 1991 by using a 5% social rate of discount. Now let us use the modified discounting method with the same discount rate on the same cost stream. This yields a discounted health cost of $683 000 000 as opposed to $9940.

As for the long-term monitoring costs, with a 10 000 years cut-off period, figures are even more dramatic. A $1.5 million annual monitoring cost, starting from year 2025, amounts to $15 000 000 000. A 5% discount rate reduces this sum to about $5.7 million in 1991. The modified discounting method yields a discounted sum of about $3.4 billion.

Table 11.8   Ordinary and modified methods for the nuclear energy example

Year	Ordinary discount factors			Modified discount factors		
	1%	2.5%	5%	1%	2.5%	5%
0	1.00	1.00	1.00	1.00	1.00	1.00
50	0.61	0.29	0.09	0.73	0.49	0.28
100	0.37	0.08	0.01	0.71	0.46	0.27
500	0.01	0.00	0.00	0.71	0.46	0.27
1 000	0.00	0.00	0.00	0.71	0.46	0.27
2 000	0.00	0.00	0.00	0.71	0.46	0.27
10 000	0.00	0.00	0.00	0.71	0.46	0.27
100 000	0.00	0.00	0.00	0.71	0.46	0.27
1 000 000	0.00	0.00	0.00	0.71	0.46	0.27

The devastating effect of ordinary discounting on future costs of nuclear projects is clear. Realizing the unacceptable consequences of ordinary discounting on distant costs created by the nuclear industry, some writers such as Logan *et al.* (1978), Cummings *et al.* (1981) and Burness *et al.* (1983) recommended further study on the discounting philosophy, especially for cases concerning future generations.

An OECD committee (OECD, 1980) contended that established and widely used economic principles are totally useless in dealing with nuclear wastes. After studying radiation exposure the committee decided not to discount future health detriments attributable to long-lasting radionuclides while noting that the difference between zero and even a very modest discount rate can be enormous. Table 11.8 shows ordinary and modified discounting methods for various interest rates and for various years.

The Waste Isolation Pilot Project is being constructed to dispose of the United States' defence wastes. In a vastly political issue like defence there may be some grounds to disregard economic criteria, although some may question this by pointing out that there are economic consequences to any political decision. However, when wastes are generated for commercial purposes such as power generation, it would be hard to dismiss economics as irrelevant in assessing the whole value of the project. As explained in Chapter 10, a good proportion of the waste inventory in the United States stems from commercial activities. No matter how high the future costs of nuclear projects may be, e.g. decommissioning power plants, monitoring wastes and health detriments, ordinary discounting makes a nuclear energy programme look economically attractive. Modified discounting, on the other hand, paints a substantially different picture.

Appendix 1
The ordinary discount factors

					Percentage						
Year	4.0	5.0	6.0	7.0	8.0	9.0	10.0	11.0	12.0	15.0
1	0.9615	0.9524	0.9434	0.9346	0.9259	0.9174	0.9091	0.9009	0.8929	0.8696
2	0.9246	0.9070	0.8900	0.8734	0.8573	0.8417	0.8264	0.8116	0.7972	0.7561
3	0.8890	0.8638	0.8396	0.8163	0.7938	0.7722	0.7513	0.7312	0.7118	0.6575
4	0.8548	0.8227	0.7921	0.7629	0.7350	0.7084	0.6830	0.6587	0.6355	0.5718
5	0.8219	0.7835	0.7473	0.7130	0.6806	0.6499	0.6209	0.5935	0.5674	0.4972
6	0.7903	0.7462	0.7050	0.6663	0.6302	0.5963	0.5645	0.5346	0.5066	0.4323
7	0.7599	0.7107	0.6651	0.6228	0.5835	0.5470	0.5132	0.4817	0.4523	0.3759
8	0.7307	0.6768	0.6274	0.5820	0.5403	0.5019	0.4665	0.4339	0.4039	0.3269
9	0.7026	0.6446	0.5919	0.5439	0.5002	0.4604	0.4241	0.3909	0.3606	0.2843
10	0.6756	0.6139	0.5584	0.5083	0.4632	0.4224	0.3855	0.3522	0.3220	0.2472
11	0.6496	0.5847	0.5268	0.4751	0.4289	0.3875	0.3505	0.3173	0.2875	0.2149
12	0.6246	0.5568	0.4970	0.4440	0.3971	0.3555	0.3186	0.2858	0.2567	0.1869
13	0.6006	0.5303	0.4688	0.4150	0.3677	0.3262	0.2897	0.2575	0.2292	0.1625
14	0.5775	0.5051	0.4423	0.3878	0.3405	0.2992	0.2633	0.2320	0.2046	0.1413
15	0.5553	0.4810	0.4173	0.3624	0.3152	0.2745	0.2394	0.2090	0.1827	0.1229
16	0.5339	0.4581	0.3936	0.3387	0.2919	0.2519	0.2176	0.1883	0.1631	0.1069

Appendix 1 (contd)

					Percentage					
Year	4.0	5.0	6.0	7.0	8.0	9.0	10.0	11.0	12.0	15.0
17	0.5134	0.4363	0.3714	0.3166	0.2703	0.2311	0.1978	0.1696	0.1456	0.0929
18	0.4936	0.4155	0.3503	0.2959	0.2502	0.2120	0.1799	0.1528	0.1300	0.0808
19	0.4746	0.3957	0.3305	0.2765	0.2317	0.1945	0.1635	0.1377	0.1161	0.0703
20	0.4564	0.3769	0.3118	0.2584	0.2145	0.1784	0.1486	0.1240	0.1037	0.0611
21	0.4388	0.3589	0.2942	0.2415	0.1987	0.1637	0.1351	0.1117	0.0926	0.0531
22	0.4219	0.3419	0.2775	0.2257	0.1839	0.1502	0.1228	0.1007	0.0826	0.0462
23	0.4057	0.3256	0.2618	0.2109	0.1703	0.1378	0.1117	0.0907	0.0738	0.0402
24	0.3901	0.3101	0.2470	0.1971	0.1577	0.1264	0.1015	0.0817	0.0659	0.0349
25	0.3751	0.2953	0.2330	0.1842	0.1460	0.1160	0.0923	0.0736	0.0588	0.0304
26	0.3607	0.2812	0.2198	0.1722	0.1352	0.1064	0.0839	0.0663	0.0525	0.0264
27	0.3468	0.2678	0.2074	0.1609	0.1252	0.0976	0.0763	0.0597	0.0469	0.0230
28	0.3335	0.2551	0.1956	0.1504	0.1159	0.0895	0.0693	0.0538	0.0419	0.0200
29	0.3207	0.2429	0.1846	0.1406	0.1073	0.0822	0.0630	0.0485	0.0374	0.0174
30	0.3083	0.2314	0.1741	0.1314	0.0994	0.0754	0.0573	0.0437	0.0334	0.0151
31	0.2965	0.2204	0.1643	0.1228	0.0920	0.0691	0.0521	0.0394	0.0298	0.0131
32	0.2851	0.2099	0.1550	0.1147	0.0852	0.0634	0.0474	0.0355	0.0266	0.0114
33	0.2741	0.1999	0.1462	0.1072	0.0789	0.0582	0.0431	0.0319	0.0238	0.0099
34	0.2636	0.1904	0.1379	0.1002	0.0730	0.0534	0.0391	0.0288	0.0212	0.0086
35	0.2534	0.1813	0.1301	0.0937	0.0676	0.0490	0.0356	0.0259	0.0189	0.0075
36	0.2437	0.1727	0.1227	0.0875	0.0626	0.0449	0.0323	0.0234	0.0169	0.0065
37	0.2343	0.1644	0.1158	0.0818	0.0580	0.0412	0.0294	0.0210	0.0151	0.0057
38	0.2253	0.1566	0.1092	0.0765	0.0537	0.0378	0.0267	0.0190	0.0135	0.0049

Appendix 1 (contd)

					Percentage					
Year	4.0	5.0	6.0	7.0	8.0	9.0	10.0	11.0	12.0	15.0
39	0.2166	0.1491	0.1031	0.0715	0.0497	0.0347	0.0243	0.0171	0.0120	0.0043
40	0.2083	0.1420	0.0972	0.0668	0.0460	0.0318	0.0221	0.0154	0.0107	0.0037
41	0.2003	0.1353	0.0917	0.0624	0.0426	0.0292	0.0201	0.0139	0.0096	0.0032
42	0.1926	0.1288	0.0865	0.0583	0.0395	0.0268	0.0183	0.0125	0.0086	0.0028
43	0.1852	0.1227	0.0816	0.0545	0.0365	0.0246	0.0166	0.0112	0.0076	0.0025
44	0.1780	0.1169	0.0770	0.0509	0.0338	0.0226	0.0151	0.0101	0.0068	0.0021
45	0.1712	0.1113	0.0727	0.0476	0.0313	0.0207	0.0137	0.0091	0.0061	0.0019
46	0.1646	0.1060	0.0685	0.0445	0.0290	0.0190	0.0125	0.0082	0.0054	0.0016
47	0.1583	0.1009	0.0647	0.0416	0.0269	0.0174	0.0113	0.0074	0.0049	0.0014
48	0.1522	0.0961	0.0610	0.0389	0.0249	0.0160	0.0103	0.0067	0.0043	0.0012
49	0.1463	0.0916	0.0575	0.0363	0.0230	0.0147	0.0094	0.0060	0.0039	0.0011
50	0.1407	0.0872	0.0543	0.0339	0.0213	0.0134	0.0085	0.0054	0.0035	0.0009

Appendix 2

Discount factors for the United Kingdom on the basis of MDM

							Discount rate, percentage								
Year (t)	1	2	3	4	5	6	7	8	9	10	11	12	13	14	15
1	0.99010	0.98039	0.97087	0.96154	0.95238	0.94340	0.93458	0.92593	0.91743	0.90909	0.90090	0.89286	0.88496	0.87719	0.86957
2	0.98043	0.96143	0.94298	0.92506	0.90765	0.89073	0.87428	0.85828	0.84272	0.82758	0.81285	0.79850	0.78454	0.77094	0.75770
3	0.97099	0.94310	0.91628	0.89048	0.86564	0.84173	0.81870	0.79631	0.77513	0.75451	0.73462	0.71543	0.69691	0.67904	0.66177
4	0.96178	0.92538	0.89072	0.85769	0.82620	0.79616	0.76749	0.74013	0.71399	0.68901	0.66514	0.64230	0.62046	0.59955	0.57953
5	0.95278	0.90826	0.86626	0.82661	0.78917	0.75378	0.72032	0.68866	0.65870	0.63032	0.60343	0.57794	0.55377	0.53082	0.50904
6	0.94401	0.89172	0.84286	0.79717	0.75441	0.71438	0.67687	0.64170	0.60871	0.57774	0.54865	0.52131	0.49560	0.47141	0.44864
7	0.93545	0.87574	0.82047	0.76927	0.72180	0.67776	0.63686	0.59886	0.56352	0.53064	0.50002	0.47149	0.44489	0.42006	0.39688
8	0.92710	0.86031	0.79906	0.74284	0.69120	0.64372	0.60003	0.55978	0.52269	0.48846	0.45687	0.42767	0.40067	0.37569	0.35255
9	0.91896	0.84541	0.77859	0.71782	0.66250	0.61210	0.56612	0.52415	0.48579	0.45070	0.41858	0.38914	0.36214	0.33735	0.31458
10	0.91102	0.83102	0.75902	0.69413	0.63559	0.58273	0.53493	0.49166	0.45246	0.41690	0.38461	0.35527	0.32857	0.30425	0.28207
11	0.90328	0.81714	0.74031	0.67171	0.61036	0.55545	0.50622	0.46205	0.42236	0.38665	0.35449	0.32550	0.29932	0.27566	0.25425
12	0.89575	0.80375	0.72244	0.65049	0.58672	0.53012	0.47983	0.43507	0.39518	0.35959	0.32779	0.29934	0.27385	0.25098	0.23044
13	0.88840	0.79083	0.70537	0.63041	0.56456	0.50662	0.45555	0.41049	0.37065	0.33539	0.30412	0.27636	0.25167	0.22968	0.21007
14	0.88125	0.77837	0.68907	0.61143	0.54380	0.48481	0.43324	0.38810	0.34852	0.31374	0.28314	0.25617	0.23236	0.21130	0.19264
15	0.87429	0.76636	0.67351	0.59348	0.52437	0.46457	0.41274	0.36772	0.32855	0.29439	0.26456	0.23846	0.21556	0.19545	0.17774
16	0.86752	0.75478	0.65865	0.57651	0.50617	0.44580	0.39390	0.34917	0.31054	0.27710	0.24811	0.22290	0.20095	0.18178	0.16501

Appendix 2 (contd)

Discount rate, percentage

Year (t)	1	2	3	4	5	6	7	8	9	10	11	12	13	14	15
17	0.86092	0.74363	0.64448	0.56047	0.48913	0.42841	0.37659	0.33228	0.29430	0.26166	0.23354	0.20926	0.18824	0.17000	0.15412
18	0.85451	0.73289	0.63096	0.54533	0.47320	0.41228	0.36071	0.31693	0.27966	0.24786	0.22064	0.19729	0.17719	0.15984	0.14483
19	0.84827	0.72254	0.61807	0.53102	0.45829	0.39734	0.34612	0.30296	0.26647	0.23554	0.20923	0.18679	0.16758	0.15109	0.13688
20	0.84221	0.71259	0.60579	0.51752	0.44435	0.38350	0.33274	0.29026	0.25459	0.22454	0.19914	0.17759	0.15924	0.14355	0.13010
21	0.83632	0.70301	0.59408	0.50477	0.43132	0.37069	0.32047	0.27872	0.24390	0.21473	0.19021	0.16952	0.15199	0.13707	0.12432
22	0.83060	0.69379	0.58292	0.49275	0.41914	0.35883	0.30921	0.26824	0.23427	0.20598	0.18233	0.16246	0.14569	0.13148	0.11938
23	0.82504	0.68493	0.57230	0.48141	0.40777	0.34785	0.29890	0.25872	0.22561	0.19818	0.17536	0.15627	0.14023	0.12668	0.11517
24	0.81965	0.67642	0.56220	0.47073	0.39715	0.33770	0.28945	0.25008	0.21781	0.19123	0.16920	0.15086	0.13549	0.12255	0.11158
25	0.81441	0.66823	0.55258	0.46065	0.38724	0.32832	0.28079	0.24224	0.21081	0.18503	0.16377	0.14612	0.13138	0.11900	0.10852
26	0.80934	0.66037	0.54343	0.45117	0.37799	0.31965	0.27286	0.23513	0.20451	0.17951	0.15897	0.14197	0.12782	0.11595	0.10592
27	0.80441	0.65283	0.53474	0.44224	0.36937	0.31164	0.26561	0.22868	0.19886	0.17460	0.15474	0.13835	0.12473	0.11333	0.10370
28	0.79964	0.64559	0.52647	0.43383	0.36133	0.30424	0.25898	0.22284	0.19378	0.17023	0.15101	0.13519	0.12206	0.11108	0.10181
29	0.79503	0.63865	0.51863	0.42592	0.35385	0.29741	0.25291	0.21755	0.18922	0.16635	0.14772	0.13242	0.11974	0.10914	0.10021
30	0.79055	0.63199	0.51118	0.41849	0.34687	0.29112	0.24737	0.21275	0.18513	0.16289	0.14482	0.13001	0.11774	0.10749	0.09884
31	0.78623	0.62562	0.50411	0.41150	0.34038	0.28531	0.24231	0.20842	0.18149	0.15982	0.14227	0.12790	0.11601	0.10607	0.09768
32	0.78204	0.61951	0.49741	0.40494	0.33435	0.27996	0.23768	0.20450	0.17818	0.15710	0.14003	0.12606	0.11451	0.10485	0.09670
33	0.77800	0.61367	0.49105	0.39879	0.32873	0.27504	0.23347	0.20095	0.17524	0.15468	0.13806	0.12446	0.11322	0.10381	0.09586
34	0.77409	0.60808	0.48504	0.39301	0.32352	0.27050	0.22962	0.19775	0.17261	0.15254	0.13632	0.12307	0.11210	0.10292	0.09515
35	0.77032	0.60274	0.47934	0.38759	0.31867	0.26633	0.22612	0.19486	0.17025	0.15063	0.13480	0.12185	0.11113	0.10216	0.09455
36	0.76668	0.59763	0.47395	0.38252	0.31418	0.26250	0.22293	0.19225	0.16815	0.14895	0.13346	0.12080	0.11030	0.10150	0.09404
37	0.76317	0.59276	0.46886	0.37777	0.31001	0.25898	0.22003	0.18990	0.16626	0.14746	0.13229	0.11988	0.10959	0.10095	0.09361
38	0.75980	0.58811	0.46405	0.37333	0.30615	0.25575	0.21739	0.18778	0.16459	0.14614	0.13126	0.11908	0.10897	0.10047	0.09324
39	0.75654	0.58368	0.45950	0.36917	0.30258	0.25278	0.21499	0.18587	0.16309	0.14498	0.13036	0.11839	0.10844	0.10007	0.09293
40	0.75341	0.57946	0.45522	0.36529	0.29927	0.25006	0.21281	0.18416	0.16175	0.14395	0.12957	0.11779	0.10799	0.09972	0.09267
41	0.75041	0.57545	0.45119	0.36167	0.29621	0.24758	0.21084	0.18262	0.16057	0.14304	0.12888	0.11727	0.10759	0.09943	0.09245
42	0.74752	0.57163	0.44739	0.35829	0.29339	0.24550	0.20905	0.18123	0.15951	0.14224	0.12828	0.11682	0.10726	0.09918	0.09227

Appendix 2 (contd)

						Discount rate, percentage									
Year (t)	1	2	3	4	5	6	7	8	9	10	11	12	13	14	15
43	0.74475	0.56801	0.44381	0.35515	0.29078	0.24322	0.20743	0.17999	0.15857	0.14153	0.12775	0.11643	0.10697	0.09897	0.09211
44	0.74210	0.56457	0.44045	0.35222	0.28838	0.24132	0.20596	0.17888	0.15773	0.14091	0.12730	0.11609	0.10672	0.09879	0.09198
45	0.73956	0.56131	0.43730	0.34950	0.28617	0.23959	0.20464	0.17788	0.15699	0.14037	0.12690	0.11580	0.10651	0.09863	0.09187
46	0.73714	0.55823	0.43435	0.34698	0.28414	0.23801	0.20344	0.17699	0.15634	0.13989	0.12655	0.11555	0.10633	0.09850	0.09177
47	0.73482	0.55531	0.43158	0.34463	0.28227	0.23658	0.20237	0.17620	0.15576	0.13947	0.12625	0.11533	0.10618	0.09839	0.09170
48	0.73261	0.55256	0.42900	0.34247	0.28056	0.23527	0.20140	0.17549	0.15525	0.13910	0.12599	0.11515	0.10605	0.09830	0.09163
49	0.73051	0.54996	0.42658	0.34046	0.27899	0.23409	0.20053	0.17486	0.15480	0.13878	0.12576	0.11499	0.10593	0.09822	0.09158
50	0.72851	0.54752	0.42433	0.33861	0.27756	0.23302	0.19975	0.17430	0.15440	0.13850	0.12557	0.11485	0.10584	0.09816	0.09153
51	0.72661	0.54522	0.42224	0.33691	0.27625	0.23205	0.19905	0.17380	0.15405	0.13826	0.12540	0.11473	0.10576	0.09810	0.09149
52	0.72481	0.54307	0.42029	0.33534	0.27506	0.23118	0.19842	0.17336	0.15374	0.13805	0.12525	0.11463	0.10569	0.09806	0.09146
53	0.72312	0.54106	0.41849	0.33390	0.27397	0.23039	0.19786	0.17297	0.15347	0.13786	0.12513	0.11455	0.10563	0.09802	0.09144
54	0.72152	0.53917	0.41683	0.33258	0.27299	0.22969	0.19737	0.17263	0.15324	0.13770	0.12502	0.11448	0.10558	0.09798	0.09141
55	0.72001	0.53742	0.41529	0.33138	0.27210	0.22905	0.19693	0.17233	0.15303	0.13757	0.12493	0.11442	0.10554	0.09796	0.09140
56	0.71860	0.53580	0.41388	0.33028	0.27130	0.22849	0.19654	0.17206	0.15285	0.13745	0.12485	0.11436	0.10551	0.09794	0.09138
57	0.71728	0.53429	0.41258	0.32928	0.27058	0.22798	0.19619	0.17183	0.15270	0.13735	0.12478	0.11432	0.10548	0.09792	0.09137
58	0.71605	0.53290	0.41140	0.32838	0.26993	0.22753	0.19589	0.17163	0.15257	0.13726	0.12472	0.11428	0.10546	0.09790	0.09136
59	0.71490	0.53162	0.41032	0.32757	0.26935	0.22714	0.19562	0.17145	0.15245	0.13718	0.12468	0.11425	0.10544	0.09789	0.09135
60	0.71385	0.53045	0.40934	0.32684	0.26884	0.22679	0.19539	0.17130	0.15235	0.13712	0.12464	0.11423	0.10542	0.09788	0.09134
61	0.71288	0.52939	0.40846	0.32619	0.26839	0.22648	0.19519	0.17117	0.15227	0.13707	0.12460	0.11421	0.10541	0.09787	0.09134
62	0.71199	0.52843	0.40767	0.32561	0.26799	0.22622	0.19502	0.17106	0.15220	0.13702	0.12457	0.11419	0.10540	0.09786	0.09133
63	0.71119	0.52756	0.40697	0.32510	0.26764	0.22599	0.19487	0.17096	0.15214	0.13699	0.12455	0.11417	0.10539	0.09786	0.09133
64	0.71046	0.52679	0.40635	0.32466	0.26734	0.22579	0.19474	0.17088	0.15209	0.13696	0.12453	0.11416	0.10538	0.09785	0.09133
65	0.70982	0.52611	0.40581	0.32427	0.26708	0.22562	0.19464	0.17082	0.15205	0.13693	0.12452	0.11415	0.10538	0.09785	0.09133
66	0.70925	0.52551	0.40534	0.32394	0.26686	0.22548	0.19455	0.17076	0.15202	0.13691	0.12450	0.11414	0.10537	0.09785	0.09133
67	0.70875	0.52501	0.40495	0.32366	0.26668	0.22536	0.19447	0.17072	0.15199	0.13689	0.12449	0.11414	0.10537	0.09784	0.09132
68	0.70834	0.52458	0.40462	0.32344	0.26653	0.22527	0.19442	0.17068	0.15197	0.13688	0.12449	0.11413	0.10537	0.09784	0.09132

Appendix 2 (contd)

Discount rate, percentage

Year (t)	1	2	3	4	5	6	7	8	9	10	11	12	13	14	15
69	0.70799	0.52423	0.40435	0.32325	0.26641	0.22520	0.19437	0.17066	0.15195	0.13687	0.12448	0.11413	0.10536	0.09784	0.09132
70	0.70772	0.52396	0.40414	0.32311	0.26632	0.22514	0.19434	0.17064	0.15194	0.13686	0.12448	0.11413	0.10536	0.09784	0.09132
71	0.70752	0.52375	0.40399	0.32301	0.26625	0.22510	0.19431	0.17062	0.15193	0.13686	0.12447	0.11413	0.10536	0.09784	0.09132
72	0.70738	0.52362	0.40389	0.32295	0.26621	0.22508	0.19430	0.17062	0.15193	0.13686	0.12447	0.11413	0.10536	0.09784	0.09132
73	0.70732	0.52356	0.40385	0.32291	0.26619	0.22507	0.19429	0.17061	0.15192	0.13686	0.12447	0.11413	0.10536	0.09784	0.09132

References

Alfred, A.M. (1968) The correct yardstick for state investment. *District Bank Review*, **166**, 21–32.

Almon, S. (1965) The distributed lag between capital appropriations and expenditures. *Econometrica*, **33**, 78–96.

Anderson, K. and Tyers, R. (1986) Distortions in the world food markets: a quantitative assessment. *World Development Report*, World Bank, Washington, DC.

Anderson, L.G. (1982) Marine fisheries, in *Current Issues in Natural Resource Policy* (ed. P.R. Portney), Resources for the Future, Washington, DC.

Anderson, M.A. (1987) A war of words: public inquiry into the designation of the North Pennines as an area of outstanding natural beauty, in *Multipurpose Agriculture and Forestry* (eds M. Merlo, G. Stellin, P. Harou and M. Whitby), Wissenschaftsverlag Vauk, Kiel, Germany.

Anderson, R. and Crocker, T. (1971) Air pollution and residential property values. *Urban Studies*, **8**, 171–80.

Arrow, K.J. (1966) Discounting and public investment criteria, in *Water Research* (eds A.V. Kneese and S.C. Smith), Johns Hopkins University Press, Baltimore, MD, pp. 13–32.

Arrow, K.J. and Fisher, A.C. (1974) Environmental preservation, uncertainty and irreversibility. *Quarterly Journal of Economics*, **89**, 312–19.

Arrow, K.J. and Kurz, M. (1970) *Public Investment, the Rate of Return and Optimal Fiscal Policy*, Johns Hopkins University Press, Baltimore, MD.

Baan, P.J.A. and Hopstaken, C.F.A.M. (1989) *Schade Vermetsting; Schade Als Gevolg Van Emissie Van Stikstaf en Foster en Batten Van Emissie Beperking*. Delft.

Bailey, R. (1989) Coal – the ultimate privatisation. *National Westminster Bank Quarterly Review*, August, 2–12.

Barbier, E. (1991) Tropical deforestation, in *Blueprint 2* (ed. D. Pearce), Earthscan, London.

Barnett, H. (1979) Scarcity and growth revisited, in *Scarcity and Growth Reconsidered* (ed. V.K. Smith), Johns Hopkins University Press, Baltimore, MD.

Barnett, H. and Morse, C. (1963) *Scarcity and Growth: the Economics of Natural Resource Availability*, Johns Hopkins University Press, Baltimore, MD.

Barnum, H.N, Tarantola, D. and Sefiady, I.F. (1980) Cost-effectiveness of an immunisation programme in Indonesia. *Bulletin WHO*, **58**(3), 499–503.

Barrett, S. (1990) *Memorandum* to House of Commons Select Committee on Energy, published as Briefing Paper, Pricing the Environment: the Economic and Environmental Consequences of a Carbon Tax, in *Economic Outlook, 1989–1993*, London Business School, February 1990, London.

Barrett, S. (1991) Global warming: economics of a carbon tax, in *Blueprint 2* (ed. D. Pearce), Earthscan, London.

Bateman, D.I. (1989) Heroes for present purposes? A look at the changing idea of land ownership in Britain. *Journal of Agricultural Economics*, **40**, 269–90.

Bateman, I. (1989) Modified discounting method: some comments. *Project Appraisal*, **4**, 104–6.

Bateman, I. (1993) Valuation of the environment, methods and techniques revealed preference methods, in *Sustainable Environmental Economics and Management* (ed. R.K. Turner), Belhaven Press, London.

Bateman, I. and Turner, R.K. (1993) Valuation of the environment, methods and techniques: the contingent valuation method, in *Sustainable Environmental Economics and Management* (ed. R.K. Turner), Belhaven Press, London.

Bator, M.F. (1958) The anatomy of market failure. *Quarterly Journal of Economics*, **72**, 351–79.

Baumol, W.J. (1952) *Welfare Economics and the Theory of State*, Harvard University Press, Cambridge, MA.

Baumol, W.J. (1968) On the social rate of discount. *American Economic Review*, **58**, 788–802.

Baumol, W.J. (1969) On the social rate of discount – comment on comments. *American Economic Review*, **59**, 930–3.

Baumol, W. and Oates, W. (1971) The use of standards and prices for the protection of the environment. *Swedish Journal of Economics*, **73**, 42–54.

Baumol, W.J. and Oates W. (1975) *The Theory of Environmental Policy*, Prentice Hall, Eaglewood Cliffs, NJ.

Baumol, W. and Oates, W. (1988) *The Theory of Environmental Policy*, Cambridge University Press, Cambridge.

Beckermann, M. (1974) *In Defence of Economic Growth*, Cape, London.

Beckerman, W. (1975) *Pricing for Pollution*, Institute of Economic Affairs, London.

Beenstock, M. (1983) Pull the plug on OPEC. *The Guardian*, 30 November.

Beesley, M.E., Coburn, T.M. and Reynolds, D.C. (1960) The London–Birmingham motorway traffic and economics. Technical paper no. 46. Road Research Laboratory, HMSO, London.

BEIR Report (1972) *Biological Effects of Ionizing Radiation, the Effects on Population Exposure to Low Levels of Ionizing Radiation*, Division of Medical Sciences, United States National Research Council, Washington, DC.

Bellinger, W.K. (1991) Multigenerational value: modifying the modified discounting method. *Project Appraisal*, **6**, 101–8.

Bishop, R.C. and Heberlein, T.A. (1979) Measuring values of extra-market goods: Are indirect measures biased? *American Journal of Agricultural Economics*, **61**, 926–30.

Bockstael, N., Strand, I. and Hanemann, W. (1987) Time and the recreational demand model. *American Journal of Agricultural Economics*, **69**, 293–302.

Bohm, P. (1972) Estimating demand for public goods: An experiment. *European Economic Review*, **3**, 11–30.

Bolin, B. (1992) Economics, energy and environment. *FEEM Newsletter*, vol. 2 July, FEEM, Milano.

Boulding, K.E. (1935) The theory of a single investment. *Quarterly Journal of Economics*, **49**, 475–94.

Boulding, K. (1966) The economics of coming spaceship earth, in *Environmental Quality in a Growing Economy* (ed. H. Harret), Johns Hopkins Press, Baltimore, MD.

Boulding, K. (1970) Fun and games with gross national product, in *Environmental Crisis* (ed. H.W. Helfrich), Yale University Press, Yale.

Bowers, J. (1990) *Economics of the Environment: the Conservationists' Response to the Pearce Report*, The British Association of Nature Conservation, London.

Braden, J.B. and Kolstad, C.D. (1991) *Measuring the Demand for Environmental Quality*, North Holland, Amsterdam.

Brent, R.J. (1990) *Project Appraisal for Developing Countries*, Harvester Wheatsheaf, Hemel Hempstead.

Brookshire, D., Ives, B. and Schulze, W. (1976) The valuation of aesthetic preferences. *Journal of Environmental Economics and Management*, **3**, 325–46.

Brookshire, D., Thayer, M.A., Schulze, W.D. and D'Arge, R.C. (1982) Valuing public goods: A comparison of survey and hedonic approaches. *American Economic Review*, **72**, 165–77.

Brookshire, D.S., Eubanks, L.S. and Randall, A. (1983) Estimating option prices and existence values for wildlife resources. *Land Economics*, **59**, 1–15.

Brookshire, D., Thayer, M.A., Tschirhart, J. and Schulze, W.D. (1984) A text of the expected utility model: Evidence from earthquake risks. Unpublished manuscript, Department of Economics, University of Wyoming, Laramie.

Broussalian, V.L. (1966) Evaluation of non-marketable investments. *Research Contribution 9*, Centre for Naval Analysis, Arlington, VA.

Broussalian, V.L. (1971) Discounting and evaluation of public investments. *Applied Economics*, **3**, 1–10.

Brown, B.J. *et al.* (1987) Global sustainability: toward definition. *Environmental Management*, **11**, 713–19.

Brundtland Report (1987) *Our Common Future*, Oxford University Press, Oxford.

Buchanan, J.M. (1969) External diseconomies, corrective taxes and market structure. *American Economic Review*, March 1960.

Buchanan, J. and Stubblebine, W. (1962) Externality. *Econometrica*, **29**, 371–84.

Buddy, L.D. and Bergstrom, J.C. (1991) Measuring environmental amenity benefits of agricultural land, in *Farming and the Countryside on Economic Analysis of External Costs and Benefits* (ed. N. Hanley), CAB International, Wallingford.

Bureau of Mines (1970) *Mineral Facts and Problems*, US Government Printing Office, Washington, DC.

Burness, H.S., Cummins, R.G., Gorman, W.D. and Lindsford, R.R. (1983) US reclamation policy and Indian water rights. *Natural Resources Journal*, **20**, 807–26.

Burnett, J. (1990) Ecology, economics and the environment. *The Royal Bank of Scotland Review*, **167**, 3–15.

Burrell, A. (1989) The demand for fertiliser in the United Kingdom. *Journal of Agricultural Economics*, **40**, 1–120.

Butler, G.D. (1959) *Introduction to Community Recreation*, McGraw-Hill, New York.

Butler, R.V. (1982) The specification of housing indexes for urban housing. *Land Economics*, **58**, 96–108.

Byerlee, D.R. (1971) Option demand and consumer surplus. *Quarterly Journal of Economics*, **85**, 523–7.

Campbell, M. (1993) Moscow in danger of acute nuclear peril. *The Sunday Times*, 9 May 1993.

Channell, J.K., Chaturvedi, L. and Neill, R.H. (1990) Human intrusion scenarios in nuclear waste repository evaluations. *Environmental Evaluation Group Research Papers*, Department of Energy, Albuquerque, New Mexico.

Clarenback (1985) Reliability of estimates of agricultural damages from floods, in US Commission on Organisation of Executive Branch of Government, in *Task Force on Water Resources and Power*, Vol. 3, US Government Printing Office, Washington, DC.

Christy, F.T. (1973) Alternative arrangements for marine fisheries: an overview, RfF/PIS-A, paper 1, Research for the Future Inc., Washington.

Church, R. (1986) *The History of the British Coal Industry, Volume 3, 1830–1913: Victorian Pre-eminence*, Clarendon Press, Oxford.

Cicchetti, C.V. and Freeman, A.M. (1971) Option demand and consumer surplus, further comment. *Quarterly Journal of Economics*, **85**, 528–39.

Clark, C.W. (1976) *Mathematical Bioeconomics*, Wiley, New York.

Clark, C.W. (1982) The carbon dioxide question: a perspective for 1982, in *Carbon Dioxide Review* (ed. C.W. Clark), Oxford University Press, Oxford, pp. 3–43.

Clark, C.W. (1986), Sustainable development of the biosphere: themes for a research programme, in *Sustainable Development of the Biosphere* (eds C.W. Clark and R.E. Munn), Cambridge University Press, Cambridge.

Clark, C.W. and Munro, G.R. (1975) The economics of fishing and modern capital theory: a simplified approach. *Journal of Environmental Economics and Management*, **2**, 92–106.

Clark, C.W., Clark, F.H. and Munro, G.R. (1979) The optimal exploitation of renewable resource stocks: problems of irreversible investment. *Econometrica*, **47**, 25–47.

Clark, I.N., Major, P.J. and Mollet, N. (1989) The development and implementation of New Zealand's ITQ management system, in *Rights Based Fishing* (eds P.A. Neher, R. Arnason and N. Mollet), Kluwer Academic Press, London, pp. 117–45.

Clawson, M. (1979) Forests in the long sweep of American history. *Science*, **204**, 1168–74.

Clawson, M. and Held, B. (1957) *The Federal Lands: Their Use and Management*, Johns Hopkins University Press, Baltimore, MD.

Clawson, M. and Knetsch, J.I. (1969) *Economics of Outdoor Recreation*, Resources for the Future Inc., Washington, DC.

Clean Air Act Amendments (1970, 1977, 1979) United States Government Printing Office, Washington DC.

Coase, R. (1960) The problem of social cost. *Journal of Law and Economics* **3**, 1–44.

Collard, D., Pearce, D. and Ulph, D. (1988) *Economics, Growth and Sustainable Environments*, Macmillan, London.

Colman, D. (1991) Land purchase as a means of providing external benefits from agricultural, in *Farming and the Countryside* (ed. N. Hanley), CAB International, Wallingford.

Commission of the European Communities (1983) *Acid Rain, A Review of the Phenomenon in the EEC and Europe*, Graham and Trotman, London.

Connor, S. (1993) Ozone depletion linked to rise in harmful radiation. *The Independent*, 23 April 1993.

Conrad, J. (1988) Nitrate debate and the nitrate policy in FR Germany. *Land Use Policy*, April, 207–18.

Conrad, J.M. (1980) Quasi-option demand value and the expected value of information. *Quarterly Journal of Economics*, **94.**

Convery, F.J. (1988) The economics of forestry in the Republic of Ireland. *The Irish Banking Review*, Autumn, 42–6.

Cooper, C. (1981) *Economic Evaluation and the Environment*, Hodder & Stoughton, London.

Cooper, J.P. (1972) Two approaches to polynomial distributed lags estimation: an expository note and comment. *American Statistician*, June, 32–5.

Copes, P. (1972) Factory rents, sale ownership and the optimum level of fisheries exploitation. *Manchester School of Social and Economics Studies*, **40,** 145–63.

Court, W.H.B. (1967) *A Concise Economic History of Britain*, Cambridge University Press, Cambridge.

Croll, B. and Hayes, C. (1988) Nitrate and water supplies in the United Kingdom. *Environmental Pollution*, **50,** 163–87.

Crutchfield, J.A. and Pontecorvo, G. (1969) *The Pacific Salmon Fisheries: A Study of Irrational Conservation*, Johns Hopkins University Press, Baltimore, MD.

Cummings, R., Schultze, W. and Meyer, A. (1978) Optimal municipal investment in boomtowns: in empirical analysis. *Journal of Environmental Economics and Management*, **5,** 252–67.

Cummings, R.G., Burness, H.S. and Norton, R.G. (1981) *The Proposed Waste Isolation Pilot Project (WIPP) and Impacts in the State of New Mexico. A Socio-economic Analysis*, EMD-2-67-1139, University of New Mexico, Department of Economics, Albuquerque.

Cummings, R., Brookshire, D. and Schultze, W. (1986) *Valuing Environmental Goods: An Assessment of the Contingent Valuation Method*, Rowman & Allenheld, Totowa, New Jersey.

Dales, J.H. (1968) *Pollution, Property and Prices*, Toronto University, Toronto.

Daly, H.E. (1990) Towards some operational principles of sustainable development. *Ecology and Economics*, **2,** 1–6.

Daly, H. and Cobb, J.B. (1989) *For the Common Good*, Green Print, London.

Dasgupta, P. (1991) Environment as a commodity, in *Economic Policy Towards the Environment* (ed. D. Helm), Blackwell, Oxford.

Davidson, F.G., Adams, F.G. and Seneca, J.S. (1966) The social value of water recreational facilities resulting from an improvement in water quality. *Water Research* (ed. K. Smith), Johns Hopkins University Press, Baltimore, MD.

Davies, B.K. (1964) The value of big game hunting in a private forest. *Transactions of the 29th North American Wildlife and Natural Resources Conference*, The Wildlife Management Institute, Washington, DC.

Davis, R. (1963) Recreation planning as an economic problem. *Natural Resources Journal*, **3**, 329–49.

De Graaff, J.V. (1957) *Theoretical Welfare Economics*, Cambridge University Press, Cambridge.

Devall, B. and Sessions, G. (1985) *Deep Ecology: Living as if Nature Mattered*, Gibbs M. Smith, Layton, Utah.

Dixon, J.A. (1991) Economic valuation of environmental resources, in *Development Research: The Environmental Challenge* (ed. J. Winpenny), Overseas Development Institute, London.

Dobb, M. (1946) *Political Economy and Capitalism*, Routledge, London.

Dobb, M. (1954) *On Economic Theory and Socialism*, Routledge, London.

Dobb, M. (1960) *An Essay on Economic Growth and Planning*, Western Printing Service, Bristol.

Dorfman, R. (1975) An estimate of the social rate of discount. *Discussion paper No. 442*, Harvard Institute of Economic Research.

Douthwaite, R. (1992) *The Growth Illusion, How Economic Growth has Enriched the Few, Impoverished the Many and Endangered the Planet*, Lilliput Press, Dublin.

Drummond, M.F. (1980) *Principles of Economic Appraisal of Health Care*, Oxford University Press, Oxford.

Drummond, M.F., Landbrook, A., Lawson, K. and Steels, A. (1986) *Studies in Economic Appraisal in Health Care*, Oxford University Press, Oxford.

Dubgaard, A. (1987) Taxation as a means to control pesticide application. Report no. 35. Institute of Agricultural Economics, Copenhagen.

Dubgaard, A. (1989) Input levies as a means of controlling the intensities of nitrogenous fertilizers and pesticides, in *Economic Aspects of Environmental Regulations in Agriculture* (eds Dubgaard, A. and Nielsen, A.), Wissenschaftsverlag Vauk Kiel, Kiel.

Dubgaard, A. (1991) Pesticide regulations in Denmark, in *Farming and the Countryside* (ed. N. Hanley), CAB International, Wallingford.

Dupuit, J. (1844) On the measurement of the utility of public works, translated from French in *International Economic Papers*, No. 2, London, 1952.

Eckstein, O. (1957) Investment criteria for economic development and the theory of intertemporal welfare economics. *Quarterly Journal of Economics*, **71**, 56–83.

Eckstein, O. (1958) *Water Resources Development, The Economics of Project Evaluation*, Cambridge University Press, Cambridge.

Eckstein, O. (1961) A survey of theory of public expenditure. *Public Finances: Needs, Sources and Utilisation* (ed. J. Buchanan), Princeton University Press, Princeton, NJ.

The Economist (1989) Coal, 2, September, p. 111.

Edwards, A. (1986) *An Agricultural Land Use Budget for the United Kingdom*, Department of Environmental Studies and Countryside Planning.

Eliot, J. (1760) *Essays upon Field Husbandry in New England*, Edes & Gill, Boston.

Energy Consultative Document (1978) *Energy Policy*, Cmnd 7101, HMSO, London.

England, R. (1986) Reducing the nitrogen input on arable farms. *Journal of Agricultural Economics*, **37**, 13–24.

EPA (1987) *Regulatory Impact Analysis of Stratospheric Ozone*, Vol. 1–111, EPA, Washington, DC.

European Community (1987) *European Community and the Environment*, Periodical 3/1987, Division IX/E.S. Co-ordination and preparation of publications, Office for Official Publications of the European Communities, Luxembourg.

European Community (1989) *A Common Agricultural Policy for the 1990's*, European Documentation, Luxembourg.

European Documentation (1985) *The European Community's Fishery Policy*, Periodical 1/1985, Office for Publications of the European Communities, L-2985 Luxembourg.

Everest, D. (1988) *The Greenhouse Effect, Issues for Policy Makers*, joint energy programme, Royal Institute of International Affairs, London.

Farrell, J. (1987), Information and the Coase theorem. *Journal of Economic Perspectives*, 1987.

Faustmann, M. (1849) Gerechnung des Wertes welchen Waldboden sowie noch nicht haubare Holzbestänce für die Waldwirtschaft besitzen, *Allgemeine Forst und Jagd-Zeitung*, **25**, 441–5.

Federal Insecticide, Fungicide and Rodenticide Act (1975) US Government Printing Office, Washington, DC.

Feldstein, M.S. (1964) The social time preference rate in cost-benefit analysis. *Economic Journal*, **74**, 360–79.

Feldstein, M.S. (1972) The inadequacy of weighted discount rates. *Cost–Benefit Analysis* (ed. R. Layard), Penguin Educational, London.

Feldstein, M.S. (1974) Financing in the evaluation of public expenditure, in *Public Finance and Stabilisation Policy* (eds W.L. Smith and J.M. Culbertson), North Holland, Amsterdam.

Fellner, W. (1967) Operational utility: the theoretical background and measurement, in *Ten Economic Studies in the Tradition of Irving Fisher* (ed. W. Fellner), John Wiley, New York.

Fischhoff, B. *et al.* (1981), *Acceptable Risk*, Cambridge University Press, Cambridge.

Fisher, A.C. and Haneman, W, M. (1983) *Endangered Species, The Economics of Irreversible Damage*, Department of Agricultural Economics, University of California.

Fisher, A., Krutilla, J.V. and Cicchetti, J. (1972) The economics of environmental preservation: a theoretical and empirical analysis, in *Economics of Natural and Environmental Resources* (ed. V.L. Smith), Gordon & Breach, New York, pp. 463–78.

Fisher, I. (1927) A statistical method for measuring utility and justice of progressive income tax, in *Ten Economic Essays Contributed in Honour of J. Bates Clarke*, Macmillan, London.

Fisher, I. (1930) *The Theory of Interest*, Macmillan, London.

Food and Agricultural Organization (1990) Sustainable management of tropical moist forests, presented for ASEAN Sub-Regional Seminar, Indonesia, by R. Schmidt, January, Rome.

Forestry Commission (1971) *Booklet 34, Forest Management Tables (metric)*, HMSO, London.

Forestry Commission (1972) *Forest Record 80*, HMSO, London.

Forestry Commission (1975) *Booklet 40*, HMSO, London.

Forestry Commission (1976) *Booklet 43*, HMSO, London.

Forestry Commission (1977a) *Wood Production Outlook in Britain*, HMSO, London.

Forestry Commission (1977b) *Leaflet 64*, HMSO, London.

Foster, C.D. and Beesley, M.E. (1963) Estimating the social benefit of constructing an underground railway in London. *Journal of the Royal Statistical Society, Series A*, 67–88.

Foster, C.D. and Beesley, M.E. (1965) Victoria line, social benefits and finances. *Journal of the Royal Statistical Society, Series A*, 67–88.

Frank, A.G. (1966) The development of underdevelopment. *Monthly Review*, **18**, 17–31.

Freeman, A.M. III (1984) The quasi-option value of irreversible development. *Journal of Environmental Economics and Management*, **11**, 292–5.

Frisch, R. (1932) *The New Methods of Measuring Marginal Utility*, Verlag von J.C.B. Mahr, Tübingen.

Gannon, C.A. (1969) Towards the strategy for conservation in a world of technological change. *Socio-economic Planning Sciences*, **3**, 158–78.

Gerasimov, I.P. and Gindin, A.M. (1977) The problem of transfering run-off from northern Siberian rivers to the arid regions of the European USSR, The Soviet Central Asia and Kazakhstan, in *Environmental Effects of Complex River Development* (ed. G.K. White), Westview Press, Boulder, Colorado.

Goldsmith, E. and Hilyard, N. (1988) T*he Earth Report*, Mitchell Beazley, London.

Gordon, H.S. (1954) The economic theory of a common property resource: the fishery. *Journal of Political Economy*, **62**, 142.

Gore, A. (1992) *Earth in the Balance, Forging a New Common Purpose*, Earthscan Publications, London.

Goundry, G.K. (1960) Forest management and the theory of capital. *Canadian Journal of Economics*, **26**, 439–51.

Gowers, A. (1984) Acid rain: now the debate moves to centre stage. *Financial Times*, 7 November 1984.

Gray, L.C. (1916) Rent under the assumption of exhaustibility. *Quarterly Journal of Economics*, **28**, 446–89.

Griffin, J.M. and Steele, H.B. (1980) E*nergy Economics and Policy*, Academic Press, New York.

Hammack, J. and Brown, G. (1974) *Water Fowl and Wetlands Toward Bioeconomic Analysis*, Johns Hopkins University Press, Baltimore, MD.

Hampson, S.F. (1972) Highland forestry: an evaluation. *Journal of Agricultural Economics*, **23**, 49–57.

Hanley, N. (1988) Valuing non-market goods using contingent valuation: A survey and synthesis. *Journal of Economic Surveys*.

Hanley, N. (1989) Valuing rural recreation sites: An empirical comparison of two approaches. *Journal of Agricultural Economics*, **40**, 361–75.

Hanley, N. (1990) The economics of nitrate pollution. *European Review of Agricultural Economics*, **17**, 129–51.

Hanley, N. (1991) *Farming and the Countryside: An Economic Analysis of External Costs and Benefits*, CAB International, Wallingford.

Hanley, N. *et al.* (1990) Why is more notice taken of economists' prescriptions for control of pollution? *Environment and Planning A*, **22**, 1421–39.

Hansard (1989) *Investment*, volume 150, No. 79, column 187, 5 April 1989, HMSO, London.

Harison, A.J. and Taylor, S.J. (1969) The value of working time in appraisal of transport expenditure – a review, Department of the Environment, Note 16, London.

Harrison, D. Jr. and Rubinfeld, D. (1978) Hedonic housing prices and the demand for clean air. *Journal of Environmental Economics and Management*, **S**, 81–102.

Hartwick, J.M. and Olewiler, N.D. (1986) *The Economics of Natural Resource Use*, Harper & Row, New York.

Harvey, D.R. (1991) Agriculture and the environment: The way ahead? in *Farming and the Countryside An Economic Analysis of External Costs and Benefits* (ed. N. Hanley), CAB International, Wallingford.

Heimlich, R.E. (1991) Soil erosion and conservation policies in the United States, in *Farming and the Countryside: an Economic Analysis of External Costs and Benefits* (ed. N. Hanley), CAB International, Wallingford.

Helliwell, D.R. (1974) Discount rates in land use planning, *Forestry*, **47**, 147–52.

Helliwell, D.R. (1975) Discount rates and environmental conservation. *Environmental Conservation*, **2**, 199–201.

Henry, C. (1974) Investment decision under uncertainty: the irreversibility effect. *American Economic Review*, **64**, 1006–12.

Hicks, J.R. (1968) *Value and Capital*, Clarendon Press, Oxford.

Hiley, W.E. (1967) *Woodland Management*, Faber & Faber, London.

Hirshleifer, J., DeHaven, J.D. and Milliman, J.M. (1960) *Water Supply, Economics, Technology and Policy*, University of Chicago Press, Chicago.

HM Treasury (1984) *Investment Appraisal in the Public Sector: A Technical Guide for Government Departments*, HMSO, London.

HMSO (1972a) *Forestry in Great Britain: An Interdepartmental Cost–benefit study*, HMSO, London.

HMSO (1972b) *Forestry Policy*, HMSO, London.

Hoch, I. and Drake, T. (1974) Wages, climate and the quality of life. *Journal of Environmental Economics and Management*, **1**, 268–95.

Hodge, I. (1991) The provision of public goods in the countryside: How should it be arranged? in *Farming and the Countryside An Economic Analysis of External Costs and Benefits* (ed. N. Hanley), CAB, International, Wallingford.

Holzman, F.D. (1958) Consumer sovereignty and the role of economic development. *Economia Internazionale*, **11**.

Hotelling, H. (1925) A general mathematical theory of depreciation. *Journal of the American Statistical Association*, **20**, 340–53.

Hotelling, H. (1931) The economics of exhaustible resources. *Journal of Political Economy*, **39**, 137–75.

Hotelling, H. (1947) Unpublished letter to the Director of the National Park Service.

Howe, C.W. (1979) *Natural Resource Economics, Issues, Analysis and Policy*, Wiley, New York.

Hueting, R. (1980) *New Scarcity and Economic Growth – More Welfare Through Less Production?*, North Holland, Amsterdam.

Hufschmidt M.M. *et al.* (1983), *Environment, Natural Systems and Development: on Economic Valuation Guide*, Johns Hopkins University Press, Baltimore, MD.

Hull, E. (1861) *The Cool Fields of Great Britain, their History, Structure and Duration with Notices of the Coal-fields of other parts of the World.*

Huppert, D.D. (1989) Comments on R. Bruce Retting's Fishery Management at a turning point? Reflections on the Evolution of rights-based fishing, in *Rights-based Fishing* (eds P.A. Neher, R. Arnason and N. Mollet), Kluwer Academic Press, London, pp. 65–8.

Hutchinson, R.W. (1989) Modified discounting method: some comments. *Project Appraisal*, **4**, 108–10.

Hutchinson, T.W. (1953) *A Review of Economic Doctrines: 1870–1923*, Clarendon Press, Oxford.

Huxley, A. (1983) *Brave New World Revisited*, Traid Grafton Books, London.

Ireland, F. (1983) Best practicable means: an interpretation, in *An Annotated Reader in Environmental Planning and Management* (eds T. O'Riordan and K. Turner), Urban and Regional Planning Series, Vol. 30, Pergamon Press, London, pp. 446–52.

Jevons, W.S. (1865) *The Coal Question: an Inquiry Concerning the Progress of the Nation and the Probable Exhaustion of our Coal* Mines, Macmillan, London.

Johansson, P.O. (1991), Valuing environmental damage, in *Economic Policy Towards the Environment* (ed. D. Helm), Blackwell, Oxford.

Joint Economic Committee of the US Congress (1968) *Economic Analysis of Public Investment Decisions: Interest Rate, Policy and Discounting Analysis*, The United States Congress, US Government Printing Office, Washington, DC.

Kahn, A.E. (1966) The tyranny of small decisions. *Kyklos*, **19**, 23–47.

Kahn, H. (1976) *The Next 200 Years*, W. Marrow, New York.

Kahn, J.R. (1991) Atrazine pollution and Chesapeake fisheries, in *Farming and Countryside an Economic Analysis of External Costs and Benefits* (ed. N. Hanley), CAB International, Wallingford.

Kapp, K.W. (1950) *The Social Cost of Private Enterprise*, Cambridge University Press, Cambridge.

Kay, J.A. (1972) Social discount rates. *Journal of Public Economics*, **1**, 359–78.

Kerr, R.A. (1979) Geological disposal of nuclear wastes: Salt's lead if challenged. *Science*, **204**, 11 May, 603.

Kirchner, G. (1990) A new hazard index for determination of risk potentials of disposed radioactive wastes. *Journal of Environmental Activity*, **11**, 71–95.

Krupnick, A., Oates, W. and Van de Verg, E. (1983) On marketable air pollution permits: the case for a system of pollution offsets. *Journal of Environmental Economics and Management*, **10**, 233–47.

Krutilla, J.V. (1967) Conservation reconsidered. *American Economic Review*, **57**, 776–86.

Krutilla, J. and Eckstein, C. (1958) *Multipurpose River Development*, Johns Hopkins University Press, Baltimore, MD.

Kuhn, T.E. (1962) *Public Enterprise Economics and Transport Problems*, University of California Press, Los Angeles, CA.

Kula, E. (1981) Future generations and discounting rules in public sector investment appraisal. *Environment and Planning A*, **13**, 899–910.

Kula, E. (1984a) Derivation of social time preference rates for the United States and Canada. *Quarterly Journal of Economics*, **99**, 873–83.

Kula, E. (1984b) Discount factors for public sector investment projects by using the sum of discounted consumption flows method. *Environment and Planning A*, **16**, 689–94.

Kula, E. (1984c) Justice and efficiency with the sum of discounted consumption flows method. *Environment and Planning A*, **16**, 835–8.

Kula, E. (1985) Derivation of social time preference rates for the United Kingdom. *Environment and Planning A*, **17**, 199–212.

Kula, E. (1986a) The analysis of social interest rate in Trinidad and Tobago. *Journal of Development Studies*, **22**, 731–9.

Kula, E. (1986b) The developing framework for the economic analysis of forestry projects in the United Kingdom. *Journal of Agricultural Economics*, **37**, 365–77.

Kula, E. (1987a) The developing framework for the economic evaluation of forestry in the United Kingdom – a reply. *Journal of Agricultural Economics*, **38**, 501–4.

Kula, E. (1987b) The social interest rate for public sector project appraisal in the UK, US and Canada. *Project Appraisal*, **2**, 1969–75.

Kula, E. (1988a) The modified discount factors for project appraisal in the public sector. *Project Appraisal*, **3**, 85–9.

Kula, E. (1988b) *The Economics of Forestry – Modern Theory and Practice*, Croom Helm, London and Timber Press, Portland, OR.

Kula, E. (1989a) The modified discounting method – comment on comments. *Project Appraisal*, **3**, 110–13.

Kula, E. (1989b) Politics, economics, agriculture and famines – the Chinese case. *Food Policy*, **14**, 13–17.

Kula, E. and McKillop, D. (1988) A planting function for private afforestation in Northern Ireland. *Journal of Agricultural Economics*, **39**, 133–41.

Land Use Study Group (1966) *Forestry, Agriculture and Multiple use of Rural Land*, HMSO, London.

Landauer, C. (1947) *The Theory of National Economic Planning*, University of California Press, Los Angeles, CA.

Landsberg, H.H., Fischman, L.L. and Fisher, J.L. (1963) *Resources in America's Future, Patterns of Requirements and Availabilities 1960–2000*, Johns Hopkins University Press, Baltimore, MD.

Lawson, R.M. (1984) *Economics of Fisheries Development*, Frances Pinters, London.

Lee, R.R. and Douglas, J.L. (1971) *Economics of Water Resources Planning*, McGraw-Hill, Bombay.

Leopold L.B. and Maddock, T. Jr. (1954) *The Flood Control Controversy*, The Ronald Press, New York.

Libecap, G.D. (1989) Comments on Anthony D. Scott's conceptual origins of rights based fishing, in *Rights Based Fishing* (eds P.A. Neher, R. Arnason and N. Mollet), Kluwer Academic Press, London, pp. 39–45.

Lind, R.C. (1964) Further comment. *Quarterly Journal of Economics,* **78,** 336–45.

Lindsay, C.M. (1967) Option demand and consumer's surplus. *Quarterly Journal of Economics,* **76,** 344–6.

Lipschutz R.D. (1981) *Radioactive Waste, Politics, Technology and Risk,* A Report of the Union of Concerned Scientists, Bellinger, Cambridge, MA.

Logan, S.E., Schulze, W.D., Ben-David, S. and Brookshire, D.S. (1978) *Development and Application of a Risk Assessment Method for Radioactive Waste Management.* Vol. III, Economic Analysis, United States Environmental Protection Agency, Office of Radiation Programs, AW-459, EPA 520/6-78-005, Washington, DC.

Lombardini, S. (1992) Transition to a market economy and environmental problems. *FEEM Newsletter,* vol. 2 July, FEEM, Milano.

Long, M.F. (1967) Collective consumption services of individual consumption goods-comment. *Quarterly Journal of Economics,* **76.**

Lönnstedt, L. (1989) Goals and cutting decisions of private small forest owners. *Scandinavian Journal of Forest Resources,* **4,** 259–65.

Lovelock, J. (1979) *Gaia: A New Look at Life on Earth,* Oxford University Press, Oxford.

L'Vovich, M.I. (1978) Turning the Siberian waters south. *New Scientist,* **79,** 834–6.

McKean, R.N. (1958) *Efficiency in Government Through System Analysis,* University of California Press, Los Angeles, CA.

McKelvey (1972) Mineral resource estimates and public policy. *American Scientist,* **60.**

McKillop, D. and Kula, E. (1987) The importance of lags in determining the parameters of a planting function for forestry in Northern Ireland. *Forestry,* **60,** 229–37.

Maass, A. (1962) *Design of Water Resources Systems,* Macmillan, New York.

Maddox, J. (1972) *The Doomsday Syndrome,* Macmillan, London.

Malone, C.R. (1990) Implications of environmental program planning for siting a nuclear repository at Yucca Mountain, Nevada. *Environmental Management,* **14,** 25–32.

Malthus, T.R. (1798) *Essay on the Principle of Population as it Affects the Future Improvement of Society,* Ward, Lock, London.

Malthus, T.R. (1815) *On the Nature and Progress of Rent,* Lord Baltimore Press, Baltimore, MD.

Marcus, A.A. (1980) *Promise and Performance: Choosing and Implementing on Environmental Policy,* Greenwood Press, Westwood, CT.

Marglin, S. (1962) Economic factors affecting system design, in *Design of Water Resource System* (ed. A. Mass), Johns Hopkins University Press, Baltimore, MD.

Marglin, S. (1963a) The opportunity cost of public investment. *Quarterly Journal of Economics,* **77,** 274–89.

Marglin, S. (1963b) The social rate of discount and the optimal rate of saving. *Quarterly Journal of Economics,* **77,** 95–111.

Markyanda, A. (1991) Economics of ozone layer, in *Blueprint 2* (ed. D. Pearce), Earthscan, London.

Markyanda, A. (1992) Environment and sustainable development: economic perspectives. *FEEM Newsletter*, vol. 3 November, FEEM, Milano.

Marshall, A. (1890) *Principles of Economics*, Macmillan, London.

Meadows, D.H., Meadows, D.L., Randers, J. and Behrens, W. III (1972) *The Limits to Growth*, Pan Books, London.

Mesarovic, M.D. and Pestel, E.C. (1974) *Mankind at the Turning Point*, Dutton, New York.

Mill, J.S. (1862) *Principles of Political Economy*, Appleton, New York.

Miller, J.R. and Lad, F. (1984), Flexibility, preserving and irreversibility in environmental decisions: a Bayesian analysis. *Journal of Environmental Economics and Management*, **11**, 161–72.

Mintzer, I.M. (1987) A matter of degrees – the potential for controlling the greenhouse effect. *Research Report 5*, World Research Institute, New York.

Mishan, E.J. (1967) *The Cost of Economic Growth*, Staples, London.

Mishan, E.J. (1971) *Cost–Benefit Analysis*, Allen & Unwin, London.

Mishan, E.J. (1975) *Cost–Benefit Analysis*, 2nd edn, Allen & Unwin, London.

Mishan, E.J. (1989) *Cost–Benefit Analysis*, 3rd edn, Unwin, London.

Mittermeir, R. (1986) Primate diversity and the tropic forest, in *Biodiversity* (ed. D.E. Wilson), National Academy, Washington, DC.

Montgomery, W. (1972) Markets in licences and efficient pollution control programs. *Journal of Economic Theory*, **5**, 395–418.

Nash, C.A. (1973) Future generations and the social rate of discount. *Environment and Planning A*, **5**, 611–17.

Neill, R.H. and Chaturved, L. (1991) *Status of the WIPP Project*, Environmental Evaluation Group Publication, Albuquerque, New Mexico.

Nelson, J. (1979) Airport noise, location rent, and the market for residential amenities. *Journal of Environmental Economics and Management*, **6**, 320–31.

Nordhaus, W.D. (1973) World dynamics: measurement without data. *Economic Journal*, **83**, 1156–83.

Nordhaus, W.D. (1989) *The Economics of the Greenhouse Effect*, Department of Economics, Yale University.

Nordhaus, W.D. and Tobin, J. (1972) *Is Growth Obsolete?* National Bureau of Economic Research, 50th anniversary colloquium, Columbia University Press, New York.

North, J. (1990) Future agricultural land use patterns, in *Agriculture in Britain: Changing Pressures and Policies*, (ed. D. Brittan), CAB International, Wallingford.

O'Byrne *et al.* (1985) Housing values, census estimates, disequilibrium and the environmental cost of airport noise; a case study of Atlanta. *Journal of Environmental Economics and Management*, **12**.

OECD Committee (1980) Radiological significance and management of tritium, carbon-14, krypton-85 and iodine-129 arising from the nuclear fuel cycle. Nuclear Energy Agency Expert Group, OECD, Paris.

Olson, M. and Zeckhauser, R. (1970) The efficient production of external economies. *American Economic Review*, **60**, 512–17.

O'Neill, C.E. (1990) An economic analysis of the contributions of freshwater angling and aquaculture to the Northern Ireland Economy. PhD thesis, The Queen's University, Belfast.

O'Neill, C.E. and Davis, J. (1991) Alternative definitions of recreational angling in Northern Ireland. *Journal of Agricultural Economics*, **42**, 174–9.

OPEC (1982) *A Comparative Statistical Analysis*, Carl Veberreuter Ges, Vienna.

OPEC (1986) *Bulletin*, Vol. 17, Carl Veberreuter Ges, Vienna.

OPEC (1987) *Bulletin*, Vol. 18, Carl Veberreuter Ges, Vienna.

OPEC (1988) *Facts and Figures*, the Secretariat, Vienna.

O'Riordan, T. (1988) The politics of sustainability, in *Sustainable Development Management* (ed. R.K. Turner), Belhaven Press, London.

O'Riordan, T. and Turner, K.R. (eds) (1983) *An Annotated Reader in Environmental Planning and Management*, Pergamon, London.

Page, T. (1977) Equitable use of resource base. *Environment and Planning A*, **9**, 15–22.

Page, T. (1983) Sharing resources with future, in *An Annotated Reader in Environmental Planning and Management* (eds T. O'Riordon and R.K. Turner), Pergamon, London.

Palmquist, R.B. (1991) Hedonic methods, in *Measuring the Demand for Environmental Quality* (eds J.B. Braden and C.D. Kalstad), North Holland, Amsterdam.

Pearce, D.W. (1977) *Environmental Economics*, Longman, London.

Pearce, D.W. (1983) *Cost–Benefit Analysis*, 2nd edn, Macmillan, London.

Pearce, D.W. (1991) *Blueprint 2 – Greening the World Economy*, Earthscan, London.

Pearce, D.W. and Markyanda, A. (1989) *The Benefits of Environmental Policies*, OECD, Paris.

Pearce, D.W. and Turner, R.K. (1990) *Economics of Natural Resources and the Environment*, Harvester Wheatsheaf, London.

Pearce, D.W., Markyanda, A. and Barbier, E.B. (1989) *Blueprint for a Green Economy*, Earthscan, London.

Pearce, D.W., Barbier, E. and Markyanda, A. (1990) *Sustainable Development – Economics and the Environment in the Third World*, Edward Elgar, Aldershot.

Pezzey, J. (1988) Market mechanisms of pollution control: 'Polluter Pays', economic and practical aspects, in *Sustainable Environmental Management: Principle and Practice* (ed. R.K. Turner), Belhaven, London, pp. 190–242.

Pezzey, J. (1989) *Economic Analysis of Sustainable Growth and Sustainable Development*, Environment Department Working Paper No. 15, World Bank, Washington, DC.

Pigou, A. (1920) *Income*, Macmillan, London.

Pigou, A. (1929) *The Economics of Welfare*, Macmillan, London.

Potter, N. and Christy, F.T. (1962) *Trends in Natural Resource Commodities: Statistics of Prices, Output, Consumption, Foreign Trade and Employment in the US, 1870–1977*, Johns Hopkins University Press, Baltimore, MD.

President's Material Commission (1952) *Resources for Freedom* (5 volumes), US Government Printing Office, Washington, DC.

Price, C. (1973) To the future – with indifference or concern? *Journal of Agricultural Economics*, **24**, 383–98.

Price, C. (1976) Blind alleys and open prospects in forestry economics. *Forestry,* **49,** 93–107.

Price, C (1978) *Landscape Economics,* Macmillan, London.

Price, C. (1981) Some economic aspects of marine management policies, the future and discount rate, *FAO Fisheries series No. 5,* FAO, Rome, p. 57–65.

Price, C. (1984a) Project appraisal and planning for underdeveloped countries – the costing of non-renewable resources. *Environmental Management,* **8,** 221–32.

Price, C. (1984b) The sum of discounted consumption flows method: equity with efficiency? *Environment and Planning A,* **16,** 829–37.

Price, C. (1987) The developing framework for the economic evaluation of forestry in the United Kingdom – a comment. *Journal of Agricultural Economics,* **38,** 497–500.

Price, C. (1989) Equity, consistency, efficiency and new rules of discounting. *Project Appraisal,* **4,** 58–65.

Quinn, E.A. (1986) Projected use, emission and banks of potential ozone depleting substances. *Rand Corporation Report,* No. 2282, EPA, Washington, DC.

Rajaraman, I. (1976) Non-renewable resources: a review of long term projects. *Futures,* 8.

Raloff, J. (1989) Study upgrades radiation risks to humans. *Science News,* **136,** 26–7.

Randall, A., Ives, B. and Eastman, C. (1974) Bidding games for valuation of aesthetic environmental improvements. *Journal of Environmental Economics and Management,* **1,** 132–49.

Randall, A. and Stoll, J.R. (1983) Existence value in a total valuation framework, in *Managing Air Quality and Scenic Resources at National Parks and Wilderness Areas* (eds R.D. Rowe and L.G. Chesthut), Westview Press, Colorado.

Raucher, R.(1975) In hearings before the US Senate Commerce Committee, 5th March 1975, US Government Printing Office, Washington, DC.

Raup, D. (1986) Diversity in crisis in the geological past, in *Biodiversity* (ed. E.O. Wilson), National Academy, Washington, DC.

Rawls, J. (1972) *A Theory of Justice,* Clarendon Press, Oxford.

Repetto, R. (1986) *World Enough and Time,* Yale University Press, New Haven.

Repetto, R. *et al.* (1989) *Wasting Assets, Natural Resources in the National Income Accounts,* World Resources Institute, Washington, DC.

Retting, R.B. (1989) Is fishery management at a turning point? Reflections on the evolution of rights based fishing, in *Rights Based Fishing* (eds P.A. Neher, P. Arnason and N. Mollet), Kluwer Academic Press, London, pp. 47–64.

Ricardo, D. (1817) *Principles of Political Economy and Taxation,* recent publication by Pelican Books, 1971, London.

Rigby, M. (1989) Modified discounting method: some comments. *Project Appraisal,* **4,** 107–8.

Roover, R.D. (1970) The concept of just price: Theory and economic policy, in *The History of Economic Theory* (ed. I.H. Rima), Holt, Rinehart and Winston, New York.

Rosen, S. (1974) Hedonic prices and implicit markets: product differentiation in pure competition. *Journal of Political Economy,* **82,** 34–55.

Rowe, R., D'Arge, R. and Brookshire, D.S. (1980) An experiment in the value of visibility. *Journal of Environmental Economics and Management,* **7,** 1–19.

Russell, N.P. (1990) Efficiency of farm conservation and output reduction policies. *Manchester Working Papers in Agricultural Economics*, WP/90-02, University of Manchester, Manchester.

Russell, N.P. (1993) Efficiency of rural conservation and supply control policy. *European Review of Agricultural Economics*, **20**, 315–27.

Ryden, J., Ball, P. and Garwood, E. (1984) Nitrate leaching from grassland. *Nature*, **311**, 50–3.

Samuelson, P. (1954) The pure theory of public expenditures. *Review of Economics and Statistics*, **36**, 387–9.

Samuelson, P.A. (1964) Principles of efficiency: discussion. *American Economic Review*.

Samuelson, P.A. (1976) Economics of forestry in an evolving society. *Economic Inquiry*, **14**, 466–92.

Sandbach, F.E. (1979) Economics of pollution control, in *Economics of Environment* (eds J. Lenihan and W.W. Fletcher), Blackie, London.

Schaefer, M.D. (1957) Some consideration of population dynamics and economics in relation to the management of marine fisheries. *Journal of the Fisheries Research Board of Canada*, **14**, 669–81.

Schmalensee, R. (1972) Option demand and consumer surplus: valuing price changes under uncertainty. *American Economic Review*, **62**.

Schnare, A. (1976) Racial and ethnic price differentials in an urban housing market. *Urban Studies*, **13**, 107–20.

Schultze, W.D., Brookshire, D.S. and Sandler, T. (1981a) The social rate of discount for nuclear waste storage: economics or ethics? *Natural Resource Journal*, **21**, 811–32.

Schultze, W., D'Arge, R. and Brookshire, D.S. (1981b) Valuing environmental commodities: Some recent experiments. *Land Economics*, **57**, 151–69.

Scott, A.D. (1955) The fishery: the objectives of sole ownership. *Journal of Political Economy*, **63**, 116–24.

Scott, A.D. (1989) Conceptual origins of rights based fishing, in *Rights Based Fishing* (eds P.A. Neher, R. Arnason and N. Mollet), Kluwer Academic Press, London, pp. 11–38.

Sen, A.K. (1961) On optimising the rate of saving. *Economic Journal*, **71**, 479–96.

Sen, A.K. (1967) The social time preference rate in relation to the market rate of interest. *Quarterly Journal of Economics*, **81**, 112–24.

Seneca, J.S. and Taussig, M.K. (1979) *Environmental Economics*, 2nd edn, Prentice-Hall, New Jersey.

Seskin, E. *et al.* (1983) An empirical analysis of economic strategies for controlling air pollution. *Journal of Environmental Economics and Management*, **10**, 112–24.

Shearman, R. (1990) The meaning and ethics of sustainability. *Environmental Management*, **14**, 1–8.

Shrader-Frechette, K.S. (1991) *Nuclear Power and Public Policy, The Social and Ethical Problems of Fission Technology*, Reidel, Boston.

Silvander, U. and Drake, L. (1991) Nitrate pollution and fishery protection in Sweden, in *Farming and the Countryside* (ed. N. Hanley), CAB International, Wallingford.

Simon, J.L. (1984) *The Resourceful Earth – A Response to Global 2000*, Blackwell, London.

Slade, M.E. (1982) Trends in natural-resource commodity prices: an analysis of the time domain. *Journal of Environmental Economics and Management*, **9**, 122–37.

Smith, L.G. (1993), *Impact Assessment and Sustainable Resource Management*, Longman Scientific and Technical, Harlow, England.

Smith, V.L. (1977) Economics of wilderness resources. *Economics of Natural and Environmental Resources* (ed. V.L. Smith), Gordon & Breach, New York, pp. 489–502.

Smith, V.K. (1979) *Scarcity and Growth Reconsidered*. Johns Hopkins University Press, Baltimore, MD.

Smith, V.K. and Krutilla, J.V. (1972) Technical change and environmental resources. *Socio-economic Planning Sciences*, **6**, 125–32.

Spiegel, H.W. (1952) *The Development of Economic Thought*, John Wiley, New York.

Sraffa, I. and Dobb, M. (1951–55) *The Works and Correspondence of David Ricardo*, Cambridge University Press, Cambridge.

Steiner, P. (1959) Choosing among alternative public investment in the water resource field. *American Economic Review*, **49**, 893–916.

Stolwijk, H.J.J. (1989) *Economische Gevolgen Voor de Veehouderiz van eeen Drietel Milieuscenario's*, CPB, Den Haag.

Stone, R. (1954) *Measurement of Consumer Expenditure and Behaviour in the United Kingdom 1920–1928*, Cambridge University Press, Cambridge.

Swanson, T. (1991) Conserving biological diversity, in *Blueprint 2* (ed. D. Pearce), Earthscan, London.

Sylvan, R. and Bennett, D. (1988) Tacism and deep ecology. *The Ecologist*, **18**, 148–59.

Tamminga, G. and Wijnands, J. (1991) Animal waste problems in the Netherlands, in *Farming and the Countryside* (ed. N. Hanley), CAB International, Wallingford.

Thaler, R. and Rosen, S. (1976) The value of saving a life: evidence form the labour market, in *Household Production and Consumption* (ed. N.J. Terlecky), National Bureau of Economic Research, New York.

Thompson, A.E. (1971) The Forestry Commission: a re-appraisal of its functions. *Three Banks Review*, September, 30–44.

Thomson, K. (1988) Future generations: the modified discounting method – a reply. *Project Appraisal*, **3**, 171–2.

Thornton, I. (1992) Report on the United Nations Conference on Environment and Development, *Globe*, Issue 8, August 1992.

Tietenberg, T. (1992), *Environmental and Natural Resource Economics*, 3rd edn, Harper-Collins, New York.

Traiforos, S., Adamontiades, A. and More, E. (1990) *The Status of Nuclear Power Technology*, World Bank, Washington, DC.

Traill, B. (1988) The rural environment; what role for Europe?, in *Land Use and the European Environment* (eds M.C. Whitby and J. Ollerenshaw), Belhaven Press, London.

Tullock, G. (1964) The social rate of discount and optimal rate of interest: comment. *Quarterly Journal of Economics*, **78**, 331–6.

Turner, R.K. (1991) Environment, economics and ethics, in *Blueprint 2* (ed. D. Pearce), Earthscan, London.

Turvey, R. (1977) Optimisation and suboptimisation in fishery regulation. *Economics of Natural and Environmental Resources* (ed. V.L. Smith), Gordon & Breach, London, pp. 175–87.

UK Energy Committee (1980–81) HC 114, HMSO, London.

UK Energy Committee (1990) *The Cost of Nuclear Power*, 7 June 1990, HMSO, London.

UK Government (1961) *The Financial and Economic Obligations of Nationalised Industries*, Cmnd 1337, HMSO, London.

UK Government (1965a) *National Plan*, Cmnd 2764, Chapter 11, HMSO, London.

UK Government (1965b) *Fuel Policy*, Cmnd 2798, HMSO, London.

UK Government (1975) *Food from our Own Resources*, HMSO, London.

UK Government (1976) *Nationalised Industries: a Review of Economic and Financial Objectives*, Cmnd 3437, HMSO, London.

UK Government (1978) *Nationalised Industries*, Cmnd 7131, HMSO, London.

Ulrich, J. (1990), 1199 Speech, *Enchanted Times*, Fall 1990, New Mexico Research and Education Enrichment Foundation, Albuquerque.

United Nations (1977) *Third Conference on the Law of the Sea*, informal composite negotiating text, UN Doc.A/Conf.62/WP.10.

US Congress (1982) Nuclear Waste Policy Act, Public Law, 97–425, January 7, 1983.

US Congress (1987) Nuclear Waste Policy Amendments Act of 1987, Public Law 100–203, December 1987, 42, USC 10101 et seq, US Government Printing Office, Washington, DC.

US Department of Energy (1978) Management of Commercially Generated Radioactive Waste, Draft Environmental Impact Statement, Department of Energy, Washington DC, DOE/EIS/0046/D, August.

US Department of Energy (1988) Managing the Nation's Nuclear Wastes, Factsheet Series, Office of Civilian Radioactive Waste Management, Washington, DC.

US Department of Energy (1990) *Preliminary Estimates of the Total System Cost for the Restructured Programme*, US Department of Energy, Office of Civilian Radioactive Waste Management, Washington.

US Environmental Protection Agency (1985) Environmental Standards for the Management and Disposal of Spent Nuclear Fuel, High Level and Transuranic Radioactive Waste, 40 CFR 191, Federal Register, 50, 38066–38089, EPA, Washington, DC.

Usher, D. (1964) The social rate of discount and optimal rate of investment: comment. *Quarterly Journal of Economics*, **78**, 641–4.

Viscusi, W.K. and Zeckhauser, R.J. (1976) Environmental policy choice under uncertainty. *Journal of Environmental Economics and Management*, **3**, 97–112.

Vogel, D. (1986) *National Styles of Regulation: Environmental Policy in Great Britain and the United States*, Cornell University Press.

Von Thunen, J.H. (1826) *Isolated State* (English edition, ed. P. Hall), Pergamon Press, London. (German edition published in 1822).

Walker, K.R. (1958) *Competition for Hill Land Between the Agriculture, Industry and Forestry Commission*, unpublished PhD thesis, Oxford University.

Wall Street Journal (1977) The waiting game: sizeable gas reserves untapped as producers await profitable prices, 23 February 1977.

Weisbrod, B.A. (1964) Collective consumption services of individual consumption goods. *Quarterly Journal of Economics*, **78,** 471–7.

West, E. (1815) *Essay on the Application of Capital*, London.

Whitby, M. (1991) The changing nature of rural land use, in *Farming and the Countryside* (ed. N. Hanley), CAB International, Wallingford.

White, L. Jr. (1967) The historical roots of ecological crisis. *Science,* **155,** 1203–7.

Winpenny, J.T. (1991), *Values for the Environment*, HMSO, London.

Winters, L.A. (1989) Agricultural policy in industrialised economies. *The Economic Review,* September, 37–41.

Wolfe Report (1973) *Some Considerations Regarding Forestry Policy in Great Britain,* Forestry Commission, Edinburgh.

World Conservation Union (1990) *Environmental Issues in Eastern Europe: Setting an Agenda,* The Royal Institute of Environmental Affairs Energy and Environmental Programme, London.

World Resources Institute (1987) A matter of degrees; the potential for controlling the greenhouse effect. Research Report 5, New York.

Wunderlich, G. (1967) Taxing and exploiting oil: the Dakota case, in *Extractive Resources and Taxation* (ed. M. Gaffney), University of Wisconsin Press, Madison, WI.

Young, A. (1804) *General View of Agriculture of Hertfordshire*, G. and W. Nicol, London.

Young, O.R. (1981) *National Resources and the State,* University of California Press, Berkeley, CA.

Zwartendyk, J. (1972) What is mineral endowment and how should we measure it? *Mineral Bulletin M.R. 126,* Canadian Government Department of Energy, Mines and Resources, Ottawa.

Author index

Davis, R.   244
DeGraff, J.V.   302
Deval, B.   41
Dixon, J.A.   238
Dobb, M.   7, 13
Dorfman, R.   60
Drummond, M.R.   233
Douthwaite, R.   10, 34, 133
Dubgaard, A.   132, 135, 136
Dupuit, J.   236
Douglas, J.L.   237

Eckstein, O.   80, 237, 305
Edwards, A.   142
Eliot, J.   32, 139
Eliot, T.S.   213
England, R.   132
Everest, D.   215, 221, 224, 225

FAO   125
Farrell, J.   182
Faustmann, M.   32, 106
Feldstein, M.   80, 302
Fellner, W.   310
Fischoff, B.   255
Fischman, L.L.   270
Fisher, A.   255, 256, 264, 268
Fisher, I.   80, 110, 310
Foster, C.D.   82, 84, 87, 88
Frisch, R.   310

Gannon, C.   264, 268
Garwood, E.   131
Gerasimov, I.P.   15
Gindin, A.M.   15
Goldsmith, W.D.   225
Gordon, H.S.   32, 42, 154
Gorman, W.D.   342
Gore, A.   16, 17
Goundrey, G.K.   110
Gowers, A.   213
Grey, L.C.   x, 145
Griffin, J.M.   172

Haberlein, T.A.   244
Hammock, J.   244
Hampson, S.F.   81
Haneman, W.   250
Hanley, N.   128, 132, 133, 186, 188,
     189, 200, 245, 246, 250
Harrison, A.J.   239
Hartwick, J.M.   18, 23, 61, 69, 206

Harvey, D.R.   130, 143
Hayes, C.   131, 133
Heimlich, R.E.   140
Held, B.   220
Helliwell, D.R.   8
Henry, C.   273
Hicks, J.R.   32
Hiley, W.E.   20, 83, 87
Hilyard, N.   225
Hirshleifer, J.   14, 305
Hoch, I.   239
Hodge, I.   xi, 128, 142
Holzman, F.D.   14
Hoopstaken, A.M.   137
Hotelling, H.   x, 110, 145, 160, 248
Howe, C.W.   30
Hueting, R.   22
Hufschmidt, M.M.   239
Huppert, D.D.   72
Hutchinson, R.W.   323, 332, 334
Hutchinson, T.W.   12
Huxley, A.   5–7

Ireland, F.   188

Kahn, A.E.   272
Kahn, H.   36
Kahn, J.R.   134
Kapp, K.W.   177
Kay, J.A.   80
Kerr, R.A.   281
Kirchner, G.   235
Knetsch, J.I.   270
Kolstad, C.D.   243
Krupnick, A.   186
Krutilla, J.V.   237, 251, 256, 264, 305
Kuhn, T.E.   305
Kula, E.   4, 60, 81, 100, 103, 127, 313,
     322, 323
Kurz, M.   80

Ladd, F.   255
Land Use Study Group   81
Landauer, C.   312
Landbrooke, A.   239
Landsberg, H.H.   220
Lawson, K.   239
Lawson, R.M.   60
Lee, R.R.   237
Leopold, L.B.   236
Libecap, G.D.   45, 72
Lind, R.C.   317

Subject index